Data, Models and Analysis

MW00964989

This volume contains the ten most cited articles that have appeared in the journal *Atmosphere-Ocean* since 1995. These articles cover a wide range of topics in meteorology, climatology and oceanography. Modelling work is represented in five papers, covering global climate model development; a cumulus parameterization scheme for global climate models; development of a regional forecast modelling system and parameterization of peatland hydraulic processes for climate models. Data rehabilitation and compilation in order to support trend analysis work on comprehensive precipitation and temperature data sets is presented in four papers. Field studies are represented by a paper on the circumpolar lead system. While the modelling studies are global in their application and applicability, the data analysis and field study papers cover environments that are specifically, but not uniquely, Canadian.

This book will be of interest to researchers, students and professionals in the various sub-fields of meteorology, oceanography and climate science.

Guoqi Han is a Research Scientist at Fisheries and Oceans Canada, and an Adjunct Professor at Memorial University, Canada. His main research field is physical oceanography, including coastal ocean modelling, satellite oceanography, as well as ocean climate variability and change. He publishes regularly in peer-reviewed scientific journals. He serves as an Editor-in-Chief of *Atmosphere-Ocean*, an Academic Editor of *PLOS One*, and an editorial member of *Ocean Modelling*. He is the Lead Scientist for ocean sciences in the Surface Water and Ocean Topography – Canada (SWOT-C) Program. He is a Vice Chair, Commission A: Space Studies of Earth's Surface, Meteorology and Climate, Committee on Space Research. He was awarded the Prize in Applied Oceanography by the Canadian Meteorological and Oceanographic Society in 1998.

Hai Lin is a Research Scientist at Environment and Climate Change Canada (ECCC), in the Meteorological Research Division. He is also an Adjunct Professor at the Department of Atmospheric and Oceanic Sciences of McGill University, Canada. His research interests include large-scale atmospheric dynamics, climate variability and numerical weather prediction. He leads the research and development of sub-seasonal and seasonal prediction systems in Recherche en Prévision Numérique of ECCC. He was the recipient of the 2010 President's Prize of the Canadian Meteorological and Oceanographic Society. He publishes regularly in international journals, served as an Associate Editor of *Monthly Weather Review* and is an Editor-in-Chief of *Atmosphere-Ocean*. He is a member of the Steering Group for Subseasonal to Seasonal Prediction of the World Weather Research Programme and World Climate Research Programme of the World Meteorological Organization.

Douw Steyn is a Professor Emeritus of Atmospheric Science at The University of British Columbia, Canada, in the Department of Earth Ocean and Atmospheric Sciences. His professional, teaching and research activities are in the field of air pollution meteorology, boundary layer meteorology, mesoscale meteorology, environmental science and interdisciplinary science. His research involves measurement and modelling studies of regional air pollution, especially in regions with complex terrain. He publishes regularly in international peer reviewed literature and has served as Editor of *Atmosphere-Ocean* and on the editorial board of *Boundary Layer Meteorology*. He is an Accredited Consulting Meteorologist and has international consultancy experience in his areas of expertise. He has been awarded the Mentorship Medal of the Canadian Federation of Earth Sciences and the Andrew Thompson Prize in Applied Meteorology by CMOS, and is a Fellow of CMOS.

Data, Models and Analysis

The Highest Impact Articles in
Atmosphere-Ocean

Edited by
Guoqi Han, Hai Lin and Douw Steyn

Routledge
Taylor & Francis Group

LONDON AND NEW YORK

CMOS
SCMO

First published 2017
by Routledge
2 Park Square, Milton Park, Abingdon, Oxon, OX14 4RN, UK

and by Routledge
711 Third Avenue, New York, NY 10017, USA

Routledge is an imprint of the Taylor & Francis Group, an informa business

© 2017 Canadian Meteorological and Oceanographic Society

All rights reserved. No part of this book may be reprinted or reproduced
or utilised in any form or by any electronic, mechanical, or other means,
now known or hereafter invented, including photocopying and recording,
or in any information storage or retrieval system, without permission in
writing from the publishers. Printed in Canada.

Trademark notice: Product or corporate names may be trademarks or
registered trademarks, and are used only for identification and
explanation without intent to infringe.

British Library Cataloguing in Publication Data
A catalogue record for this book is available from the British Library

ISBN13: 978-1-138-04765-5

Typeset in Times New Roman
by RefineCatch Limited, Bungay, Suffolk

Publisher's Note
The publisher accepts responsibility for any inconsistencies that may have
arisen during the conversion of this book from journal articles to book chapters,
namely the possible inclusion of journal terminology.

Disclaimer
Every effort has been made to contact copyright holders for their permission to
reprint material in this book. The publishers would be grateful to hear from any
copyright holder who is not here acknowledged and will undertake to rectify
any errors or omissions in future editions of this book.

Contents

Citation Information   vii

Notes on Contributors   ix

Foreword   xv
David Grimes

1. Introduction   1
   Guoqi Han, Hai Lin and Douw Steyn

2. The International Polar Year (IPY) Circumpolar Flaw Lead (CFL) System Study: Overview and the Physical System   3
   D. G. Barber, M. G. Asplin, Y. Gratton, J. V. Lukovich, R. J. Galley, R. L. Raddatz and D. Leitch

3. Rehabilitation and Analysis of Canadian Daily Precipitation Time Series   22
   Éva Mekis and William D. Hogg

4. Temperature and Precipitation Trends in Canada During the 20th Century   40
   Xuebin Zhang, Lucie A. Vincent, W. D. Hogg and Ain Niitsoo

5. Changes in Daily and Extreme Temperature and Precipitation Indices for Canada over the Twentieth Century   61
   Lucie A. Vincent and Éva Mekis

6. An Overview of the Second Generation Adjusted Daily Precipitation Dataset for Trend Analysis in Canada   78
   Éva Mekis and Lucie A. Vincent

7. Parametrization of Peatland Hydraulic Properties for the Canadian Land Surface Scheme   93
   Matthew G. Letts, Nigel T. Roulet, Neil T. Comer, Michael R. Skarupa and Diana L. Verseghy

8. The 15-km Version of the Canadian Regional Forecast System   106
   Jocelyn Mailhot, Stéphane Bélair, Louis Lefaivre, Bernard Bilodeau, Michel Desgagné, Claude Girard, Anna Glazer, Anne-Marie Leduc, André Méthot, Alain Patoine, André Plante, Alan Rahill, Tom Robinson, Donald Talbot, André Tremblay, Paul Vaillancourt, Ayrton Zadra and Abdessamad Qaddouri

9. The Canadian Fourth Generation Atmospheric Global Climate Model (CanAM4). Part I: Representation of Physical Processes   123
   Knut von Salzen, John F. Scinocca, Norman A. McFarlane, Jiangnan Li, Jason N. S. Cole, David Plummer, Diana Verseghy, M. Cathy Reader, Xiaoyan Ma, Michael Lazare and Larry Solheim

10. Sensitivity of Climate Simulations to the Parameterization of Cumulus Convection in the Canadian Climate Center General Circulation Model   145
    G.J. Zhang and Norman A. McFarlane

11. The UVic Earth System Climate Model: Model Description, Climatology, and Applications to Past, Present and Future Climates   169
    Andrew J. Weaver, Michael Eby, Edward C. Wiebe, Cecilia M. Bitz, Phil B. Duffy, Tracy L. Ewen, Augustus F. Fanning, Marika M. Holland, Amy MacFadyen, H. Damon Matthews, Katrin J. Meissner, Oleg Saenko, Andreas Schmittner, Huaxiao Wang and Masakazu Yoshimori

Index   237

Citation Information

The following chapters were originally published in various issues of *Atmosphere-Ocean*. When citing this material, please use the original page numbering for each article, as follows:

Chapter 2

D. G. Barber, M. G. Asplin, Y. Gratton, J. V. Lukovich, R. J. Galley, R. L. Raddatz & D. Leitch (2010) The International polar year (IPY) circumpolar flaw lead (CFL) system study: Overview and the physical system, Atmosphere-Ocean, 48:4, 225–243, DOI: 10.3137/OC317.2010

Chapter 3

Eva Mekis & William D. Hogg (1999) Rehabilitation and analysis of Canadian daily precipitation time series, Atmosphere-Ocean, 37:1, 53–85, DOI: 10.1080/07055900.1999.9649621

Chapter 4

Xuebin Zhang, Lucie A. Vincent, W. D. Hogg & Ain Niitsoo (2000) Temperature and precipitation trends in Canada during the 20th century, Atmosphere-Ocean, 38:3, 395–429, DOI: 10.1080/07055900.2000.9649654

Chapter 5

Lucie A. Vincent & Éva Mekis (2006) Changes in Daily and Extreme Temperature and Precipitation Indices for Canada over the Twentieth Century, Atmosphere-Ocean, 44:2, 177–193, DOI: 10.3137/ao.440205

Chapter 6

Éva Mekis & Lucie A. Vincent (2011) An Overview of the Second Generation Adjusted Daily Precipitation Dataset for Trend Analysis in Canada, Atmosphere-Ocean, 49:2, 163–177, DOI: 10.1080/07055900.2011.583910

Chapter 7

Matthew G. Letts, Nigel T. Roulet, Neil T. Comer, Michael R. Skarupa & Diana L. Verseghy (2000) Parametrization of peatland hydraulic properties for the Canadian land surface scheme, Atmosphere-Ocean, 38:1, 141–160, DOI: 10.1080/07055900.2000.9649643

Chapter 8

Jocelyn Mailhot, Stéphane Bélair, Louis Lefaivre, Bernard Bilodeau, Michel Desgagné, Claude Girard, Anna Glazer, Anne-Marie Leduc, André Méthot, Alain Patoine, André Plante, Alan Rahill, Tom Robinson, Donald Talbot, André Tremblay, Paul Vaillancourt, Ayrton Zadra & Abdessamad Qaddouri (2006) The 15-km version of the Canadian regional forecast system, Atmosphere-Ocean, 44:2, 133–149, DOI: 10.3137/ao.440202

Chapter 9

Knut von Salzen, John F. Scinocca, Norman A. McFarlane, Jiangnan Li, Jason N. S. Cole, David Plummer, Diana Verseghy, M. Cathy Reader, Xiaoyan Ma, Michael Lazare & Larry Solheim (2013) The Canadian Fourth Generation Atmospheric Global Climate Model (CanAM4). Part I: Representation of Physical Processes, Atmosphere-Ocean, 51:1, 104–125, DOI: 10.1080/07055900.2012.755610

Chapter 10

G. J. Zhang & Norman A. McFarlane (1995) Sensitivity of climate simulations to the parameterization of cumulus convection in the Canadian climate centre general circulation model, Atmosphere-Ocean, 33:3, 407–446, DOI: 10.1080/07055900.1995.9649539

Chapter 11

Andrew J. Weaver, Michael Eby, Edward C. Wiebe, Cecilia M. Bitz, Phil B. Duffy, Tracy L. Ewen, Augustus F. Fanning, Marika M. Holland, Amy MacFadyen, H. Damon Matthews, Katrin J. Meissner, Oleg Saenko, Andreas Schmittner, Huaxiao Wang & Masakazu Yoshimori (2001) The UVic earth system climate model: Model description, climatology, and applications to past, present and future climates, Atmosphere-Ocean, 39:4, 361–428, DOI: 10.1080/07055900.2001.9649686

For any permission-related enquiries please visit:
http://www.tandfonline.com/page/help/permissions

Notes on Contributors

David Barber writes with regards to the International Polar Year (IPY) Circumpolar Flaw Lead (CFL) system study. The IPY-CFL program was conducted between 2007 and 2011. The team was able to keep a fully outfitted research icebreaker (CCGS Amundsen) mobile throughout the winter of the IPY within the Cape Bathurst Flaw lead in the Southern Beaufort Sea. The project brought together over 450 investigators from 27 different countries in a first ever project designed to examine winter processes on ocean-sea ice-atmosphere coupling. Results from the overall study have resulted in hundreds of journal papers documenting the various processes which drive the coupling in this unique marine environment. The study clarified a number of key processes and in particular brought focus on the fact that winter processes are highly sensitive to high frequency events such as storms. An Inuit-led team brought Inuit knowledge to the study and fully integrated this with 9 science teams allowing for full system examination. The study was also testament to the fact that 20 years ago a study like this would not have been possible due to extreme ice features in the southern limb of the Beaufort Sea ice gyre. Climate change effects were clearly evident and a key enabling feature of this unique study.

Éva Mekis and William Hogg describe the development of the adjusted daily rainfall and snowfall measurements which are used for trends analysis in Canada. The adjustment procedures removed systematic biases due to changes in the measurement program so climate change detection became achievable. The rain measurement errors are corrected for wind undercatch, evaporation losses and rain gauge specific wetting losses. Snow density corrections based upon coincident ruler and Nipher measurements were applied to all snow ruler measurements to estimate the proper water equivalent of snowfall. The adjustments for trace flags also bring observations closer to real amounts. Where necessary, records from neighboring stations were joined creating longer records. The adjustments described here not only remove the inconsistencies for purposes of generating a more homogeneous time series, they also greatly reduce, or ideally, eliminate bias within the full range of the time series. This makes these data much more appropriate for use in water balance studies, runoff analyses and for general model verification purposes. This first generation of adjusted rainfall and snowfall has been used by hundreds of scientists from different government agencies and universities for the analysis of climate change and climate change impacts in Canada.

Zhang et al. analyse trends in Canadian temperature and precipitation during the 20th century using, for the first time, homogenized temperatures and adjusted precipitation data which include adjustments for site relocation, changes in observing programs and corrections for known instrument changes and measurement program deficiencies. The data were first interpolated to a 50 × 50 km grid to take into account the uneven distribution of the station data which is also sparse in northern Canada (north of 60°N). From 1950–1998, a distinct pattern of warming in the south and west, and cooling in the northeast regions of the country was found in winter and spring. Across Canada, precipitation increased by 5% to 35%, with significant negative trends in the southern regions during the winter. The ratio of snowfall to total precipitation increased, with negative trends in southern regions during the spring. The causes of the different spatial and temporal trends are not discussed but there is some evidence of agreement between the observed trends in Canadian climate and those predicted by the Global Climate Models incorporating an increase in atmospheric greenhouse gases.

Lucie Vincent and Éva Mekis analyse the trends and variations in several indices of daily and extreme temperature and precipitation in Canada. The indices are based on homogenized daily temperature and adjusted daily precipitation. Previous studies on climate change in Canada have shown changes in long-term climatic means and it became pertinent to investigate if the warming in Canada was accompanied by detectable changes in temperature and precipitation extremes. Extreme climate events have also the greatest and most direct impact on our everyday lives, community and environment. The analysis of the temperature indices indicate fewer cold nights, cold days and frost days and, conversely, more warm night, warm days and summer days across the country for 1950–2003. The analysis of the precipitation indices reveals more days with precipitation, a decrease in the precipitation intensity and a decrease in the maximum number of consecutive dry days. No consistent changes are found in most of the indices of extreme precipitation.

Éva Mekis and Lucie Vincent present the second generation of adjusted daily precipitation for trend analysis in Canada. Twelve years after the creation of the first generation (described in the preceding article by Mékis and Hogg), it became vital to describe the latest version of the adjusted rainfall and snowfall to properly document the newer methodologies and the impact of changes for the large user community. This article provides the step-by-step procedures to produce the adjusted precipitation and the trend analysis of annual and seasonal rainfall and snowfall for 1950–2009 and 1900–2009. Several new locations were added; the increased coverage and enhanced metadata improved the overall data quality. The trace flag and snow water equivalent adjustment were also updated. The impact of adjustments was examined in detail, the rainfall amount increased by 5 to 20% in the Canadian Arctic with the adjustments, while the effect of snow adjustments displayed an even bigger range and variability (from none to up to 50%) throughout the country. By providing adjusted rainfall and snowfall separately, the study of the change in precipitation phase was made possible. Analysis of the trends indicate an increase in rainfall across the country while a mix of increasing and decreasing trends was found during the summer in the Canadian Prairies. Snowfall increased mainly in the north and decreased in the southwestern regions of the country. This dataset is now available at the Government of Canada Open Data portal (http://open.canada.ca/en/open-data) and it is widely used by scientists working in government agencies, universities, and by the public in general.

Matthew Letts et al. write on the development of the organic soil parameters for the Canadian Land Surface Scheme (CLASS). Most land surface packages for climate models assume the soils comprise sand, silt and clay, but much of the northern latitudes are covered with soils that are largely made up of organic material. For example twelve percent of the land surface of Canada, and similar areas of Fennoscandinavia and Russia, are covered with peatlands. The thermal and hydraulic properties of organic matter is very different than that of mineral soils largely due to the high porosity of organic matter. In this paper, Letts et al. compile data on the thermal and hydraulic properties of peat because of the degree of decomposition, or humification, of the peat. These parameters were incorporated into CLASS along with an algorithm to simulate the water table and methane emissions. At the time this was the first inclusion of organic matter into a land surface package, but many climate models now use a similar parameter set. Without the inclusion of organic matter in climate models the presence and absence of permafrost and the annual thermal cycle cannot be simulated with any degree of confidence.

Jocelyn Mailhot and Stéphane Bélair describe a mesoscale version of the Global Environmental Multiscale (GEM) model implemented operationally at the Canadian Meteorological Centre in May 2004. The major upgrades include increased vertical and horizontal resolution (15 km) and improvements to the physics parameterization package (boundary layer clouds, shallow and deep convection, gravity wave drag and low-level blocking due to subgrid-scale orography). Various aspects of the improved performance of the new system are documented to provide useful guidance to the Canadian operational forecasters community. The development of the GEM 15 km forecast system has also been instrumental for subsequent developments toward kilometer- and sub-kilometer-scale forecast systems, such as the experimental systems used during the Vancouver 2010 and Sochi 2014 Winter Olympics, the Toronto 2015 Pan American Games and the recently implemented pan-Canadian 2.5 km High Resolution Deterministic Prediction System.

Knut von Salzen provides a summary of physical processes in the Canadian fourth Generation Atmospheric Global Climate model (CanAM4), which is the latest in a long series of models, starting with the Canadian Climate Centre Spectral Atmospheric General Circulation model in the early 1980s. CanAM4 continues to represent the backbone of the Canadian Earth System Model developed by the Canadian Centre for Climate Modelling and Analysis, Environment and Climate Change Canada. Substantially improved parameterizations of clouds, aerosols and radiation distinguish CanAM4 from earlier versions of the model. Important applications of CanAM4 include projections of future global climate for CMIP5 (the Coupled Model Intercomparison Project, Phase 5) in support of the Intergovernmental Panel on Climate Change (IPCC) 5th Assessment report (AR5) and its operational use for seasonal forecasts of temperature and precipitation for Canada. CanAM4 has also been widely used to study changes in Arctic climate and short-lived climate pollutants. In the near future, CanAM4 will be replaced by an updated version of the model which includes further improved representation of atmospheric physical processes.

Guang Zhang and Norman McFarlane document a parameterization scheme utilizing the entraining plume conceptual framework pioneered by Arakawa and Schubert in 1974 (AS) to represent the effects of deep moist convection on the larger scale circulation in global models. A major simplification from the original AS formulation was the adoption of a spectrum of convective elements whose cloud-base mass fluxes are uniformly distributed as a function of their fractional entrainment rates. The corresponding plume ensemble is defined in terms of a range of entrainment rates that are determined from the prognostic large-scale temperature and humidity fields of the GCM. The scheme is easily implemented, robust, and general enough to be used in a wide range of modeling applications. The "Zhang-McFarlane" (ZM) cumulus parameterization has been adopted in a wide range of modeling applications including the operational global climate models of the Canadian Centre for Climate Modelling and Analysis (CCCma), the Community Atmospheric Model of the National Centre for Atmospheric Research (NCAR CAM), and a number of global climate models in use within the international climate research community. Since publication of this paper in *Atmosphere-Ocean*, Dr Guang Zhang has developed additional refinements, linking onset of convection and cloud-base mass flux to production of convective instability by large-scale flow. A comprehensive microphysics parameterization has also been incorporated into the ZM scheme recently to better describe the microphysical processes inside convection. These changes have substantially improved the performance of the ZM scheme.

Andrew Weaver, together with his students, postdoctoral fellows and research associates document the development of the UVic Earth System Climate Model (ESCM). Now used by more than 130 external researchers at more than 75 locations around the world, the UVic ESCM consists of a three-dimensional ocean general circulation model coupled to a thermodynamic/dynamic sea-ice model, an energy-moisture balance atmospheric model with dynamical feedbacks and a thermo-mechanical land-ice model. The model is used to examine future projections of climate change under increasing atmospheric CO_2 as well as the climate of the Last Glacial Maximum. A long 6000-year integration is also conducted to test the assumption often used in ocean model evaluation that the present-day ocean observations are in equilibrium with the present-day radiative forcing.

Notes sur les contributeurs

David Barber écrit dans le contexte de l'Étude sur le chenal de séparation circumpolaire de l'année polaire internationale (API) réalisée entre 2007 et 2011. L'équipe a pu maintenir en mouvement un brise-glace de recherche entièrement équipé (le NGCC Amundsen) pendant la période hivernale de l'API à l'intérieur du chenal de séparation circumpolaire du cap Bathurst dans le sud de la mer de Beaufort. L'étude a permis de réunir plus de 450 chercheurs venant de 27 pays, dans un projet innovateur visant à examiner les processus hivernaux de couplage océan-glace de mer-atmosphère. Les résultats de l'étude ont abouti à des centaines d'articles scientifiques documentant les divers processus entraînant les couplages dans ce milieu marin unique. L'étude a documenté un nombre de processus clés et a particulièrement mis en évidence la sensibilité des processus hivernaux aux phénomènes fréquemment rencontrés comme les tempêtes. Une équipe menée par des Inuits a apporté des connaissances inuites qui ont été pleinement intégrées par neuf équipes scientifiques, permettant l'examen complet du système. L'étude a prouvé que 20 ans plus tôt, une telle étude n'aurait pas été possible en raison des caractéristiques des formations glacielles extrêmes dans le flanc sud du tourbillon de glace de la mer de Beaufort. Les effets du changement climatique étaient manifestement évidents et ont joué un rôle puissant dans cette étude unique.

Éva Mekis et William Hogg décrivent l'élaboration des mesures ajustées de précipitations pluviales et neigeuses quotidiennes nécessaires à l'analyse des tendances pour le Canada. Ces procédures de réglage ont permis de supprimer des écarts systématiques causés par les changements dans le programme de mesure, ce qui a rendu la détection des variations climatiques réalisable. Les erreurs de mesure de la pluie ont été corrigées pour la diminution de capture due au vent, les pertes par évaporation et les pertes de mouillage propres au pluviomètre. Des corrections de densité de neige basées sur des mesures coïncidentes avec une règle et un nivomètre de Nipher ont été appliquées à toutes les mesures de neige pour estimer l'équivalence appropriée en eau de la neige. Des ajustements pour les indications de trace amènent aussi les observations plus près des quantités réelles. Au besoin, les données provenant de stations voisines ont été jointes pour allonger la série temporelle. Les réglages décrits ici ne supprime pas seulement les incohérences, permettant de générer une série chronologique plus homogène, ils réduisent aussi considérablement ou, idéalement, éliminent les biais dans toute l'étendue de la série temporelle. Ceci rend ces données beaucoup plus appropriées pour les études de bilan hydrique, analyses de ruissellement et vérification des modèles en général. Cette première génération de précipitations liquides et neigeuses ajustées a été utilisée par des centaines de scientifiques de différents organismes gouvernementaux et universitaires pour analyser les changements climatiques et leurs impacts au Canada.

Zhang et coll. analysent les tendances des températures et des précipitations canadiennes au cours du XXe siècle, en utilisant, pour la première fois, des températures homogénéisées et des données de précipitations ajustées incluant des ajustements pour la relocalisation des sites, les changements de programme d'observation, les corrections pour les modifications connues des instruments et les faiblesses du programme de mesure. Tout d'abord, les données ont été interpolées à une maille de 50 × 50 km pour tenir compte de la répartition inégale des données des stations, qui en plus, sont rares dans le Nord canadien (au nord du 60e parallèle). De 1950 à 1998, on a trouvé de nettes tendances au réchauffement dans le sud et l'ouest ainsi qu'au refroidissement dans les régions du nord-est du pays en hiver et au printemps. Partout au Canada, les précipitations ont augmenté de 5 à 35 %, mais avec des tendances négatives significatives dans les régions méridionales en hiver. Le ratio des chutes de neige par rapport aux précipitations totales a augmenté, mais avec des tendances négatives pour les régions méridionales au printemps. Les causes des différentes tendances spatiales et temporelles ne sont pas discutées, mais il semble y avoir compatibilité entre l'évolution observée du climat canadien et celle prédite par les modèles climatiques globaux qui intègrent une augmentation des gaz à effet de serre.

Lucie Vincent et Éva Mekis analysent les tendances et les variations de plusieurs indices de température et de précipitation quotidienne et extrême au Canada. Les indices sont basés sur les températures quotidiennes homogénéisées et les précipitations quotidiennes ajustées. De précédentes études sur les changements climatiques au Canada ont révélé des variations dans les moyennes climatiques à long terme. Il est donc devenu pertinent d'investiguer si le réchauffement climatique au Canada est accompagné de changements décelables dans les extrêmes de température et de précipitation, car les phénomènes climatiques extrêmes ont un impact plus important et direct sur notre vie quotidienne, la collectivité et l'environnement. L'analyse des indices de température indique moins de nuits et de journées froides et moins de jours de gel, et à l'inverse, plus de nuits et de journées chaudes et de journées d'été partout au Canada de 1950 à 2003. L'analyse des indices de précipitation révèle plus de jours pluvieux, une diminution de l'intensité des précipitations et une diminution du nombre maximal de jours consécutifs de temps sec. On ne trouve aucune modification importante dans la majorité des indices de précipitation extrême.

Éva Mekis et Lucie Vincent présentent la deuxième génération de précipitations quotidiennes ajustées pour l'analyse des tendances au Canada. Douze ans après la création de la première génération (décrite dans l'article précédent de Mékis et Hogg), il est devenu vital de décrire la dernière version des précipitations pluvieuses et neigeuses ajustées afin de bien documenter les nouvelles méthodologies et l'impact des changements pour la vaste communauté d'utilisateurs. Cet article fournit toutes les étapes de la procédure à suivre pour produire les précipitations ajustées et pour analyser les tendances des précipitations et des chutes de neige annuelles et saisonnières de 1950 à 2009 et de 1900 à 2009. Plusieurs nouveaux sites ont été ajoutés ; la couverture accrue et des métadonnées plus élaborées ont amélioré la qualité générale des données. Les réglages pour l'indicateur de trace et les équivalents en eau de la neige ont également été mis à date. L'impact des réglages a été examiné en détail : la quantité de pluie a augmenté de 5 à 20 % dans l'Arctique canadien avec les réglages, tandis que l'effet des réglages pour la neige a affiché un éventail encore plus large et plus variable (de 0 à 50 %) à l'échelle du pays. En fournissant séparément les précipitations pluviales et neigeuses ajustées, l'étude du changement de phase des précipitations est devenue possible. L'analyse des tendances indique une augmentation des précipitations dans tout le pays alors qu'un mélange d'augmentation et de réduction au cours de l'été a été trouvé pour les Prairies canadiennes. Les chutes de neige ont augmenté principalement dans le nord et diminué dans les régions du sud-ouest du pays. Cet ensemble de données est maintenant disponible sur le portail des données ouvertes du gouvernement du Canada (http://ouvert.canada.ca/fr/open-data) et est largement utilisé par les scientifiques des organismes gouvernementaux, des universités, et par le public en général.

Matthew Letts et coll. écrit sur le développement du paramétrage des sols organiques pour le schéma canadien de la surface terrestre (CLASS). La majorité des représentations de la surface terrestre utilisées dans les modèles climatiques suppose que les sols sont constitués de limon, de sable et d'argile alors qu'une grande partie des latitudes septentrionales soient recouvertes de sols principalement constitués de matières organiques. Par exemple, 12 % des terres émergées du Canada et d'endroits similaires de Fenno-Scandinavie et de Russie sont recouverts de tourbières. Les propriétés thermiques et hydrauliques de la matière organique diffèrent grandement de celles des sols minéraux, principalement à cause de la grande porosité de la matière organique. Dans cet article, Letts et coll. compilent des données sur les propriétés thermiques et hydrauliques de la tourbe en fonction du degré de décomposition ou d'humification de la tourbe. Ces paramètres ont été intégrés au CLASS, accompagnés d'un algorithme permettant de simuler la nappe phréatique et les émissions de méthane. À l'époque, il s'agissait de la première inclusion de la matière organique dans un ensemble de surface terrestre, mais plusieurs modèles climatiques utilisent désormais un ensemble de paramètres similaires. Sans l'inclusion de la matière organique dans les modèles climatiques, la présence ou l'absence du pergélisol et le cycle thermique annuel ne peuvent pas être simulées avec un degré quelconque de confiance.

Jocelyn Mailhot et Stéphane Bélair décrivent une version à méso-échelle du système régional de prévision, devenue opérationnelle au Centre Météorologique Canadien en mai 2004. Les principales modifications incluent une augmentation de la résolution verticale et horizontale (15 km) et des améliorations majeures à la physique du modèle (couche limite nuageuse, convection restreinte et convection profonde, effets orographiques sous-maille des ondes de gravité déferlantes et blocage des vents de surface). On y souligne les diverses améliorations du nouveau système afin de fournir des informations utiles aux prévisionnistes d'opérations canadiens. Le développement du système de prévision à 15 km a aussi représenté une étape importante pour les développements subséquents de systèmes de prévision aux échelles kilométriques et sous-kilométriques, tels que les systèmes expérimentaux utilisés durant les Jeux olympiques d'hiver de Vancouver en 2010 et de Sochi en 2014, les Jeux panaméricains de Toronto en 2015, et la nouvelle version opérationnelle pancanadienne à 2.5 km du Système de Prévision Déterministe à Haute Résolution.

Knut von Salzen fournit un aperçu des processus physiques dans le modèle climatique atmosphérique canadien de quatrième génération (CanAM4), le dernier d'une longue série de modèles commençant avec le modèle spectral de la circulation générale du Centre climatique canadien du début des années 1980. Le CanAM4 demeure la base du modèle canadien du système terrestre mis au point par le Centre canadien de la modélisation et de l'analyse climatique de Environnement et Changement climatique Canada. Des paramétrisations substantiellement améliorées des nuages, des aérosols et du rayonnement distinguent le CanAM4 des versions précédentes du modèle. Parmi les applications importantes du CanAM4 on note les projections du changement climatique mondial pour le projet d'inter-comparaison de modèles couplés n°5 (CMIP5) en soutien au cinquième rapport d'évaluation (AR5) du groupe d'experts intergouvernemental sur l'évolution du climat (GIEC), et son utilisation opérationnelle pour les prévisions saisonnières de température et de précipitation au Canada. Le CanAM4 a également été fréquemment utilisé pour étudier les changements climatiques de l'Arctique et le sort des polluants climatiques éphémères. Dans un proche avenir, le CanAM4 sera remplacé par une nouvelle version offrant des représentations améliorées des processus physiques atmosphériques.

Guang Zhang et Norman McFarlane documentent un schéma de paramétrisation basé sur le modèle conceptuel du panache d'entraînement mis au point par Arakawa et Schubert en 1974 pour représenter les effets de la convection humide profonde sur la circulation à plus grande échelle dans les modèles planétaires. Une simplification majeure de la formulation Arakawa-Schubert d'origine est l'adoption d'un éventail d'éléments convectifs, dont les flux de masse à la base des nuages sont uniformément répartis en fonction de leur taux d'entraînement fractionné. L'ensemble des panaches correspondants est défini en termes d'une gamme de débits d'entraînement déterminés à partir des champs de température et d'humidité pronostiques à grande échelle du modèle de circulation générale. Le système est facilement mis en place, robuste et assez général pour être utilisé dans un large éventail d'applications de modèles. La paramétrisation des cumulus de Zhang-McFarlane a été adoptée dans un grand nombre d'applications de modélisation, y compris les modèles opérationnels du climat mondial du Centre canadien de modélisation et d'analyse du climat (CCmaC), le modèle atmosphérique communautaire du centre national de recherche atmosphérique américain (*National Centre for Atmospheric Research* (NCAR CAM)), et plusieurs modèles climatiques globaux utilisés internationalement dans le monde de la recherche sur le climat. Depuis la publication de cet article dans *Atmosphere-Ocean*, le Dr Guang Zhang a mis au point d'autres raffinements qui relient l'apparition de la convection et le flux de masse à la base des nuages à la production d'instabilité convective par l'écoulement à grande échelle. Une paramétrisation microphysique complète a aussi été incorporée récemment au schème de Zhang-McFarlane pour mieux décrire les processus microphysiques à l'intérieur de la convection. Ces modifications ont considérablement amélioré les performance du système Zhang-McFarlane.

Andrew Weaver, aidé de ses étudiants, de stagiaires postdoctoraux et d'associés de recherche ont documenté le développement du modèle du système climatique terrestre de l'Université de Victoria (ESCM). Maintenant utilisé par plus de 130 chercheurs externes, dans plus de 75 sites autour du monde, l'ESCM de l'Université de Victoria se compose d'un modèle tridimensionnel de circulation océanique générale couplé à un modèle de glace marine thermodynamique/dynamique, à un modèle atmosphérique à bilans d'énergie et d'humidité avec rétroactions dynamiques et à un modèle thermomécanique des glaces sur les terres émergées. Le modèle est utilisé pour examiner les prévisions des changements climatiques à venir sous une augmentation du CO_2 atmosphérique ainsi qu'en fonction du climat du dernier maximum glaciaire. Une intégration sur 6000 ans est également réalisée pour tester l'hypothèse, souvent utilisée lors de l'évaluation des modèles océaniques, que les observations actuelles de l'océan sont en équilibre avec le forçage radiatif actuel.

Foreword – for CMOS 50th Anniversary Publication

Today, the Canadian Meteorological and Oceanographic Society (CMOS) is celebrating its 50th anniversary. From its humble beginnings, the Society played a central role in connecting academia, government and private sector interests in meteorology, and by 1977, in oceanography as well. The *Atmosphere-Ocean* journal served as an important conduit for sharing our proud Canadian moments and scientific accomplishments. Science, applications and analytical methods have underpinned our success in understanding and predicting changes that we see in weather, climate, oceanography and the atmospheric environment.

This Special Edition of *Atmosphere-Ocean* brings together some of the best works presented over the years. While this is but a taste of the wide breadth of scientific accomplishments from all over Canada by both members and other contributors of the Society, this edition celebrates the richness of our achievements. No doubt as you turn the pages of this book, many will relate to the important advances we have made over these years, names that have become iconic or achievements that you have seen celebrated at the highest level in Canada and internationally. Some of your own advances and achievements may even be based on the great works of past Canadian scientists showcased in this compilation.

I hope that you share, as I do, our pride in all that has been accomplished over the past 50 years. I would like to personally thank everyone who worked on producing *Atmosphere-Ocean* over the years; their dedication has made this special 50th anniversary compilation possible.

David Grimes
Assistant Deputy Minister
Meteorological Service of Canada
And
President of World Meteorological Organization since 2011

Préface — concernant la publication du cinquantième anniversaire de la SCMO

Aujourd'hui, la Société canadienne de météorologie et d'océanographie (SCMO) célèbre son cinquantième anniversaire. Depuis ses modestes débuts, la société a joué un rôle essentiel permettant de relier le milieu universitaire, le gouvernement et le secteur privé dans les domaines de la météorologie, et depuis 1977, de l'océanographie. La revue Atmosphere-Ocean nous a permis de partager nos moments de fierté et nos réalisations scientifiques en tant que Canadiens. Notre succès à comprendre et à prédire les changements que nous observons sur le plan de la météorologie, du climat, de l'océanographie et de l'environnement atmosphérique s'appuie sur la science, les applications et les méthodes d'analyse.

Cette édition spéciale d'Atmosphere-Ocean rassemble quelques-unes des meilleures contributions publiées au fil des ans. Alors qu'il ne s'agit que d'un aperçu de la vaste gamme de réalisations scientifiques provenant de partout au Canada par les membres de la société et leurs collaborateurs, cette édition célèbre la richesse de nos accomplissements. Sans doute qu'au fil de votre lecture, plusieurs d'entre vous s'identifieront aux progrès importants réalisés au cours des années, aux noms devenus emblématiques ou aux réalisations célébrées de façon prestigieuse au Canada et à l'étranger. Certaines de vos avancées et réalisations sont peut-être même issues des grandes contributions scientifiques canadiennes présentées dans cette compilation.

J'espère que vous partagerez aussi bien que moi, cette fierté par rapport au travail accompli au cours des cinquante dernières années. Je tiens à remercier personnellement tous ceux qui ont œuvré à la production d'Atmosphere-Ocean au fil des ans ; leur dévouement a rendu possible cette compilation spéciale qui souligne ce cinquantième anniversaire.

David Grimes
Sous-ministre adjoint
Service météorologique du Canada
et
Président de l'Organisation météorologique mondiale depuis
2011

Introduction

This book celebrates the 50th anniversary of the Canadian Meteorological and Oceanographic Society (CMOS) by republishing a compilation of papers published in the CMOS flagship journal *Atmosphere-Ocean*. The republished papers were selected from all papers published in *Atmosphere-Ocean*, and its predecessor *Atmosphere* on the basis of being the most-cited papers in the past five years, with some attention paid to papers highly cited in 2015. This was done in order to capture very recently published papers of note.

Readers of this compilation will notice a number of features common to the selection. All works address either meteorological or oceanographic phenomena of importance to Canada, or important advances made by Canadian scientists. While *Atmosphere-Ocean* is a venue for publication on meteorological and oceanographic phenomena wherever they might occur, it is not surprising that Canadian authors publish their best work on Canadian phenomena in a Canadian Journal. While Canadian authors do publish their work in non-Canadian journals, it is gratifying that they publish some of their highly cited works in *Atmosphere-Ocean*. Readers might also notice that the most highly cited papers in the journal span meteorology, climate science and oceanography, the three editorial "departments" of *Atmosphere-Ocean*. This is a satisfying reflection of the strength of all branches of the sciences represented in CMOS.

CMOS is a bilingual society, and *Atmosphere-Ocean* contains papers written in both French and English, though French language papers form a small minority. It is therefore not surprising, though disappointing, that there are no French language papers in this book. While this is in part due to the small number of French language papers in *Atmosphere-Ocean*, it is also due to the fact that English has become the *lingua franca* of science. Scientists whose first language is not English often choose to publish their work in English in order to maximize their readership.

The papers in this book generally do not represent the last word in their topic, but rather, like all scientific works, represent a significant point of progress in a journey of discovery. Each of the papers has built on preceding work, and leads to subsequent work. The high frequency of citation clearly demonstrates their impact on their respective fields. In order to give readers a sense of the background work underlying the papers, and to provide some insight into the influence the papers have had on their field, we have asked corresponding authors of each of the papers to provide a short description. Papers are ordered as follows: A description of a field study; four papers on climate data sets and their analysis; and five papers on numerical model development, roughly in order of increasing scale.

Introduction

Ce livre célèbre le cinquantième anniversaire de la Société canadienne de météorologie et d'océanographie (SCMO) en publiant une compilation d'articles déjà parus dans *Atmosphere-Ocean,* la revue phare de la SCMO. Les articles republiés ont été sélectionnés parmi tous ceux parus dans *Atmosphere-Ocean* et son prédécesseur *Atmosphere* en fonction de leur fréquence de citation au cours des cinq dernières années, et particulièrement en 2015. Cette approche permet de promouvoir les articles éminents publiés récemment.

Les lecteurs de cette compilation remarqueront un certain nombre de caractéristiques communes à la sélection. Tous les travaux abordent des phénomènes météorologiques ou océanographiques d'importance pour le Canada, ou des progrès importants réalisés par les scientifiques canadiens. Bien que *Atmosphere-Ocean* publie des articles touchant à des phénomènes météorologiques et océanographiques où qu'ils puissent se produire, il n'est pas étonnant que les auteurs canadiens publient leurs meilleurs travaux concernant des phénomènes canadiens dans une revue canadienne. Bien que les auteurs canadiens publient une partie de leurs travaux dans des revues étrangères, nous sommes fiers que certains de leurs articles fréquemment cités paraissent dans *Atmosphere-Ocean.* Les lecteurs pourraient aussi remarquer que les articles hautement cités dans la revue englobent les domaines de la météorologie, de la climatologie et de l'océanographie, les trois « départements » éditoriaux d'*Atmosphere-Ocean.* Il s'agit d'une réflexion satisfaisante de la force de tous les domaines scientifiques représentés par la SCMO.

La SCMO est une société bilingue et *Atmosphere-Ocean* publie des articles rédigés en français et en anglais, bien que les articles en langue française s'y retrouvent en minorité. Il n'est donc pas étonnant, bien que décevant, qu'aucun article rédigé en français ne soit inclus dans ce livre. Ceci est en partie dû au fait que seul un petit nombre d'articles sont publiés en français dans *Atmosphere-Ocean,* mais aussi que l'anglais est devenu la *lingua franca* de la science. Les scientifiques dont la langue maternelle diffère de l'anglais choisissent régulièrement de publier leurs résultats en anglais afin de maximiser leur lectorat.

En général, les articles publiés dans ce livre ne représentent guère le dernier mot sur un sujet, mais plutôt, comme tout ouvrage scientifique, un progrès important sur le parcours de la découverte. Chaque article est fondé sur des travaux précédents et contribue à l'avancement de la recherche. La fréquence élevée des citations démontre clairement l'impact des travaux dans leurs domaines respectifs. Afin d'offrir aux lecteurs une vue d'ensemble du travail que sous-tendent les articles et de fournir un aperçu de l'influence qu'ils ont dans leur domaine, nous avons demandé aux auteurs correspondants de fournir une brève synthèse de leur article. Les articles sont classés comme suit : la description d'une étude sur le terrain ; quatre articles sur des ensembles de données climatiques et leur analyse, et enfin cinq articles sur l'élaboration de modèles numériques, en ordre de l'échelle du projet.

The International Polar Year (IPY) Circumpolar Flaw Lead (CFL) System Study: Overview and the Physical System

D.G. Barber[1], M.G. Asplin[1], Y. Gratton[2], J. V. Lukovich[1], R. J. Galley[1], R. L. Raddatz[1] and D. Leitch[1]

[1]*Centre for Earth Observation Science, Department of Environment and Geography*
University of Manitoba, Winnipeg, Manitoba
[2]*Institut National de Recherche Scientifique, Centre Eau, Terre et Environnement, Québec, Quebec*

ABSTRACT   *The Circumpolar Flaw Lead (CFL) system study is a Canadian-led International Polar Year (IPY) initiative with over 350 participants from 27 countries. The study is multidisciplinary in nature, integrating physical sciences, biological sciences and Inuvialuit traditional knowledge. The CFL study is designed to investigate the importance of changing climate processes in the flaw lead system of the northern hemisphere on the physical, biogeochemical and biological components of the Arctic marine system. The circumpolar flaw lead is a perennial characteristic of the Arctic throughout the winter season and forms when the mobile multi-year (MY) pack ice moves away from coastal fast ice, creating recurrent and interconnected polynyas in the Norwegian, Icelandic, North American and Siberian sectors of the Arctic. The CFL study was 293 days in duration and involved the overwintering of the Canadian research icebreaker CCGS Amundsen in the Cape Bathurst flaw lead throughout the annual sea-ice cycle of 2007–2008.*

In this paper we provide an introduction to the CFL project and then use preliminary data from the field season to describe the physical flaw lead system, as observed during the CFL overwintering project. Preliminary data show that ocean circulation is affected by eddy propagation into Amundsen Gulf (AG). Upwelling features arising along the ice edge and along abrupt topography are also detected and identified as important processes that bring nutrient rich waters up to the euphotic zone. Analysis of sea-ice relative vorticity and sea-ice area by ice type in the AG during the CFL study illustrates increased variability in ice vorticity in late autumn 2007 and an increase in new and young ice areas in the AG during winter. Analysis of atmospheric data show that a strong northeast–southwest pressure gradient present over the AG in autumn may be a synoptic-scale atmospheric response to sensible and latent heat fluxes arising from areas of open water persisting into late November 2007. The median atmospheric boundary layer temperature profile over the Cape Bathurst flaw lead during the winter season was stable but much less so when compared to Russian ice island stations.

RÉSUMÉ   *[Traduit par la rédaction] L'étude du système du chenal de séparation circumpolaire (CSC) est une initiative de l'Année polaire internationale (API) menée par le Canada et à laquelle 350 participants provenant de 27 pays ont pris part. L'étude, de nature multidisciplinaire, fait appel aux sciences physiques et biologiques ainsi qu'au savoir traditionnel Inuvialuit. L'étude du CSC vise à examiner les répercussions des processus climatiques changeants dans le système du chenal de séparation de l'hémisphère Nord sur les composantes physiques, biogéochimiques et biologiques du système marin arctique. Le CSC est une caractéristique permanente de l'Arctique durant la saison d'hiver et se forme quand la banquise mobile de glace de plusieurs années s'éloigne de la banquise côtière fixe en créant des polynies récurrentes et interconnectées dans les secteurs norvégien, islandais, nord-américain et sibérien de l'Arctique. Pour mener l'étude du CSC, qui a duré 293 jours, le brise-glace de recherche canadien NGCC Amundsen est demeuré dans le chenal de séparation du cap Bathurst tout l'hiver, c'est-à-dire pendant tout le cycle annuel des glaces de mer de 2007–2008.*

Dans cet article, nous fournissons une introduction au projet du CSC, puis nous utilisons les données préliminaires de la saison sur le terrain pour décrire le système physique du chenal de séparation, tel qu'observé durant l'hiver du projet du CSC. Les données préliminaires montrent que la circulation océanique est influencée par la propagation de tourbillon dans le golfe d'Amundsen (GA). Des caractéristiques de remontée d'eau le long de la lisière des glaces et le long d'éléments topographiques abrupts ont aussi été observées et identifiées comme des processus importants apportant des eaux riches en nutriments jusqu'à la zone euphotique. L'analyse de la vorticité relative de la glace de mer et de la superficie de la glace de mer par type dans le GA durant l'étude du CSC révèle une variabilité accrue dans la vorticité de la glace à la fin de l'automne 2007 et un accroissement des superficies de nouvelle et de jeune glace dans le GA durant l'hiver. L'analyse des données atmosphériques montre que la présence d'un fort gradient de pression nord est – sud ouest au-dessus du GA en automne peut être une réponse atmosphérique d'échelle synoptique aux flux de chaleur sensible et latente à partir des zones d'eaux libres qui ont persisté jusque vers la fin de novembre 2007. Le profil thermique de la couche limite atmosphérique médiane au-dessus du chenal de séparation du cap Bathurst durant la saison hivernale était stable, mais beaucoup moins que les profils observés aux stations des îles de glace russes.

1 Introduction

The world expects confirmation of global warming to appear first and be strongest in the polar regions of our planet (IPCC, 2007). Climate change has immediate implications for the sustainability of northern communities and the health and well-being of individuals and their economies. Creative responses based on sound research, shared knowledge and the engagement of people at all levels are required to meet this critical challenge. Observations indicate that the Arctic Ocean and its peripheral seas are presently warming. The extent of Arctic sea ice has shrunk at an average annual rate of about 45,100 km^2 (±4600 km^2) per year from 1979 to 2006 (Parkinson and Cavalieri, 2008). Five of the lowest summer sea-ice extents have occurred since 1998, with 2007 having the lowest minimum on instrumental record. The thickness of the multi-year (MY) ice has also decreased by about 40% over the past 30 years (Rothrock et al., 1999; Liu et al., 2004). Recent studies have documented variations in the Northern Annular Mode and associated surface atmospheric pressure fields (Thompson and Wallace, 1998). The resulting strengthening of westerly winds has deflected the freshwater plumes of several large rivers eastward (Thompson and Wallace, 2001) and has increased the export of sea ice through the transpolar drift (Kwok, 2008). The fresh water on the continental shelves normally forms a shield between the ice and the underlying warm Atlantic water. The eastward advection of this shield has allowed contact between the ice and increasingly warmer invading Atlantic waters (Polyakov et al., 2005), thus enhancing sea-ice melt. In the Canada Basin, the Beaufort Gyre (BG) also affects the reduction of sea ice and the formation of the flaw lead system. Recent results (Lukovich and Barber, 2006) show that the reversal of the BG, triggered by increased cyclogenesis over the Canada Basin (Zhang et al., 2004), has increased in frequency since 1990, thereby affecting both sea-ice extent and dynamics.

Flaw leads and polynyas are important physical features in the Arctic climate system and provide a unique environment from which we can gain insights into the changing polar marine ecosystem. Oceanographically, the high ice production in the flaw lead system contributes significantly to brine fluxes from the continental shelves into the deep basins (Martin et al., 1995). These fluxes drive biogeochemical fluxes on and off the continental shelves and control many aspects of gas and mass fluxes across the ocean-sea-ice-atmosphere (OSA) interface. Meteorologically, we expect the flaw lead system to play a central role in the steering of cyclones within the area; we also expect that the connection to the central pack portends a large-scale teleconnection to hemispheric scale pressure patterns such as the Arctic Oscillation (e.g., Dmitrenko et al., 2005).

Biologically, the circumpolar flaw lead preconditions the shelves to be productive portions of the marine ecosystem with the early availability of light and increased availability of nutrients through advection and upwelling at the shelf break. We expect ecosystem-wide enhancements to productivity in these areas, sustained for longer periods throughout the annual cycle. These expectations are supported by early use of the flaw lead by apex predators such as birds, beluga and bowhead whales, and polar bears. The 2003–2004 Canadian Arctic Shelf Exchange Study (CASES) overwintering program revealed a surprisingly active zooplankton-fish-mammal community on the inner shelf during the fall and winter months. Cluster analysis and non-metric multi-dimensional scaling of the zooplankton collected during the fall of 2002 indicate distinct assemblages in the Mackenzie Shelf–Franklin Bay region, the Cape Bathurst polynya and the Beaufort Slope (Darnis et al., 2008). The ongoing analysis of the 2003 CASES collections will enable us to test that these ice-dynamics-dependent assemblages also exist in spring and summer; however, a critical missing piece of the puzzle is the winter distribution of the zooplankton, Arctic cod and mammals on the shelf and at the shelf slope.

The complexities of the response of sea ice to changing oceanic and atmospheric forcing, and the subsequent response of the marine ecosystem to these changes, are key motivating principles for the International Polar Year (IPY) and the Circumpolar Flaw Lead (CFL) system study (CEOS, 2007). The CFL program is structured around three integrated components: a field study, an observatory and a modelling study. This triumvirate is designed to examine the importance of climate processes that are changing the nature of the flaw lead system in the northern hemisphere and the effect of these changes on marine ecosystem processes, contaminant transport, fluxes of nutrients and biogenic carbon, and exchange of greenhouse gases across the OSA interface. The CFL project contrasts the early opening and late closing of the flaw lead area against that of the adjacent fast ice. This contrast focuses on the oceanic and atmospheric forcing of the ice cover in these two regions and describes how these physical processes moderate biogeochemical processes within the Arctic marine ecosystem. In total, over 350 scientists from 27 countries were organized around ten science teams, addressing a wide range of research questions spanning local to regional scales:

1) How will climate variability and/or change affect the timing and extent of the flaw lead system through predictable ocean and atmospheric forcing?
2) How will reduced sea-ice cover affect seasonal OSA coupling in the flaw lead system?
3) How will ecosystem productivity and carbon cycling be altered by a change in the timing of the flaw lead system?
4) How will climate variability and/or change affect the adjacent fast ice ecosystem?
5) How will changes in the timing of precipitation and the formation and decay of sea ice affect the contributions of sympagic versus pelagic production to carbon cycling?

The overarching objectives of this paper are to introduce the IPY-CFL project and then provide an overview of key physical processes within the flaw lead system. Examples of ocean, sea-ice and atmospheric data collected during the CFL

4

overwintering field program in 2007–2008 are placed in the context of a 30-year climatology, spanning the last 30 years (1979–2008), of these variables. We examine various aspects of the physical oceanography of the study area, with a particular emphasis on ocean circulation, eddy advection and surface circulation. We present the sea-ice climatology for the CFL study area and examine sea-ice dynamic and thermodynamic processes throughout the annual cycle. We also present examples of atmospheric data and summarize key atmospheric processes. We conclude with a discussion of key coupling mechanisms across the OSA interface with a particular emphasis on how this coupling controls the spatial and temporal evolution of the Cape Bathurst flaw lead polynya complex.

2 The Circumpolar flaw lead system study

The CFL field study employed the Canadian Coast Guard Ship (CCGS) *Amundsen* (hereinafter *Amundsen*) as the primary field research platform. This represents the first time that a research icebreaker overwintered and stayed mobile in the circumpolar flaw lead of the northern hemisphere. The short-term objectives of the CFL project were to collect a dataset that describes the physical controls of marine ecosystem productivity in the open water and thin ice of the winter CFL system; to use these data to improve physically based models of atmospheric, sea-ice and oceanic processes; and to develop improved modelling approaches to predict biophysical processes within the Arctic marine and flaw lead physical systems. The long-term objectives were to use these findings to inform policy decisions regarding ocean management, climate impacts and adaptation in this region of Canada's Arctic. Our results will eventually be fully integrated with other IPY-funded projects that focused on other sectors of the CFL system in the European, Russian and Greenlandic sectors of the Arctic.

The CFL field study program began 18 October 2007. We commenced sampling in "transect mode" sampling 74 different open-water sites arranged in transects within the Amundsen Gulf (AG) and the southern Beaufort Sea between 18 October 2007 and 27 November 2007. The transects ran along the Mackenzie Shelf, northwards along the west side of Banks Island, northward from the Mackenzie River, and along the pack ice edge (Fig. 1a). Multiple moorings were installed during open-water sampling operations throughout the AG (Figs 1 and 3) in collaboration with ArcticNet. These moorings monitor chemical, physical and biological variables within the water column over an annual cycle and are a continuation of an ongoing marine observation program that began during the CASES overwintering project (2003–2004). On 28 November 2007, the ship entered a "drift mode" where the ship parked in a suitable ice floe until ice conditions or movement necessitated a move to another location. A "suitable" piece of ice consisted of an ice pan that was sufficiently large, thick, homogeneous and undisturbed, allowing for researchers to set up equipment and collect samples on, over

and under the ice. A total of 44 drift stations were sampled between 28 November 2007 and 31 May 2008 in the northern AG, south of Banks Island (Fig. 1b). Prevailing winds forced the sea ice (and ship) to drift along westerly vectors, in accordance with the Beaufort Sea Ice Gyre (hereinafter referred to as the Beaufort Gyre (BG). Time spent at each drift site averaged 3±4 days, with the maximum duration being 22 consecutive days stationed on one floe.

The initial CFL field study plan called for the establishment of a semi-permanent ice camp along a fast ice edge, which typically forms between Banks Island and Cape Perry (at approximately 70°45'N, 123°10'W); however, this was not possible as high ice mobility, driven by strong easterly winds, prevented the formation of fast ice throughout much of the AG. The lack of a long-term ice camp was compensated for by using several fast ice sampling sites throughout the ice melt season of May and June. The majority of these sites were located on the south side of the AG at the entrance to two coastal bays (Franklin Bay and Darnley Bay) where a scuba diving program aided in sample collection (Fig. 1b). Fast ice was also sampled in Prince of Wales Strait and near the north end of Banks Island and along the western edge of M'Clure Strait. A total of 17 fast ice stations were sampled averaging 1.3 days in duration. The maximum duration was 9 consecutive days. Open-water sampling along the established transects resumed at the end of June 2008. Between 1 June 2008 and 7 August 2008 (the end of the field season), a total of 96 sites were sampled (Fig. 1a). Several transects from open water into fast ice or mobile pack ice were also conducted (Fig. 1b), and many sites from the previous open-water season were revisited. In July 2008, multiple moorings were again collected, serviced and redeployed (Fig. 3).

The *Amundsen* was specially outfitted to support multidisciplinary polar marine science directly (Fig. 2). Equipment included a rosette equipped with a conductivity-temperature-depth (CTD) sensor, zooplankton nets, meteorological equipment, box coring equipment, a remotely operated vehicle (ROV) and an ensemble of moorings. Much of this equipment can be deployed off the side of the ship or from the ship's moonpool, a specially designed hole in the bottom of the ship that can be opened (only when the ship is stationary) to allow for the deployment of equipment in winter conditions. The ship also has wet, dry and several other specialized laboratories (i.e., cold laboratories, freezer laboratory, and radioactive laboratory). The ship also has the Portable In-situ Laboratory for Mercury Speciation (PILMS), which contains a class-100 clean room allowing trace metal analysis. Various vehicles on the *Amundsen* facilitated sampling in the different seasons including a zodiac, air-ice boats, snowmobiles, an all terrain vehicle (ATV) with snow plough, a half-track and a B0-105 helicopter.

An ensemble of physical sampling activities collected data from the three main components of the flaw lead physical system – the ocean, the sea ice and the atmosphere. The samplings were conducted regularly throughout the duration of the field study. The physical oceanography was assessed by

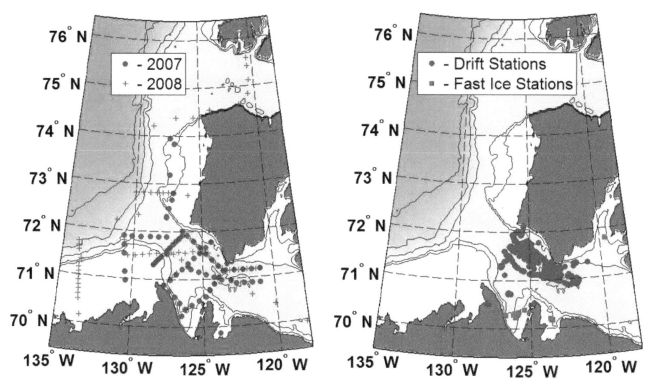

Fig. 1   a) Distributed open-water sample sites sampled in the fall of 2007 and the summer of 2008; b) drift stations and fast ice stations sampled in the winter and spring of 2007–2008.

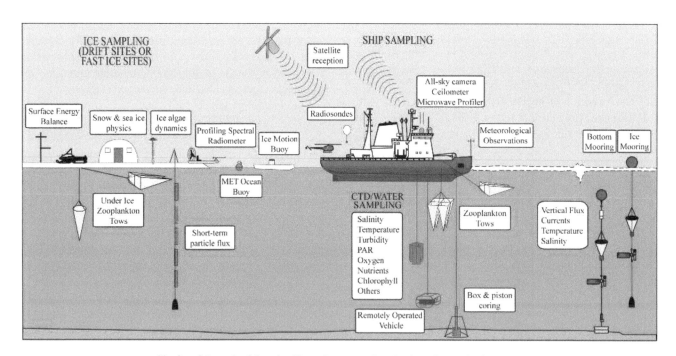

Fig. 2   Schematic of the scientific equipment used on the *Amundsen* and at ice camps.

collecting salinity, temperature and density profiles of the water column and profiling vertical and horizontal ocean currents (Fig. 4). Sea-ice and snow sampling assessed thermodynamics (temperature and salinity profiles), microstructure, snow grains, sea-ice dynamics (motion) and sea-ice thicknesses. Ship-based and upper-level meteorological programs

assessed surface meteorology, cloud cover, and upper-level profiles of temperature and water vapour. Automated sensors were augmented with meteorological balloon launches throughout the program. Examples of these field data are presented later in this paper to describe the physical system of the flaw lead and place observations made during the

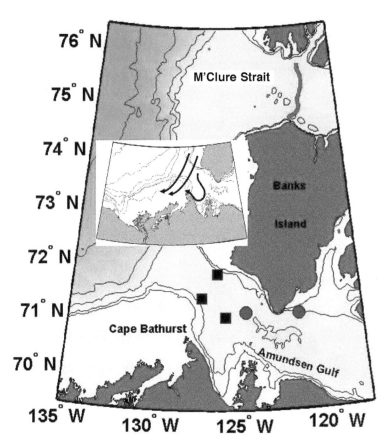

Fig. 3   Location of the three ArcticNet moorings (black squares) and two drift stations (blue circles) where eddies were observed. The thick red line across the mouth of M'Clure Strait shows where the MVP section was obtained.

2007–2008 annual cycle in the context of a 30-year climatology. A physical sciences special issue is planned for the Journal of Geophysical Research (oceans and atmospheres) expected in 2011.

The CFL study also incorporated a comprehensive biological sampling program to assess how changes in the physical flaw lead system, e.g., sea-ice cover reduction and increasing sea temperatures, could affect biogeochemical processes, biological productivity, fisheries resources and marine mammal populations of the coastal Canadian Arctic. The fragmented, thin and often absent ice cover in the flaw lead allows solar radiation to reach the surface layer of the ocean where it triggers photosynthesis by microscopic algae. Researchers investigating the pelagic and benthic food web examined how and to what extent microalgae growing in the flaw lead are consumed by the zooplankton and the benthos. Relative to adjacent ice-covered regions, enhanced algal production in the flaw lead may translate into biological hot spots where higher zooplankton and benthos production and abundance will prevail. We also investigated how Arctic cod, a keystone species in the Arctic food web, uses the flaw lead for feeding, overwintering, reproduction and as a nursery ground for their young. Results from the biological sampling component of the CFL study are beyond the scope of this paper, and therefore detailed sampling equipment, methods employed and results will be described in a forthcoming special issue on the

IPY-CFL biological program in the journal Polar Biology (expected in 2011).

In addition to the extensive science sampling program, the CFL field program also included outreach programs, including three Schools on Board field programs (national, international and circum-Arctic Inuit), an Artists on Board program, national and international media, an Inuit policy workshop, a visit by the British High Commissioner, as well as community visits to the local hamlets of Sachs Harbour, Paulatuk and Ulukhaktok. Members of the hunters and trappers committees (Sachs Harbour HTC, Paulatuk HTC and Olokhaktomiut HTC), community corporations (Sachs Harbour Community Corporation, Paulatuk Community Corporation and Ulukhaktok Community Corporation), and Elders' committees (Sachs Harbour Elders Committee, Paulatuk Elders Committee and Ulukhaktok Elders Committee) were brought on-board the *Amundsen* for an exchange of knowledge. More information on these programs is available on the project website (CEOS, 2007).

3 Data and methods

On station, ocean vertical profiles of pressure, temperature, salinity and fluorescence were obtained with a CTD-rosette system through the ship's moonpool (Fig. 2). The rosette was equipped with a SeaBird 911+ CTD with a sampling rate of

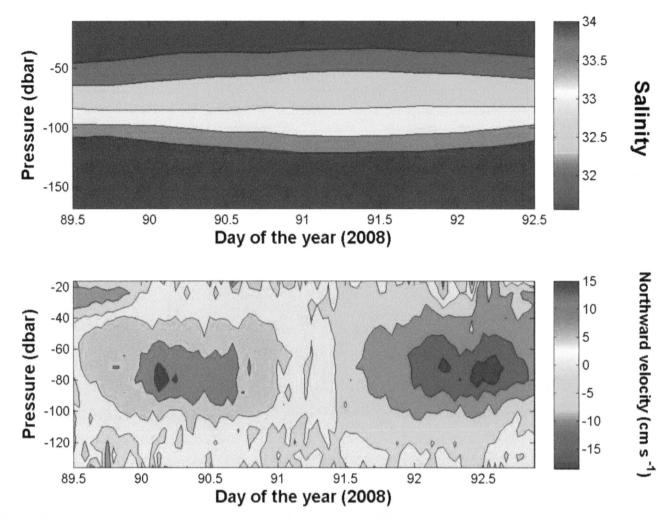

Fig. 4   Salinity (upper panel) and northward velocity (lower panel) across the March 2008 eddy near Banks Island (easternmost blue circle on Fig. 3). The velocity core is centred at 80 m and the velocity reverses at mid-core (day 91.5 or 31 March 2008).

24 Hz. The descent rate was 1 m s^{-1} and all data were averaged every decibar. The profiles typically range in depth from 10 m to the bottom. Data shallower than 10 m were contaminated by the ship and were discarded. A high resolution CTD profiler, the Self Contained Autonomous MicroProfiler (SCAMP; Stevens et al., 1999) provided a 1 mm resolution for the same temperature, salinity and fluorescence between the surface and 60–80 m, at a number of selected stations. SCAMP data are used to estimate the dissipation and the turbulent eddy diffusion in the surface mixed layer. A Moving Vessel Profiler (MVP) enabled the acquisition of vertical profiles of the same variables again while the ship was underway (Figs 3 and 5). The sensors, the same as on our CTD-rosette, are housed in a small "fish" that is allowed to free fall behind the ship. The MVP is used only in light ice or open water. The MVP horizontal resolution varies with the depth range being sampled and the ship's velocity, from 0.5 km at 4 kt to 10–15 km at 12 kt. Vertical profiles of horizontal velocities were measured with a ship-mounted 150 kHz Acoustic Doppler Current Profiler (ADCP). Three moorings were deployed in the CFL region in collaboration with ArcticNet, and their locations are shown on Fig. 3 (black squares). The mooring lines included current meters (Aanderaa RCM 11 and Nortek Aquadopp), ADCPs (RD Instrument ADCP and Nortek Continental) and various CTDs (SeaBird 37, RBR XR and Rockland's ACL). The two moorings between Cape Bathurst and Sachs Harbor included two McLane Moored profilers (MMP) that carried pressure, temperature, salinity and fluorescence sensors. They moved up and down the water column between depths of 40 and 140 m every two hours between October 2007 and July 2008.

The formation of the flaw lead results from both the formation and timing of fast ice edges and the motion of the mobile offshore pack ice (particularly the BG). Properties of sea-ice dynamics were investigated using weekly Special Sensor Microwave Imager (SSM/I) ice motion data from the National Snow and Ice Data Center (NSIDC), calculated from daily Advanced Very High Resolution Radiometer (AVHRR), Scanning Multichannel Microwave Radiometer (SMMR), SSM/I and International Arctic Buoy Program (IABP) data (Fowler, 2003). Mobility of the MY ice pack is characterized by ice motion anomalies relative to the

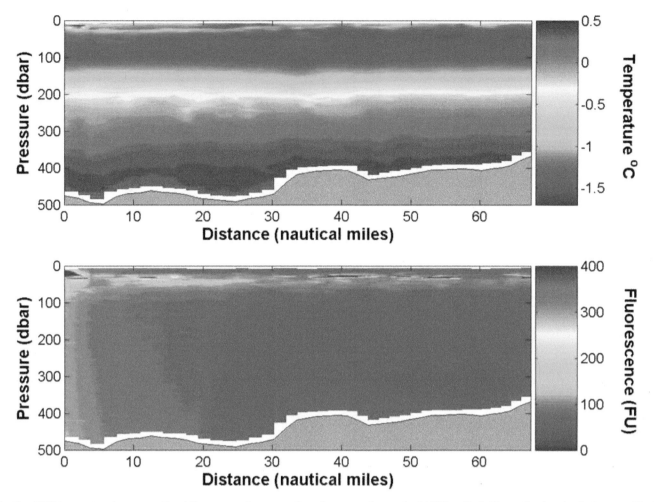

Fig. 5   MVP temperature (upper panel) and fluorescence (lower panel) section across the mouth of M'Clure Strait. The section location is shown on Fig. 3.
The section was visited on 5 July 2008.

1979–2008 climatological mean for September 2007, the month during which the record minimum in sea-ice extent in the Arctic occurred (Fig. 6a). Shaded contours depict the square root of the sum of the square of zonal and meridional ice motion anomalies (Fig. 6b). Sea-ice dynamics in the Beaufort Sea and thus the CFL study region are governed by circulation of the BG (Proshutinsky et al., 2002). Relative vorticity ($\omega = \partial v/\partial x - \partial u/\partial y$ for u (v) the zonal (meridional) component of ice motion) is used to monitor the orientation and magnitude of the BG, namely the cyclonic (anticyclonic) circulation associated with divergence (convergence) of the ice pack. Mean relative vorticity is computed from weekly sea-ice motion vectors with a nominal spacing of 25 km, using a finite-differencing scheme, as described in Lukovich and Barber (2006), and further details may be found therein. Year-week plots of relative vorticity spatially averaged over the CFL study region from 165°W to 125°W and 70°N to 85°N show circulation in the Beaufort Sea region (BSR) from fall (approximately week 36) in 1979 until fall (week 35) in 2008 (Fig. 6c). Twenty-two ice motion buoys (Global Positioning System position only) were also deployed during the CFL field program to provide information on smaller-

scale features of the BG circulation in the southern BSR and these data will be used in future analyses of sea-ice motion.

Sea-ice concentration by type in the CFL study area is examined using weekly digital ice chart data for the western Arctic region from the Canadian Ice Service (CIS) digital archive which were gridded (one pixel = 4 km²) (Fig. 7). For each digital ice chart, the fraction of total, old, first-year (a superset of thick first-year (>120 cm), medium first-year (70–120 cm) and thin first-year (30–70 cm), young (10–30 cm) and new ice (< 10 cm) in each pixel was multiplied by four to obtain the coverage (in square kilometres) of each ice type in each pixel. Coverage by type in each pixel ($n = 159685$) was summed and divided by the total study area (638,740 km²) yielding the fractional coverage of the study area by sea-ice type for each weekly chart. CIS digital data are based on manual interpretation of RADARSAT-1 (a primary data source since 1996), National Oceanic and Atmospheric Administration (NOAA)-AVHRR and Envisat Advanced Synthetic Aperture Radar (ASAR) and in situ observations, aerial and marine surveys (Fequet, 2002). The CIS Archive Documentation Series contains a substantial

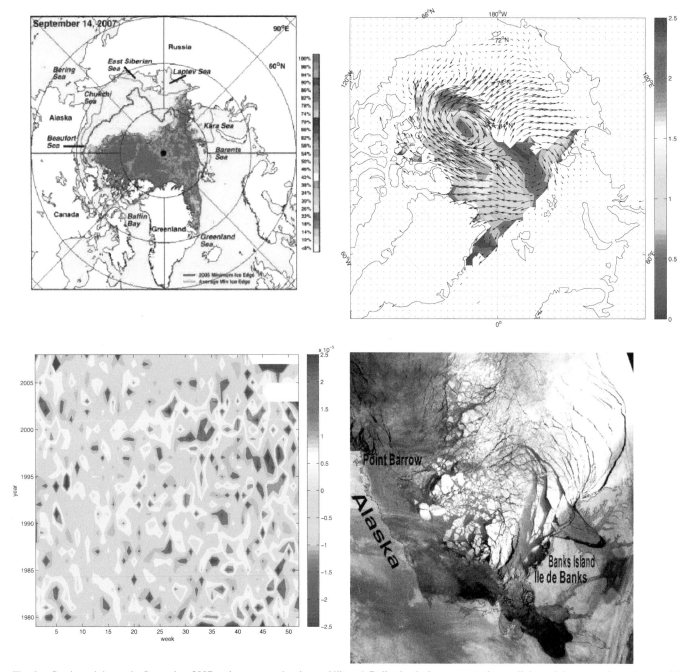

Fig. 6.   Sea-ice minimum in September 2007 and accompanying ice mobility. a) Daily Arctic ice concentration outlining minimum sea-ice extent on 14 September 2007 (reproduced from Fig. 2 in Comiso et al. (2008) by permission of the American Geophysical Union). b) Ice motion anomalies (cm s^{-1}) for September 2007. c) Year-week plot of ice relative vorticity anomalies for 1979–2008, spatially averaged from 70°N to 85°N. d) Implications of increased ice mobility for MY pack ice: NOAA image of the flaw lead to the west of Banks Island in January 2008 (downloaded 25 November 2008 from Environment Canada, Canadian Ice Service, http://www.ice.ec.gc.ca/app/WsvPageDsp.cfm?id=11892&Lang=eng; image:NOAA).

treatise on the observational and mapping accuracy of the CIS digital data, including quality indices for each region through time (Canadian Ice Service, 2006).

The climatology of the CFL study period is compared to the 1979–2006 climatology for the region, the time period for which satellite-based sea-ice data products are available. Seasonal anomalies for sea level pressure (SLP), surface vector winds and surface air temperature are calculated for the CFL study period by subtracting the 1979–2006 seasonal

mean values from the 2007–2008 seasonal mean values (Fig. 8). Daily and six-hourly gridded atmospheric data were retrieved from archives maintained by the NOAA-Cooperative Institute for Research in Environmental Sciences (CIRES) Climate Diagnostics Centers which originate from the National Centers for Environmental Prediction (NCEP) Reanalysis I Project (Kalnay et al., 1996). The spatial resolution of these data is a 2.5° × 2.5° (latitude and longitude) grid.

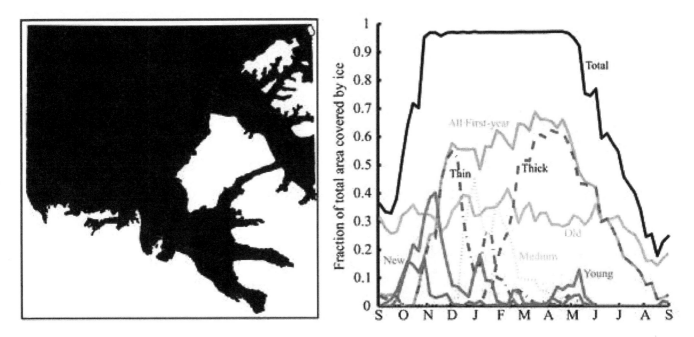

Fig. 7 (a) Map of the CFL study area and (b) the fraction of the total area (638,740 km^2, in black in (a)) that is covered by each sea-ice type through the annual cycle between September 2007 and September 2008.

An ensemble of meteorological data including SLP, air temperature, relative humidity, and wind speed and direction were collected with an AXYS Technology Inc. Automated Voluntary Observation System (AVOS), located on the roof of the ship's wheelhouse (approximately 18 m above the sea surface). The AVOS obtains atmospheric pressure data using a Vaisala PTB210 sensor, with 0.01 mb resolution and an accuracy of 0.15 mb, and is corrected to sea level by the AVOS software. Wind speed and direction were collected from an RM Young 05103 anemometer, with a directional accuracy of 0.3°, and magnitude accuracy within ±0.3 m s^{-1}. Temperature and relative humidity were measured with a Rotronics MP 101A temperature and relative humidity sensor, with a resolution of 0.1°C and an accuracy of ±0.3°C. Daily averaged SLP, air temperature and relative humidity are presented from 18 October 2007 to 04 August 2008. Daily wind vectors are represented by hourly averages at 00:00 UTC (Fig. 9).

Gridded six-hourly mean sea level pressure (MSLP) data were used to derive cyclone tracks and frequencies for the study region using the cyclone tracking algorithm of Serreze (1995; Fig. 10). The cyclone tracking exercise was conducted to detect cyclones that passed through the area bounded by 55°N–85°N and 160°E–105°W. The cyclone tracking exercise boundaries represent an area larger than the CFL study region, extending further west and north. This captures cyclones that may not necessarily pass directly over the CFL study area but may still influence the OSA interface within the CFL study region via winds, waves or coupling of heat and water vapour exchange (an example of this is presented in Fig. 11).

From December 2007 to July 2008, a Radiometrics TP/WVP 3000 microwave radiometer profiler (MWRP) measured atmospheric temperature and absolute humidity to a height of 10 km throughout the CFL study area. Temperature and humidity values were derived from microwave brightness temperatures using the manufacturer's neural network retrievals that had been trained using radiosonde measurements and a radiative transfer model (Solheim et al., 1998). Detailed information about this instrument is available in Solheim et al. (1998), Gulder and Spankuch, (2001), Ware et al. (2003) and Gafford et al. (2008). High temporal resolution (approximately 1 minute) atmospheric soundings were averaged to generate hourly profiles (Figs 11 and 12). The lower portions (surface to 2 km) of the 00:00 and 12:00 UTC temperature profiles were analyzed to characterize the thermodynamic structure of the atmospheric boundary layer (ABL) through the 2008 winter (January–March) and spring (April–June).

4 The physical system

Very little is known about the physical oceanography of the AG. The latest reviews (McLaughlin et al., 2006; Ingram et al., 2009) sketched a residual, cyclonic circulation in the AG. Kulikov et al. (1998) showed, however, that this situation may reverse at times. Peterson et al. (2008) used ice beacons, as well as RADARSAT Synthetic Aperture Radar (SAR) and ENVISAT ASAR data to estimate ice motion in late April and early May 2004. They found that ice motion was intermittent, often negligible, and that it could be explained by wind events. This means a weak surface residual circulation dominated by wind-driven events. It confirms the conclusions

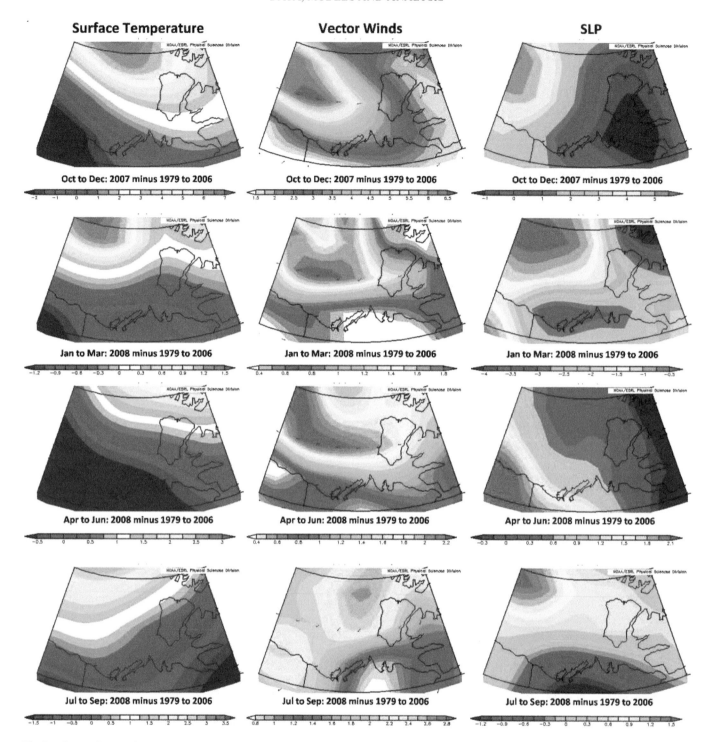

Fig. 8   Seasonal mean air temperature, vector wind speed and sea level pressure anomalies calculated for the study region for OND, JFM, AMJ and JAS. Anomalies for 1 September 2007 to 31 August 2008 are calculated against the climatology of 1 September 1979 to 31 August 2006 (images: NOAA/Earth System Research Laboratory (ESRL) Physical Sciences Division, Boulder Colorado; from   http://www.esrl.noaa.gov/psd/cgi-bin/data/composites/printpage.pl (26 November 2008)).

deduced from mooring data (Kulikov et al., 1998; Ingram et al., 2009; Lanos 2009) that the current variability is much larger than the mean flow. According to Lanos (2009), the residual circulation was anticyclonic in 2003–2004. The residual circulation is defined as the circulation remaining after a 25 h filter has been applied to the time series. Lanos

(2009) found that the surface water generally enters the AG near Banks Island and exits near Cape Bathurst (see the inset in Fig. 3); the circulation is reversed at depth. How and where the circulation loop is closed inside the AG is still a matter of debate. The data obtained during the CFL project, once the 1065 CTD profiles (over 1500 if the MVP profiles are

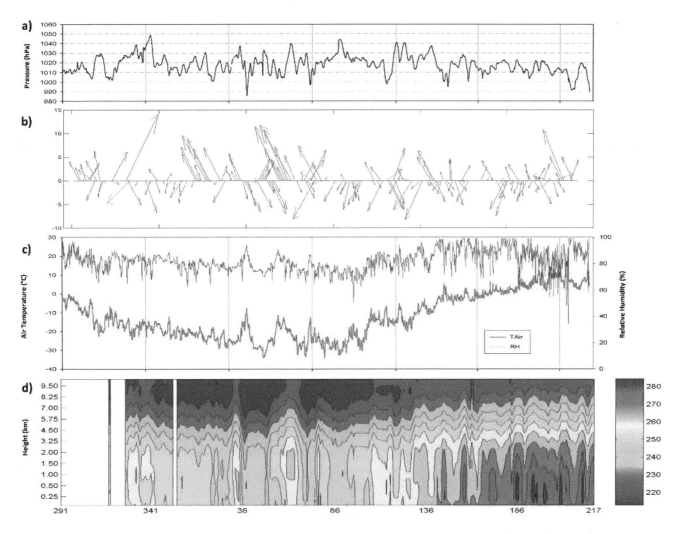

Fig. 9   Time series of a) sea level pressure, b) mean 00Z wind vectors, c) surface air temperature and relative humidity and d) mean daily upper level air temperatures from a Radiometrics MVP-3000A microwave profiling radiometer for the entire CFL study period (18 October 2007–08 August 2008).

included) have been processed, will help us answer these questions.

The AG is also the location of numerous wind-driven and tidally driven upwelling events. Williams and Carmack (2008) discuss the wind and tidally driven upwelling in the Cape Bathurst vicinity, and they show that these events bring nutrient-rich Pacific waters to the surface. These nutrients support large biological production in the water column and benthos. A major ice-edge upwelling was also observed in Franklin and Darnley bays in early June 2008. The 32.5 salinity isoline is displaced from its equilibrium depth of 42 m to the surface. It took seven days to return to its equilibrium depth (Mundy et al., 2009). The major difference between coastal upwelling and ice-edge upwelling is the fact that along-ice currents can be coupled with under-ice circulation and can bring nutrients to the ice bottom.

Lanos (2009) investigated the seasonal and interannual exchanges between the southern Beaufort Sea and the AG for the period 2002–2006 and found large interannual variability. He also introduced the concept of new, local water masses

created at the surface by brine rejection due to sea-ice formation. These water masses may be observed on the temperature-salinity diagrams obtained from moored instruments. Some of these freezing events were also recorded by the MMP profiles (not shown) between November 2007 and April 2008, when the flaw lead was open.

Early results show two interesting, first-time observations made during the CFL field study. The first one is the detection of two eddies in the AG (blue dots in Fig. 3) in the winter of 2008. The first one was observed between 24 and 31 January 2008 in the middle of the AG while the second one was observed between 29 March and 1 April 2008 east of Nelson Head (southern tip of Banks Island). Both eddies were centred at 80 m depth. The salinity structure and the northward velocity across the second eddy are shown on Fig. 4. Many eddies were observed in the Canada Basin over the years (Timmerman et al., 2008). The typical diameter of eddies observed on the Alaskan Shelf (Kadko et al., 2008) is 16 km. Since the *Amundsen* was virtually immobile during this period, it can be concluded that the eddy drift velocity

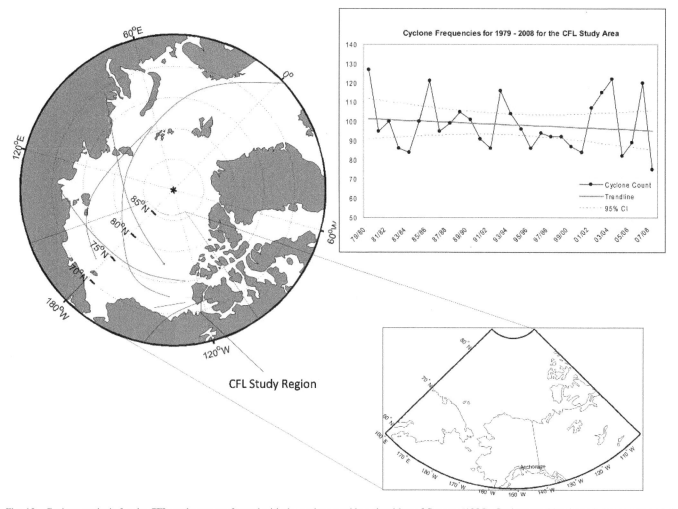

CFL Study Region

Fig. 10 Cyclone analysis for the CFL study area performed with the cyclone tracking algorithm of Serreze (1995). Cyclone tracking boundaries are expanded to capture variability in large-scale synoptic circulation. Typical summer (red) and winter (blue) cyclone paths of storms that affected the CFL study area. Cyclone frequencies calculated for 1 September 1979 to 31 August 2008. A non-significant ($p = 0.2328$) trend of -0.21 ± 5.42 (95% CI) cyclone counts per year is identified for the defined region.

was of the order of 3 km d^{-1} (16 km in 5 days). The lens shape structure can be seen in the salinity contours on Fig. 4 (upper panel). The velocity reversal from +15 cm s^{-1} (northward) to −15 cm s^{-1} (southward) is also clearly visible on the lower panel. Almost all eddies observed in the Beaufort Sea are anticyclonic (Timmermans et al., 2008) whether they are generated by baroclinc instabilities (Spall et al., 2008) or by brine rejection (Matsumura and Hasumi, 2008). Assuming an anti-cyclonic eddy (brine rejection process), the eddy drifted west-ward under the ship. Eddies are known to be generated on the Alaskan Shelf (Pickart, 2004). Many dynamic processes can generate these eddies. Some of these processes, mostly related to current instabilities, are discussed by McLaughlin et al. (2006). The March eddy was a mature eddy in the sense that the lens was well formed and isolated from the surface by waters of different properties. Another very good local eddy formation candidate is shallow convection (Marshall, 1999) during freeze-up. Released brine will sink to its equilibrium depth, approximately 80–120 m in our study region. The January eddy was believed to be still linked to the surface and

in the process of being formed. More work is clearly needed to identify and describe the generating mechanisms of these two eddies.

In June to July 2008, the *Amundsen* was able to reach the mouth of M'Clure Strait near 75°N because of the record low ice concentrations along the west coast of Banks Island. For the first time, water samples and CTD-rosette profiles were obtained at five stations (Fig. 1a) across the mouth of the strait. An MVP section (Fig. 5) was performed on the return trip. A coherent structure centred at 80 m was observed in mid-strait in the temperature data (Fig. 5, upper panel). This structure is also recognizable on the salinity section (not shown). The most interesting features are the fluorescence patches (Fig. 5, lower panel) between 20 and 40 m. These patches would have been impossible to identify with only the data from the five traditional CTD-rosette stations. The later rosette data will complete a long and unique section across M'Clure Strait, Barrow Strait and Lancaster Sound that will enable studies of the mixing of Atlantic and Pacific waters across the Canadian Archipelago.

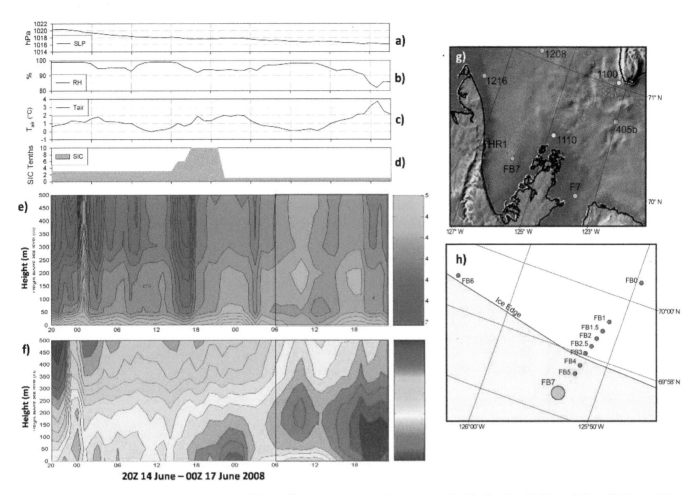

Fig. 11 Atmospheric coupling: measurements taken while the ship performed an ice-edge transect in Franklin Bay from 20:00 UTC 14 June–00:00 UTC 17 June 2008. a) sea level pressure b) relative humidity (10 m), c) air temperature (10 m), d) sea-ice concentration, e) ABL vapour densities from 0–500 m at 50 m resolution, f) ABL temperatures from 0–500 m at 50 m resolution, g) Franklin Bay area and stations and h) stations visited on transect between fast ice and open water. An approaching low pressure system enhances air-sea coupling from 06:00 UTC 16 June–00:00 UTC 17 June 2008 (black boxes in panels (e) and (f)).

From the perspective of sea ice, the CFL field project was preceded by the dramatic record summer minimum extent recorded in September 2007 (Comiso et al., 2008). This loss of seasonal and perennial ice cover in 2007 is a consequence of both dynamic and thermodynamic forcing. Investigation of dynamic forcing responsible for the record reduction in ice cover in 2007 showed that a persistent SLP high over the BSR during the preceding summer (June–August) resulted in anticyclonic circulation in ice motion and favoured increased advection from the Pacific to the Atlantic sector of the Arctic (Kwok, 2008). SLP anomalies over the BSR in the summer of 2007 were also shown to induce Ekman ice drift towards the central Arctic (or convergence due to anticyclonic circulation and concomitant upwelling), resulting in increased ice-edge retreat (Ogi et al., 2008). This influence of atmospheric forcing on ice dynamics is observed in ice motion anomalies in the BSR for September 2007 (Fig. 6b), which exhibit maxima in the Canada Basin characteristic of predominantly anticyclonic circulation (as depicted by ice motion vectors). Year-week plots of ice relative vorticity anomalies from 1979 to

2008 further highlight the effect of atmospheric forcing on sea ice. Enhanced anticyclonic ice activity induced by surface winds from the SLP high is characterized by negative sea-ice relative vorticity anomalies from week 40 in 2007 to approximately week 20 in 2008 (blue shading in Fig. 6c).

Investigation of thermodynamic forcing mechanisms responsible for the record reduction in ice cover in September 2007 demonstrates a significant increase in bottom ice melt due to an increase in open water fraction and thus solar heat absorption in the southern Beaufort Sea (Perovich et al., 2008). More recent studies attribute both the anomalous atmospheric circulation and bottom ice melt in the BSR to a dipole anomaly pattern in the northern hemisphere that drives ice from the western to the eastern Arctic and gives rise to anomalous oceanic heat flux to the Arctic Ocean via Bering Strait (Wang et al., 2009). Reinforcement of ice-albedo feedback mechanisms (namely an increase in solar heating of the upper ocean with increases in open water and heat advection from the Pacific) will give rise to continued thinning of sea ice in the western Arctic, the consequences of which are now

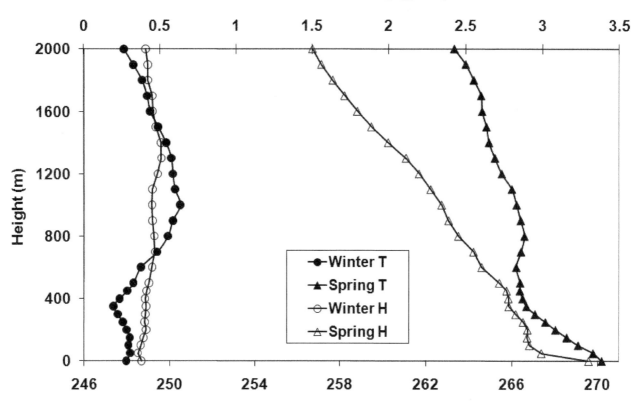

Fig. 12 Median (00:00 and 12:00 UTC) temperature (*T*) and absolute humidity (*H*) profiles for winter (January–March) and spring (April–June) 2008. The near surface (0–350 m) layer of the median temperature profile had a weakly stable lapse rate in winter and a near-neutral lapse rate in spring.

being observed. Indeed, the unprecedented formation of a flaw lead spanning approximately 100 km to the west of Banks Island in January 2008 (Fig. 6d) highlights the implications of a thinner and more mobile ice pack for ice cover in the western Arctic in particular, in addition to the potential for increased influence of atmospheric forcing on sea ice.

At the beginning of September 2007, much of the sea ice present in the study area was old ice, with very small amounts of thinner, seasonal sea ice types. By the third week of September, the region began to freeze up, signalled by rapid increases in the fraction of new and young ice (Fig. 7). On the first day of October in a given year, all first-year sea ice grown in the previous winter is reclassified as old (MY) ice, having survived a summer's melt. Near the end of October, the study region was almost completely ice covered, and by the first week of November much of the new and young ice present in October had thickened into thin first-year sea ice (Fig. 7). By the middle of December much of the thin first-year sea ice had thickened into medium first-year ice. The occurrence of thick first-year sea ice in the CFL study area began increasing rapidly at the beginning of January 2008 (Fig. 6d), constituting nearly all the seasonal sea ice in the region by the beginning of April (Fig. 7). Sea ice in the study area began to break up in the first week of May, declining to

a minimum of less than 20% coverage in the second week of August 2008 (Fig. 7).

While the evolution of the sea-ice cover described above is controlled by thermodynamic processes, episodes of sea-ice motion intermittently affect the icescape of the region. These processes are best illustrated by observing the variability of old sea ice over the annual cycle and perturbations caused in the fractions of the other stages of development (Fig. 7). It is obvious that lead formation in the CFL region occurs throughout the winter as evidenced by rapid and substantial increases in fractional coverage after initial fall freeze-up in the time series of new and young sea ice (Fig. 7) and by the fact that the *Amundsen* stayed mobile throughout the winter. Substantial lead formation occurred in the first week of January 2008, the middle of February and for much of April and May, all denoted by a spike in the young-ice fraction, and to a lesser degree, increases in new-ice fraction (Fig. 7). Each of these instances of lead formation coincided with reductions in the fraction of old ice and first-year sea ice in the CFL study area (Fig. 7).

From the perspective of the atmosphere, it is clear that momentum exchange between the atmosphere and ice surface accounts for a significant amount of variability in ice motion. The MSLP anomalies for October to December (OND) show an anomalously strong SLP gradient running northeast to southwest (Fig. 8). The strong pressure gradient corresponds with anomalously strong easterly winds over the Beaufort Sea. A dipole in surface temperature anomalies exists for OND, with negative (positive) temperature anomalies occurring over the AG (Beaufort Sea north of Alaska). This pattern likely reflects increased advection of cold, continental air from the Canadian Archipelago via strong easterly winds that were commonly observed during the CFL field program (Fig. 9). These winds likely arise from the large temperature gradient present over the region at this time. This gradient represents an area of strong baroclinicity, an important element in cyclogenesis. Furthermore, strong westward advection of sea ice was monitored using ice drift buoys and was likely generated by the strong easterly winds. This will be analyzed further for cyclogenesis and wind-forcing of sea-ice motion in future studies.

January to March (JFM) MSLP anomalies show a strong south–north gradient in pressure anomalies and correspond to anomalously strong northeasterly winds centred over the southern Beaufort Sea (Fig. 8). Negative anomalies are identified for surface temperatures and for JFM, with the strongest negative anomalies occurring northeast and north of Banks Island, respectively. The MSLP pattern for April to June (AMJ) is characterized by negative pressure anomalies over the coastal zone and positive anomalies over M'Clure Strait and Banks Island. This pattern is accompanied by strong easterly wind anomalies over the southern Beaufort Sea. AMJ surface air temperature anomalies follow an east–west gradient, with positive (weak negative) anomalies present over the coastal regions of the southern Beaufort Sea (Canadian Archipelago). July to September (JAS) MSLP anomalies are characterized by a southeast to northwest pressure gradient with strong northerly wind anomalies. Temperature anomalies for JAS reveal negative anomalies over the coastal region of the Northwest Territories and positive anomalies over most of the southern Beaufort Sea. Both anomalies are possibly caused by increased cloud cover arising from an increased area of open water occurring early in the season.

A summary of the meteorological conditions observed during the CFL study is presented in Fig. 9. The surface air temperature time series shows a typical seasonal cycle with surface air temperatures ranging from −33.5°C in the winter to as high as +21°C in the summer (recorded at a location close to shore). Surface temperatures had a downward trend from the start of the CFL study into early 2008. Surface air temperatures then increased slowly into the summer months and typically hovered around or above freezing. Vertical temperature profiles from the MWRP show a similar seasonal trend with cold temperatures present aloft during winter and a striking shift towards warmer temperatures in mid-April at all levels in the atmosphere. The winter temperature profiles are characterized by temperature inversions aloft, which are primarily attributed to the negative surface net radiation budget. Short periods of notably intense warm-air inversions are also identifiable at and above the 850 mb level (approximately

1.5 km) in the MWRP temperature profile throughout the study period and are attributed to warm-air advection arising from approaching cyclones (decreasing SLP and southerly winds). There is a notable seasonal shift from cold to relatively warmer temperatures in mid-April at all levels in the atmosphere.

Variability in SLP throughout the study period reflects the passage of cyclones and anticyclones. Wind vectors reveal strong southeasterly winds dominating the region from mid-November 2007 to late February 2008 and are attributable to early winter storms arising from an increasing temperature gradient between persisting areas of open water and declining surface temperatures over the pack ice and landmasses. The wind regime switched to a predominantly northerly regime as the sea ice approached its maximum ice extent in late March and the Beaufort high dominated the atmosphere. The wind pattern remains fairly stable until the summer, when wind direction becomes more variable (Fig. 9). The cyclones and anticyclones are further characterized by the corresponding day-to-day variability in the surface air and relative humidity data. Periods of cold and warm advection aloft associated with passing cyclones are identifiable in the MWRP atmospheric temperature profile throughout the study period (not shown). Two notable midwinter warm advection events are noted in early February and late March with temperatures reaching as high as −10°C and −15°C respectively and are immediately followed by cold air advection (not shown).

Cyclone-tracking analysis detected 75 separate cyclones that passed within the cyclone tracking boundaries from 1 September 2007 to 31 August 2008 (Fig. 10). This represents an annual cyclone count less than the average cyclone climatology for the region and is attributable to a seasonally strengthened Beaufort High. The cyclone climatology for the region for 1979–2008 reveals a non-significant ($p = 0.2328$) trend of −0.21 ±5.42 (95% confidence interval (CI)) cyclone counts per year for the defined area and is consistent with the findings of Simmonds and Keay (2009). The highest cyclone frequencies were observed in OND when a strong temperature gradient existed between the open ocean and the rapidly cooling ice and landmasses. This corresponds to the October cyclone maximum identified by Hudak and Young (2002) and also with the increased autumn Arctic cyclone activity identified by Zhang et al. (2004). Cyclone tracks were composited to create typical summer and winter storm tracks and correspond with those identified by other recent Arctic cyclone climatologies (Sorteberg and Walsh, 2008). Storm origins resembled those identified by Hudak and Young (2002), who classified 58% of southern Beaufort Sea storms as Arctic in origin, 27% as Pacific and 15% as irregular. The cyclone tracks identified also correspond well with recent studies of cyclone variability. Sorteberg and Walsh (2008) have shown that cyclones generated along the coast of Alaska travel into the east Siberian Sea via the Beaufort Sea during the winter, while during the summer, a convergence of cyclones is observed, in keeping with earlier studies by Serreze and Barry (1988) and Zhang et al. (2004).

We believe that cyclones may play a key role in the development of the sea-ice regime in the flaw lead. A cyclone can fracture the underlying sea ice and promote sea-ice divergence during and immediately following the passage of the cyclone (Brummer et al., 2003). This will increase coupling between the OSA system where increased upward energy and moisture fluxes from the ocean may lead to a more unstable atmospheric boundary layer and the intensification of the passing cyclone. This process would be most prevalent during the autumn and early winter seasons when strong atmospheric baroclinicity is produced by the rapidly strengthening temperature gradient between the ocean and atmosphere and when thin first-year ice can easily be fractured by divergence under cyclone forcing. Intensification of cyclones, driven by heat fluxes, may increase if sea-ice concentration and extent continue to decline. Simmonds and Keay (2009) suggest that the reduction in sea ice is leading to an increase in cyclone intensity rather than cyclone frequency, and this will be a topic of future analysis.

Our sampling during the CFL project allows us to investigate both spatial and temporal variability in OSA coupling and, in particular, the influence that this coupling has on the atmospheric boundary layer (ABL). During the CFL study, the *Amundsen* traversed a variety of sea-ice types and concentrations, and thus there were many opportunities to observe OSA coupling in areas of mixed sea-ice concentration. By way of illustration, we provide an example of OSA coupling when the *Amundsen* followed a transect from an area of nearly open water (3/10 sea-ice concentration) to an area of fast ice (10/10) and then back to open water (1/10) from 15 June to 17 June 2008 (Fig. 11). Examination of the transition from 10/10 back to 1/10 reveals a strong contrast in vertical temperatures up to approximately 150 m in altitude. This effect is also identifiable (to a lesser extent) in the vapour density profile for the same period. The influences of an approaching low-pressure system on OSA coupling are notable in temperature and vapour density profiles from 06:00 UTC 16 June to 00:00 UTC 17 June 2008.

An analysis of the ABL vertical profiles obtained during the CFL project provides a wealth of data on surface-atmosphere coupling. We present here only the winter and spring season summaries. As expected, the temporal variability was also high throughout the CFL study period and we are still working through much of the data. In summary, however, the average winter season temperature profiles at 00:00 and 12:00 UTC were analyzed together. The median temperature profile (Fig. 12) had a shallow surface-based inversion (150 m; lapse rate = 0.12 K 100 m^{-1}), below a weakly stable layer (150 to 350 m; lapse rate = –0.38 K 100 m^{-1}) which was topped by an elevated inversion (350 to 1000 m; lapse rate = 0.48 K 100 m^{-1}). The overall surface to 1000 m lapse rate was 0.26 K 100 m^{-1}; the surface temperature, T_{sfc}, was 248.0 K (–25.2°C), and ΔT to the top of the inversion was 2.6 K. For purposes of contrast, winter statistics for six station years (in the 1980s) of 00:00 and 12:00 UTC soundings from three Russian drifting ice islands near the North Pole (Serreze

et al., 1992) indicated that, over thick ice, inversions occurred 100% of the time (76% surface based and 24% elevated). The median inversion depth was 1200 m, and the median temperature difference across the inversion, ΔT, was 11.8 K.

For the spring season, the temperature profiles at 00:00 and 12:00 UTC were, once again, grouped together even though a weak diurnal cycle was apparent (Fig. 12). The median ABL temperature profile had a near-neutral surface layer (350 m; lapse rate = –1.00 K 100 m^{-1} with T_{sfc} = 270.2 K or –3.0°C) under a weakly stable layer from 350 to 600 m and an inversion from 600 to 800 m. Overall, the elevated stable layer (350–800 m) had a nearly isothermal lapse rate of –0.02 K 100 m^{-1}, and ΔT was –0.09K. Again, for comparison, the drifting ice island climatology for the spring season had an inversion 97% of the time (37% surface based and 60% elevated). The median inversion height was 643 m, and the median ΔT was 5.4K (Serreze et al., 1992).

While radiative cooling of the surface tends to induce a downward sensible heat flux which cools the ABL, creating and maintaining an inversion during the Arctic winter, the weak stability of the median winter ABL profile over the flaw lead, with its thin ice cover and open leads, suggested that upward sensible heat flux occurred locally which reduced the magnitude of the area-average downward heat flux. This interpretation is consistent with the modelling results of Andreas et al. (2002) and Lupkes et al. (2008) which indicated that leads strongly influence the area-average sensible heat flux during the Arctic night and thus the depth and temperature lapse rate of the ABL. During the transition to open water in the spring season, in spite of the melting of snow and ice and the cold surface temperature of the Arctic Ocean, which both tend to induce a downward sensible heat flux and an inversion, mechanical turbulence and the upward flux of sensible heat through the thin ice and from the open leads appears to have generated enough mixing for the median lapse rate of the surface layer of the ABL to exhibit neutral stability.

The median absolute humidity profiles, based on the 00:00 and 12:00 UTC soundings, for the winter (0.01 g m^{-3} per 100 m for ABL depth = 1000 m), and spring (–0.11g m^{-3} per 100 m for ABL depth = 800 m) were also determined (Fig. 12). The moisture lapse rates, weakly positive in winter and negative in spring, were consistent with thermal damping of the upward flux of water vapour due to the stable temperature profile in the winter and with the vertical mixing of neutrally buoyant surface air parcels in the spring.

5 Summary

We have presented an introduction to the CFL study and an overview of the ocean, sea-ice and atmospheric processes which are identified as important in the formation, maintenance and dissolution of the Cape Bathurst flaw lead. Data collected from the 2007–2008 IPY-CFL overwintering project is placed in the context of a 30-year climatology, and a select subset of physical processes observed during the period are described using preliminary data. The physical system is characterized by highly dynamic ice conditions, is prone to physical forcing from both the atmosphere and ocean and is an excellent region from which to gain multidisciplinary insights into the changing polar marine ecosystem.

The physical oceanography of the CFL study region results from the interplay of deep basin-shelf coupling, especially on the Mackenzie shelf, and the role the atmosphere plays in ocean circulation in the presence and absence of sea ice. Upwelling features, both topographically generated in the absence of ice and those associated with ice edges, appear to be very important processes that bring nutrient-rich waters from depth into the euphotic zone. In the presence of ice, remote atmospheric forcing from as far as Bering Strait can modify the local circulation and water masses. Storms in Bering Strait will generate Kelvin waves propagating eastward under the ice into the CFL region. The local residual circulation is weak in the absence of ice when local wind-driven events dominate the surface circulation. Early results have shown the presence (and persistence) of eddies that propagated into the AG and that are generated locally by brine rejection events. These eddies have the potential to affect both circulation and entrainment of nutrients in the bottom of the surface mixed layer. Furthermore, ice conditions were favourable for the collection of the first CTD and MVP profiles at the western entrance to M'Clure Strait.

Analysis of BG ice relative vorticity highlights the predominance of anticyclonic activity associated with a strengthened and persistent Beaufort High over the Canada Basin noted in previous studies (Kwok, 2008; Ogi et al., 2008). In particular, extrema in sea-ice motion anomalies relative to the 1979–2008 climatological mean are observed in the Canada Basin in September 2007, while ice relative vorticity anomalies demonstrate increased anticyclonic activity in MY ice in fall 2007.

The CFL project provided a first-hand opportunity to observe OSA coupling, particularly how heat fluxes from large areas of open water in the AG influenced the surrounding atmosphere. During OND, a strong northeast–southwest pressure gradient yielded strong easterly winds and westward advection of cold air from the Canadian Archipelago. It appeared that this may be a synoptic-scale response in the atmosphere towards areas of positive temperature anomalies, likely due to sensible and latent heat fluxes arising from areas of open water persisting into late November, but more analysis is required. This is further detailed in the meteorological data record from the CFL study period where OND is characterized by long durations of strong easterly surface winds.

Another interesting and complex example of OSA coupling was observed during an ice-edge transect during the period 15–17 June 2008 that also warrants further study.

When the atmosphere forces divergence of the sea ice, a dramatic increase in the fluxes of mass, momentum and energy are observed across the OSA interface. The median ABL temperature profile over the Cape Bathurst flaw lead during the winter season was stable but much less so than the climatological profiles for Russian ice island sites in the Arctic Basin. During the spring, the median lapse rate of the surface layer of the ABL exhibited neutral stability. This apparent influence of sensible heat flux, from open leads and through thin ice, on the stability of the ABL, and thus on OSA, should grow in importance as climate change reduces the relative proportion of MY ice in the BG and increases the spatial extent of the Cape Bathurst flaw lead.

The CFL study represents the first time that a research icebreaker has stayed mobile throughout an annual cycle anywhere in the northern hemisphere CFL system. The IPY provided us with an opportunity to conduct an internationally collaborative, large-scale, system level study of all aspects of the physical, biological and biogeochemical coupling within this complex system. The findings expected from this research will have direct relevance for the modelling community, industry, policymakers and northern residents; they also represent critical comprehensive knowledge that will provide important information to understand better how future climate change will affect the Arctic. This paper provides an overview of the study and some context for the physical processes (both state variables and climatology) of relevance to the more detailed analyses forthcoming from the IPY-CFL project.

Acknowledgements

The National Centers for Environmental Prediction (NCEP) provided the reanalysis data, and the National Snow and Ice Data Center (NSIDC) provided the passive microwave sea-ice concentration data. Thanks to B. Else for processing the wind data. Funding for the CFL project was provided by the Canadian International Polar Year (IPY) program office, the Natural Sciences and Engineering Research Council (NSERC), the Canada Research Chairs (CRC) Program, the Canada Foundation for Innovation (CFI), and numerous international partner organizations. Special thanks to the officers and crew of the CCGS *Amundsen*, whose dedication and excellence made the CFL project a unique milestone in polar marine science.

References

ANDREAS, E. L.; P. S. GUEST, P. O. G. PERSSON, C. W. CHRISTOPHER, T. W. HORST, R. E. MORITZ and S. R. SEMMER. 2002. Near-surface water vapor over polar sea ice is always near ice saturation. *J. Geophys. Res.* **107**: 8.1–8.15.

BRÚMMER, B.; G. MÚLLER and H. HOEBER. 2003. A Fram Strait cyclone: properties and impact on ice drift as measured by aircraft and buoys. *J. Geophys. Res.* **108**: (D7): 4217, doi: 10.1029/2002JD002638.

CANADIAN ICE SERVICE. 2006. *Canadian Ice Service Digital Archive–Regional charts: History, accuracy, and caveats.* Retrieved 16 December 2008 from http://ice.ec.gc.ca/IA_DOC/cisads_no_001_e.pdf.

CEOS (CENTRE FOR EARTH OBSERVATION SCIENCE). 2007. *Arctic Climate Change: Circumpolar Flaw Lead Study.* Retrieved from http://www.ipy-cfl.ca.

COMISO, J.C.; C.L. PARKINSON, R. GERSTEN and L. STOCK. 2008. Accelerated decline in the Arctic sea ice cover. *Geophys. Res. Lett.* : L01703, doi:10.1029/2007GL031972.

DARNIS, G.; D.G. BARBER and L. FORTIER. 2008. Sea ice and the onshore-offshore gradient in pre-winter zooplankton assemblages in the southeastern Beaufort Sea. *J. Mar. Syst.* (3-4): 994–1011.

DMITRENKO, I.; K. TYSHKO, S. KIRILLOV, J. HÖLEMANN, H. EICKEN and H. KASSENS. 2005. Impact of flaw polynyas on the hydrography of the Laptev Sea. *Global Planet. Change*, : 9–27, doi:10.1016/j.gloplacha. 2004.12.016.

FEQUET, D. (Ed.). 2002. *MANICE: Manual of standard procedures for observing and reporting ice conditions* (9th ed.) Ottawa, ON: Canadian Ice Service, Environment Canada.

FOWLER, C. 2003. *Polar Pathfinder daily 25 km EASE-grid sea ice motion vectors.* Natl. Snow and Ice Data Center, Retrieved 2 December 2008 from http://arcss.colorado.edu/data/nsidc-0116.html.

GAFFARD, C.; J. NASH, E. WALKER, T. J. HEWISON, J. JONES and E. G. NORTON. 2008. High time resolution boundary layer description using combined remote sensing instruments. *Ann. Geophys.* 26: 2597–2612.

GULDER, J. and D. SPANKUTH. 2001. Remote sensing of the thermodynamic state of the atmospheric boundary layer by ground-based microwave radiometry. *J. Atmos. Oceanic Technol.* 19: 925–933.

HUDAK, D.R. and J.M.C. YOUNG. 2002. Storm climatology of the southern Beaufort Sea. ATMOSPHERE-OCEAN, 40 (2): 145–158.

INGRAM, R. G.; W. J. WILLIAMS, B. VAN HARDENBERG, J. T. DAWE and E. C. CARMACK. 2008. Seasonal circulation over the Canadian Beaufort Shelf. In: D. Barber, L. Fortier, J. Michaud (Eds). *On Thin Ice: a synthesis of the Canadian Arctic Shelf Exchange Study (CASES)* (pp. 14–35). Winnipeg, MB: Aboriginal Issues Press.

IPCC (INTERGOVERNMENTAL PANEL ON CLIMATE CHANGE). 2007. *Climate Change 2007: The Physical Science Basis: Contribution of Working Group I to the Fourth Assessment.* S. Solomon, D. Quin, M. Manning, Z. Chen, M. Marquis, K.B. Averyt, M. Tignor and H.L. Miller (Eds). New York, NY: Cambridge University Press.

KADKO, D.; R.S. PICKART and J. MATHIS. 2008. Age characteristics of a shelf-break eddy in the western Arctic and implications for shelf-basin exchange. *J. Geophys. Res.* 113: C02018, 1–15, doi: 10/1029/2007JC004429.

KALNAY, E.; M. KANAMITSU, R. KISTLER, W. COLLINS, D. DEAVEN, L. GANDIN, M. IREDELL, S. SAHA, G. WHITE, J. WOOLLEN, Y. ZHU, M. CHELLIAH, W. EBISUZAKI, W. HIGGINS, J. JANOWIAK, K. C. MO, C. ROPELEWSKI, J. WANG, A. LEETMAA, R. REYNOLDS, R. JENNE and D. JOSEPH. 1996. The NCEP/NCAR 40-year reanalysis project. *Bull. Am. Meteorol. Soc.* : 431–437.

KULIKOV, E. A.; E. C. CARMACK and R. W. MACDONALD. 1998. Flow variability at the continental shelf of the Mackenzie Shelf in the Beaufort Sea. *J. Geophys. Res.* 103: 12725–12741.

KWOK, R. 2008. Summer sea ice motion from the 18 GHz channel of AMSR-E and the exchange of sea ice between the Pacific and Atlantic sectors. *Geophys. Res. Lett.* : L03504, doi:10.1029/2007GL032692.

LANOS, R. 2009. Circulation régionale, masses d'eau, cycles d'évolution et de transports entre la mer de Beaufort et le Golfe d'Amundsen. Ph.D. thesis, INRS-Eau, terre et environnnement, Québec, Quebec.

LIU, J.P.; J.A. CURRY and Y.Y. HU. 2004. Recernt Arctic sea ice variability: Connections to the Arctic Oscillation and the ENSO. *Geophys. Res. Lett.* 31 (9): L09211, doi: 10.1029/2004GL019858.

LUKOVICH, J. and D. BARBER. 2006. Atmospheric controls on sea ice motion in the southern Beaufort Sea. *J. Geophys. Res.* 111: doi:10.1029/2005JD006408.

LÜPKES, C.; T. VIHMA, G. BIRNBAUM and U. WACKER. 2008. Influence of leads in sea ice on the temperature of the atmospheric boundary layer during polar night. *Geophys. Res. Lett.* : L03805, doi: 10.1029/2007GL032461.

MCLAUGHLIN, F.A.; E.C. CARMACK, R. G. INGRAM, W. G. WILLIAMS and C. MICHEL. 2006. Oceanography of the Northwest Passage. In A.L. Robinson and K. Brink (Eds). *The Sea* (Vol. 14B pp. 1213–1244). New York, NY: Harvard University Press.

MARSHALL, J. 1999. Open-ocean convection: observations, theory, and models. *Rev. Geophys.* (1): 1–64.

MARTIN, S.; R. DRUCKER and M. FORT. 1995. A laboratory study of frost flower growth on the surface of young sea ice. *J. Geophys. Res.* 100(C4): 7027–7036.

MATSUMURA, Y. and H. HASUMI. 2008. Brine-driven eddies under sea ice leads and their impact on the Arctic Ocean mixed layer. *J. Phys. Oceanogr.* : 146–163.

MUNDY, C. J.; M. GOSSELIN, J. EHN, Y. GRATTON, A. ROSSNAGEL, D.G. BARBER, J. MARTIN, J.-É. TREMBLAY, M. PALMER, K. ARRIGO, G. DARNIS, L. FORTIER, B. ELSE and T. PAPAKYRIAKOU. 2009. Contribution of under-ice primary production to an ice-edge upwelling phytoplankton bloom in the Canadian Beaufort Sea. *Geophys. Res. Lett.* 36: L17601, doi:10.1029/2009GL038837.

OGI, M.; I.G. RIGOR, M.G. MCPHEE and J.M. WALLACE. 2008. Summer retreat of Arctic sea ice: Role of summer winds. *Geophys. Res. Lett.* : L24701, doi: 10.1029/2008GL035672.

PARKINSON, C.L. and D.J. CAVALIERI. 2008. Arctic sea ice variability and trends, 1979–2006. *J. Geophys. Res.* 113: C07003, doi:10.1029/2007JC004558.

PEROVICH, D.K.; J. RICHTER-MENGE, K. JONES and B. LIGHT. 2008. Sunlight, water, and ice: Extreme Arctic sea ice melt during the summer of 2007. *Geophys. Res. Lett.* : L11501, doi:10.1029/2008GL034007.

PETERSON, I.K.; S.J. PRINSENBERG and J.S. HOLLIDAY. 2008. Observations of sea ice thickness surface roughness and ice motion in Amundsen Gulf. *J. Geophys. Res.* 113: C06016, 1–24.

PICKART, R. S. 2004. Shelfbreak circulation in the Alaskan Beaufort Sea: Mean structure and variability. *J. Geophys. Res.* 109: doi: 10.1029/2003JC001912.

POLYAKOV, I.V.; A. BESZCZYNSKA, E.C. CARMACK, I.A. DMITRENKO, E. FAHRBACH, I.E. FROLOV, R. GERDES, E. HANSEN, J. HOLFORT, V.V. IVANOV, M.A. JOHNSON, M. KARCHER, F. KAUKER, J. MORISON, K.A. ORYIK, U. SCHAUER, H.L. SIMMONS, O. SKAGSETH, V.T. SOKOLOV, M. STEELE, L.A. TOMOKHOV, D. WALSH and J.E. WALSH. 2005. One more step toward a warmer Arctic. *Geophys. Res. Lett.* 32 (17): L17605, doi:10.1029/2005GL023740.

PROSHUTINSKY, A.; R.H. BOURKE and F.A. MCLAUGHLIN. 2002. The role of the Beaufort Gyre in Arctic climate variability: Seasonal to decadal timescales. *Geophys. Res. Lett.* 29(23): 2100, doi: 10.1029/2002GL015847.

ROTHROCK, D.A.; Y. YU and G.A. MAYKUT. 1999. Thinning of the Arctic sea-ice cover. *Geophys. Res. Lett.* 26(23): 3469–3472.

SERREZE, M.C. 1995. Climatological aspects of cyclone development and decay in the Arctic. ATMOSPHERE-OCEAN, (1): 1–23.

SERREZE, M.C. and R.G. BARRY. 1988. Synoptic activity in the Arctic basin. 1979–85. *J. Clim.* 1: 1276–1295.

SERREZE, M. C.; KAHL, J. D. and R.C. SCHNELL. 1992. Low-level temperature inversion of the Eurasian Arctic and comparisons with Soviet drifting station data. *J. Clim.* : 615–629.

SIMMONDS, I. and K. KEAY. 2009. Extraordinary September Arctic sea ice reductions and their relationships with storm behavior over 1979–2008. *Geophys. Res. Lett.* 36: L19715, doi:10.1029/2009GL039810.

SOLHEIM, F.; J. R. GODWIN, E. R. WESTWATER, Y. HAN, S. J. KEIHM, K. MARCH and R. WARE. 1998. Radiometric profiling of temperature, water vapor, and cloud liquid water using various inversion methods. *Radio Sci.* : 393–404.

SORTEBERG, A. and J.E. WALSH. 2008. Seasonal cyclone variability at 70N and its impact on moisture transport in the Arctic. *Tellus*, 60A: 570–586.

SPALL, M.A.; R.S. PICKART, P.S. FRATANTONI and A.J. PLUEDDEMANN. 2008. Western Arctic shelf break eddies: formation and transport. *J. Phys. Oceanogr.* : 1644–1668.

STEVENS, C.; M. SMITH and A. ROSS. 1999. SCAMP: measuring turbulence in estuaries, lakes and coastal waters. *Water Atmos.* (2): 20–21.

THOMPSON, D.W.J. and J.M. WALLACE. 1998. The Arctic Oscillation signature in the wintertime geopotential height and temperature fields. *Geophys. Res. Lett.* 25 (9): 1297–1300.

THOMPSON, D.W.J. and J.M. WALLACE. 2001. Regional climate impacts of the Northern Hemisphere Annular Mode. *Science*, 293 (5527): 85–89.

TIMMERMANS, M.-L.; J. TOOLE, A. PROSHUTINSKY, R. KRISHFIELD and A.

PLUEDDEMANN. 2008. Eddies in the Canada Basin, Arctic Ocean, observed ice-tethered profilers. *J. Phys. Oceangr.* : 133–145.

WANG, J.; J. ZHANG, E. WATANABE, M. IKEDA, K. MIZOBATA, J. E. WALSH, X. BAI and B. WU. 2009. Is the dipole anomaly a major driver to record lows in Arctic summer sea ice extent? *Geophys. Res. Lett.* : L05706, doi:10.1029/2008GL036706.

WARE, R.; R. CARPENDER, J. GULDNER, J. LILJEGREN, T. NEHRKORN, F. SOLHELM and F. VANDENBERGHE. 2003. A multichannel profiler of temperature, humidity, and cloud liquid. *Radio Sci.* : 44.1–44.13.

WILLIAMS, J.W. and E.J. CARMACK. 2008. Combined effect of wind-forcing and isobaths divergence on upwelling at Cape Bathurst, Beaufort Sea. *J. Mar. Res.* **66** (5): 645–663.

ZHANG, X.; J. E. WALSH, J. ZHANG, U. S. BHATT and M. IKEDA. 2004. Climatology and interannual variability of Arctic cyclone activity, 1948–2002. *J. Clim.* **17**: 2300–2317.

Rehabilitation and Analysis of Canadian Daily Precipitation Time Series

Eva Mekis and William D. Hogg

Atmospheric Environment Service, Climate Research Branch, Downsview, Ontario

ABSTRACT *The goal of this project was to develop adjustment procedures to use daily resolution data to generate high quality time series of precipitation and to perform regional trend analyses on the resulting datasets. A total of 69 locations, most with data covering the period 1900–96 were used. Data availability in much of the Canadian Arctic was restricted to 1948–96. By using daily data, improved corrections to precipitation data, not practical with monthly data, could be implemented. For each of three rain gauge types, corrections to account for wind undercatch and evaporation were implemented. Gauge specific wetting loss corrections were applied for each rainfall event. For snowfall, ruler measurements were used throughout the time series, to minimize potential discontinuities introduced by the adoption of Nipher shielded snow gauge measurements in the mid-1960s. Density corrections based upon coincident ruler and Nipher measurements were applied to all ruler measurements. Where necessary, records from neighbouring stations were joined employing a technique based on a simple ratio of observations. The adjustment procedures used remove systematic biases due to changes in the measurement program but do not account for inhomogeneities related to local site changes, etc. It is assumed that such local changes introduce random inhomogeneities which are smoothed by combining the results from numerous stations. Work to adjust data from about 500 stations and generate monthly grids or maps is well underway but preliminary trend results were examined for this project by grouping stations by region. Regional time series of normalized anomalies are computed as the arithmetic mean of stations within the region. Annual and seasonal graphs of national and regional time series are presented. The national time series shows an increase in precipitation of 1.7% of mean/ decade over 1948–95. The greatest increase is in the autumn. For the same period, Canada north of 55°N showed an increase of 2.3% of mean/decade, more than the south, but much less than the 4–5% of mean/decade suggested by Groisman and Easterling (1994). The ratio of liquid to solid precipitation for Canada has declined slightly over the 1948–95 period.*

RÉSUMÉ *Le but de ce projet était de développer des procédures d'ajustement afin de pouvoir utiliser les données à résolution journalière pour générer des séries temporelles de pluviométrie de haute qualité et pour exécuter des analyses régionales de tendances à partir des ensembles de données ainsi obtenus. Au total, 69 stations furent utilisées, la plupart ayant des données couvrant la période 1900–96. La disponibilité des données dans l'Arctique canadien était limitée à la période 1948–96. En utilisant les données journalières, on pouvait effectuer des corrections améliorées aux données de pluviométrie qui ne sont pas utilisables avec des données mensuelles. Pour chacun des trois types de pluviomètre, on a introduit des corrections pour compenser la sous-capture due au vent et à l'évaporation. Des corrections pour perte par mouillage propres à chaque pluviomètre furent appliquées pour chaque occurrence de pluie. Pour la neige, les mesures au moyen d'une règle furent utilisées pour toute la série temporelle, afin de minimiser les discontinuités de mesures qui auraient pu être introduites #par l'adoption du capteur de neige protégé de Nipher au milieu des années 1960. Des corrections de densité fondées sur les mesures obtenues simultanément au moyen de la règle et du capteur Nipher furent appliquées à chaque mesure par règle. Lorsque nécessaire, les dossiers de stations voisines furent combinés au moyen d'une technique basée sur le rapport simple entre les observations. Ces méthodes d'ajustement enlèvent les tendances systématiques dans le programme de mesure, mais ne tiennent pas compte des hétérogénéités dues aux déplacements locaux des sites, etc. On suppose que de tels changements locaux induisent des hétérogénéités aléatoires qui sont aplanies en combinant les résultats de plusieurs stations. La tâche d'ajuster les données d'environ 500 stations et de produire des grilles mensuelles ou des cartes est bien engagée, mais dans ce projet les premiers calculs de tendances ont été examinés en groupant les stations par région. Les séries temporelles régionales d'anomalies normalisées sont constituées de la moyenne arithmétique des stations dans la région. Des graphiques annuels et saisonniers des séries temporelles nationales et régionales sont présentés. La série temporelle nationale montre une augmentation de la précipitation égale à 1,7% de la moyenne par décade entre 1948 et 1995. L'augmentation maximum est en automne. Pour la même période, le Canada au nord de 55°N a subi une augmentation égale à 2,3% de la moyenne par décade, plus qu'au sud, mais beaucoup moins que le 4–5% de la moyenne par décade suggéré par Groisman et Easterling (1994). Le rapport entre la précipitation liquide et solide pour le Canada a diminué un peu au cours de la période 1948–95.*

1 Introduction

The identification of local, regional and global climate change has become the key issue in climatology. Precipitation is a key element of climate. It is of primary importance for society since changes of climate to wetter or drier conditions could have large social and economic consequences. Similar to other climate elements, precipitation time series often contain inhomogeneities from several sources. The availability of a long term, continuous and homogeneous time series for precipitation provides tremendous advantages for climate researchers. Previous efforts in Canada to rehabilitate climate time series have concentrated on the Historical Canadian Climate Database (HCCD) containing monthly climate data for ≈130 locations (Gullett et al., 1992). The present work draws from some aspects of the HCCD, but is fundamentally different from it. The primary goal of the study has been to use daily resolution rainfall and snowfall observations separately and to adjust for the known inhomogeneities (such as change of site location, change of observing procedure, instrument deficiencies, etc.). The observed increasing frequency of trace measurements in northern Canada can also be handled at a daily resolution. In addition, valuable analyses of extremes, variability, etc. are possible with daily data.

A further benefit to the approach used here is that the adjusted data are a much closer approximation to the true amount of precipitation participating in the hydrologic cycle. Throughout the climate data collection period in Canada, techniques for measuring precipitation have varied in their ability to estimate true amounts, but virtually all have generated underestimates. The adjustments described here not only remove the inconsistencies for purposes of generating a more homogeneous time series, they also greatly reduce, or ideally, eliminate bias within the full range of the time series. This makes these data much more appropriate for use in water balance studies, runoff analyses, etc.

Most methodologies used in climate change studies require continuous long time series. Missing value estimation can be extremely difficult for locations and parameters with high spatial variance and without valuable surrounding station information. In this approach, the mixed two parameter gamma distribution method was applied (Stern and Coe, 1984). Occasionally, station joining is required due to station relocation. Overlapping periods are used to minimize possible inhomogeneities.

Recent studies have attempted to develop a generalized way of adjusting precipitation data for North America, including Canada. Groisman and Easterling (1994) carried out climate change analyses for North America based on these "corrected" precipitation data. Since they used a generalized correction methodology (multiplication factor for a given year) and did not need specific metadata information for each station, the authors were able to work with a large number of stations. The focus of the present work is the precise adjustment, to the extent possible, of known inhomogeneities, which required an extended station history information search. This metadata requirement limited the number of stations which could be used, necessitating careful station selection for the study.

The selected station network was distributed through 11 regions as depicted in Fig. 1 (Environment Canada, 1992), representing the major climate characteristics across Canada. The resulting regional time series are consistent and suitable for use in various local and global climate change

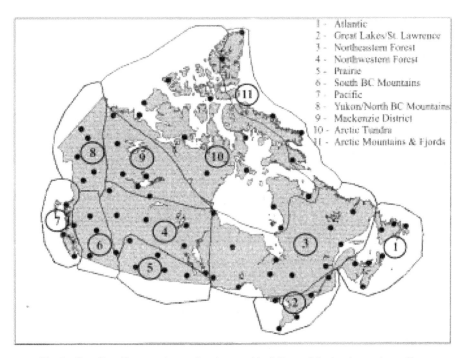

Fig. 1 Canadian climate regions and stations used in daily precipitation time series studies.

investigations, studies and analyses. When additional adjusted data are available, mapping or gridding of the results will be more desirable. Because Global Circulation Model (GCM) predictions magnify change in high latitudes, there is special interest in describing and understanding changes in precipitation in northern Canada (>55°N) (Bradley and England, 1978; Skinner and Maxwell, 1994; Walsh, 1995). The adjustment problem is especially complex in the North, where blowing snow and frequent ice crystal events play an important role in the final precipitation totals. A subset of all stations at latitudes higher than 55°N was identified and used for a preliminary analysis of northern Canada.

The depth of freshly fallen snow, measured by ruler, has been the standard climate measurement of snowfall since Canadian climate measurements began. These depth measurements have been converted to snow water equivalent by assuming a density of 100 kg m^{-3} (10:1 depth conversion) for fresh snow. The Nipher shielded snow gauge, introduced in the 1960s, provided a more accurate estimate of snow water equivalent (Metcalfe et al., 1994). However, the combination of ruler and Nipher gauge data would create a serious inhomogeneity in a time series. Ruler measurements have undergone fewer changes over time and are available at a much larger number of stations, but the superior accuracy of snow water equivalence measurements of the Nipher are extremely desirable. To maximize homogeneity in time, daily snow ruler measurements were used. To improve absolute accuracy of snow water equivalent estimates, and reduce the effect of known regional variability in the density of fresh snow, average densities calculated using Nipher measurements to determine snow water equivalent were used for this study.

The creation and development of a reliable, homogeneous, high resolution (in both time and space) precipitation dataset is an ongoing task. The completion of one step almost always identifies new possible problems. Much research remains to be done (homogeneity assessment, optimal station density, inclusion of surrounding station information in missing value estimation, use of more sophisticated methods for areal average computation, etc.). This report summarizes the data adjustment procedures completed so far and gives the results of trend analyses, based on the selected 11 climate regions and for Canada as a whole. Results are further separated according to annual and seasonal total precipitation and for rain and snow separately. Due to the special interest in the High Arctic, an analysis was also performed using the subset of stations at latitudes higher than 55°N.

2 Region and Station Selection

The selection of stations and regions was based on the existing HCCD (Gullett et al., 1992). Region boundaries separate climatologically and physiographically different characteristics of Canada. The major objectives of station selection were to select the longest and most reliable records possible, to minimize the number of station combinations and to cover all of Canada with a similar station distribution.

An attempt was made to create daily precipitation datasets for all stations covering all years available since 1895. Rain and snow (liquid and solid states of precipitation) were handled independently during the entire adjustment procedure. These precipitation data are to be used to examine trends in variability and extremes; ratio of liquid to solid precipitation and its relationship to temperature change; change of snow cover over time; and the relationship between trends in temperature and precipitation. The methodology is designed to be easy to update in future years.

The geographical distribution of the regions and the stations is shown in Fig. 1. Despite all efforts, the selected network is more dense in the south. Only a few stations are located in the Arctic and the majority of measurements there were not started until the 1940s. For this study 92 stations representing 69 locations were selected (23 stations are combined). Updating and expansion of this dataset is ongoing, but due to station closures and data delays, 1996 could not be included in the present study.

In addition to the daily rainfall gauge and snowfall ruler data for each location, paper station history files were searched thoroughly for the following metadata information:

- date of any relocation
- installation date of all precipitation gauges (date$n$)
- introduction date of 6-hourly measurement program
- introduction date of hourly weather-type measurement program.

For selected synoptic stations the following additional data were extracted from the National Climate Data Archive (NCDA) for the longest available time interval:

- daily total precipitation
- 6-hourly total precipitation (rainfall and/or snowfall water equivalent)
- hourly weather-type data (rain, rain showers, drizzle, freezing rain, freezing drizzle, ice crystals, snow, snow grains, etc.).

Further information (wind speed at the height of the gauge orifice, height of gauge orifice above ground, roughness length, height of the wind speed measuring instrument above ground, wind speed measured by instrument, average vertical angle of obstacles around the gauge, etc.) was required for an additional set of stations which were used for the determination of density of fresh snowfall. Table 1 contains an annotated list of the stations.

3 Developing Homogeneous Precipitation Time Series

This section reviews the steps which were implemented to make Canadian precipitation time series more homogeneous. As is the case elsewhere, Canadian datasets have inhomogeneities associated with change in measurement techniques, station relocation, gauge wetting losses, amounts too small to

TABLE 1.   List of stations

Station Name	Station Id.	From-To	Joined	Latitude	Longitude	Elev [m]	Region – Period	
Charlottetown	8300400	1910–1991		46.28	−63.13	48		
Chatham	8101000	1895–1996	1943	47.02	−65.45	34		
Deer Lake	8401500	1933–1996		49.22	−57.40	22		
Gander	8401700	1937–1996		48.95	−54.57	151	1	1895–1996
St. John's	8403506	1895–1996	1942	47.62	−52.73	132		
Sydney	8205700	1895–1996	1941	46.17	−60.05	56		
Yarmouth	8206500	1895–1996	1941	43.83	−66.08	43		
Gore Bay	6092925	1916–1993	1948	45.88	−82.57	193		
Harrow	6133360	1917–1991		42.03	−82.90	191		
Ottawa	6105976	1895–1996		45.32	−75.67	116	2	1895–1996
Quebec	7016294	1895–1995	1952	46.80	−71.38	73		
Welland	6139445	1895–1996		43.00	−79.27	177		
Amos	7090120	1913–1996		48.57	−78.13	310		
Bagotville	7060400	1895–1996	1944	48.33	−71.00	159		
Big Trout Lake	6010738	1939–1992		53.83	−89.87	220		
Churchill	5060600	1932–1996	1943	58.75	−94.07	35		
Goose	8501900	1942–1996		53.32	−60.42	46		
Kapuskasing	6073960	1918–1996		49.42	−82.47	227	3	1918–1994
Kenora	6034075	1895–1996	1939	49.78	−94.37	407		
Moosonee	6075425	1895–1993	1933	51.27	−80.65	10		
Sept–Iles	7047910	1944–1996		50.22	−66.27	55		
Thunder Bay	6048261	1909–1993	1942	48.37	−89.32	199		
Wabush Lake	8504175	1943–1996	1961	52.93	−66.87	551		
Fort McMurray	3062693	1944–1995		56.65	−111.22	369		
Fort Nelson	1192940	1938–1996		58.83	−122.58	382		
Great Falls	5031200	1923–1996		50.47	−96.00	249		
Lynn Lake	5061646	1949–1996	1969	56.87	−101.08	357	4	1938–1994
Fairview	3072520	1931–1991		56.07	−118.38	670		
The Pas	5052880	1911–1996	1943	53.97	−101.10	271		
Brandon	5010485	1895–1996		49.92	−99.95	409		
Edmonton	3012208	1895–1995	1938	53.30	−113.58	715		
Medicine Hat	3034480	1895–1996		50.02	−110.72	717	5	1895–1995
Regina	4016560	1898–1995		50.43	−104.67	577		
Saskatoon	4057120	1895–1992		52.17	−106.68	501		
Banff	3050520	1895–1994		51.18	−115.57	1397		
Barkerville	1090660	1895–1995		53.07	−121.52	1265		
Fort St. James	1092970	1895–1995		54.45	−124.25	686	6	1895–1995
Princeton	1126510	1895–1995	1939	49.47	−120.52	700		
Bella Coola	1060840	1895–1995		52.38	−126.58	35		
Quatsino	1036570	1895–1995		50.53	−127.62	2	7	1895–1995
Prince Rupert	1066481	1908–1995	1963	54.30	−130.43	34		
Victoria	1018620	1898–1996	1941	48.65	−123.43	20		
Dawson	2100402	1901–1995	1977	64.05	−139.13	369		
Dease Lake	1192340	1945–1995		58.42	−130.00	816		
Mayo A	2100700	1925–1996		63.62	−135.87	504	8	1939–1995
Watson Lake	2101200	1939–1995		60.12	−128.82	690		
Whitehorse	2101300	1942–1995		60.72	−135.07	703		
Fort Simpson	2202101	1897–1996	1964	61.75	−121.23	168		
Hay River	2202400	1909–1996	1944	60.83	−115.78	166		
Inuvik	2202570	1927–1995	1958	68.30	−133.48	68		
Norman Wells	2202800	1904–1996	1944	65.28	−126.80	67	9	1909–1996
Fort Smith	2202200	1946–1996		60.02	−111.95	203		
Yellowknife	2204100	1948–1996		62.47	−114.45	205		
Baker Lake	2300500	1943–1996		64.30	−96.08	18		
Cambridge Bay	2500600	1942–1996		69.10	−105.12	27		
Coral Harbour	2301000	1949–1996		64.20	−83.37	64		
Hall Beach	2402350	1929–1995		68.78	−81.25	8		
Inukjuak	7103282	1945–1996		58.45	−78.12	3		
Kuujjuaq	7113534	1957–1996		58.10	−68.42	34	10	1945–1996
Kuujjuarapik	7103536	1922–1993		55.28	−77.77	21		
Mould Bay	2502700	1947–1996		76.23	−119.33	15		
Resolute	2403500	1926–1996		74.72	−94.98	67		
Sachs Harbour	2503650	1948–1996		71.98	−125.28	86		
Cape Dyer	2400654	1946–1996		66.58	−61.62	393		
Alert	2400300	1948–1996		82.50	−62.33	63		
Clyde	2400800	1955–1995		70.48	−68.52	25		
Eureka	2401200	1959–1992		79.98	−85.93	10	11	1955–1995
Iqaluit	2402590	1950–1996		63.75	−68.53	34		
Cape Parry	2200675	1957–1992		70.17	−124.68	17		

25

be quantified (trace values) and errors associated with wind. Missing data had to be dealt with in a way which preserved the major characteristics (the distribution, mean, extremes, etc.) of the original data. Techniques to solve all of these problems had to be developed and implemented. Station history files were searched and electronic metadata files created for each station to aid in this task.

Earlier gauge adjustment methodologies employed systematic error correction on monthly and/or annual time series (Sevruk, 1982; Legates, 1987; Legates and Willmott, 1990; Groisman and Legates, 1994). There are specific problems in the Canadian data, which would be difficult (if not impossible) to correct on a monthly level (e.g., wetting loss and trace measurements). The methodology for correction of systematic biases can be improved when performed on daily rain gauge and snow ruler data. Part of the daily rain and snow adjustment methodology was based on procedures developed for 6-hourly synoptic station data by Metcalfe et al. (1994). Since their methodology was designed for a different time-step and purpose, several modifications had to be implemented.

a Adjustments Associated with Daily Rain Gauge Measurements

The daily rain gauge data contain inhomogeneities associated with instrument changes, (wind effects, wetting losses during pour out, evaporation losses, etc.) and trace measurements.

Canadian rain measurement methodologies have been modified several times over the years. The official rain gauge in Canada is currently the Type-B, which was introduced at most locations during the 1970s. Prior to this the Meteorological Service of Canada (MSC) gauge was used. The MSC gauge was originally manufactured totally from copper but the inside container was modified to a soft plastic material with different wetting characteristics around 1965. Metcalfe et al. (1997) give a more complete description of these gauges.

Different gauges behave differently due to wind effects, wetting loss, etc. Both MSC and Type-B gauges are mounted relatively low to the ground to reduce the undercatch due to wind. Ongoing intercomparison of Environment Canada rain gauges confirms that systematic differences between the MSC gauge and a pit gauge (gauge with orifice at ground level – wind speed is at a minimum) is about 4% (Goodison and Louie, 1986). The systematic difference between the Type-B gauge and a pit gauge is ≈2% at an open, windy site.

The wetting loss has two basic components: the water subject to evaporation from the surface of the funnel and the inner walls of the precipitation gauge after a precipitation event before emptying, and the water retained on the walls of the gauge and on the funnel after emptying (Metcalfe et al., 1997; Routledge, 1997). The MSC copper and plastic inserts have different wetting loss characteristics, as is also the case for the all plastic Type-B gauge, with its different thermal characteristics and direct-reading design.

The necessary adjustments associated with instrument changes for daily rain gauge measurements can be summarized as:

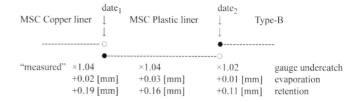

b Adjustments Associated with Trace Measurements

The trace adjustment is very important in Canada especially in the northern part of the country, where precipitation amounts are relatively low and many trace events are recorded. In these conditions, the sum of all trace amounts becomes a significant fraction of the total precipitation. The definition of trace as given by the Manual of Surface Weather Observations (MANOBS) (Environment Canada, 1996): "The vertical depth of water or water equivalent is normally expressed to the nearest 0.2 mm. Precipitation amounts up to 0.2 mm are exceedingly difficult to measure. Less than 0.2 mm is called a Trace."

Trace amounts are identified by a "flag" in the archive data files but are assigned an absolute value of zero when monthly precipitation totals are calculated. This leads to an underestimation of monthly and annual precipitation totals. The present study makes an attempt to estimate the actual amount of daily rain and snow accumulated in trace events, for each station, in a systematic and consistent way.

The problem of trace adjustment has not been addressed outside Canada, since trace is rarely recorded in other countries and has relatively little significance except in arid climates such as found in the Canadian Arctic. In the Arctic, the combination of very low annual precipitation amounts; extremely low evaporation throughout the long dark winter; plus frequent occurrences of trace observations, makes the accumulation of these minuscule amounts significant to the annual total precipitation and the water balance of the region. Metcalfe et al. (1994) report that trace amounts can account for 80% of precipitation occurrences at some locations in northern Canada. Because of the long winters in the Arctic and the high frequency of trace observations during frozen precipitation events, it is important to adjust for traces observed during snowfall or ice crystal events. For completeness, even though rainfall amounts are small in the Arctic, adjustments for rainfall trace were also implemented.

A few local experiments have been performed in Canada to assign a definite amount to trace events for both rain and snow. For rainfall, Woo and Steer (1979) estimated the trace adjustment amount for 6-hourly values as 0.03 mm. Although a very thorough experiment, it was performed for only one location (Resolute) and for only 5 trace events in the same year and was considered inadequate to provide the basis for rainfall trace adjustment across the country. Instead, all amounts for the interval 0.0–0.2 mm were considered equally probable for events classified as trace and the interval average of 0.1 mm was considered representative of a trace measurement amount. Since rain trace is also exposed to wetting, evaporation and retention losses prior to measurement, use of

0.1 mm for the rain only trace adjustment should be conservatively low (for details see Routledge, 1997). As well, rain trace events in the North have relatively little effect on the final total precipitation accumulation, so approximation procedures were considered adequate.

The practice of measuring precipitation every 6 hours at synoptic stations but archiving daily total precipitation may lead to a further underestimation. There could be as many as four observations of trace contributing to a single archived daily trace. Adjustments could not be made exclusively on the synoptic archive data because they have only been available since the 1950s and don't separately record rain and snow. Since the present adjustment is performed on daily data, it is important to determine the average number of 6-hourly trace measurements included in one daily trace flag. In order to be able to properly adjust for trace at different locations, the original 6-hourly archive information had to be classified based on the type of precipitation occurring when trace was recorded. For the selected 29 northern stations (most of them above 55°N) three major classes of trace were determined – rain, ice crystal and snow. The actual classification of trace associated events (the determination of the precipitation type) was performed through the comparison of the hourly archive weather-type information and the 6-hourly total precipitation measurements. Assuming distinct classes (every trace belongs to one and only one of the three classes) and based on all information, the count of different types of trace events can be computed for each year. The summary results can be found in Table 2.

For relating the daily and 6-hourly trace flag observations the Trace Occurrence Ratio (T_{or}) was computed for each station, each month and year for the period of overlapping synoptic and daily archived data:

$$\text{Trace Occurrence Ratio } (T_{or}) = \frac{\text{\# of trace flags in 6-hourly archive}}{\text{\# of trace flags in daily (rain and snow) archive}}.$$

The T_{or} values include all 6-hourly trace counts, even for those cases when measurable precipitation and trace events happened on the same day (if precipitation and trace occur on the same day, the daily observation file will only contain the precipitation amount without the effect of trace). Thus T_{or} inherently includes an allowance for adjustment of trace on days with measured precipitation in other synoptic periods. Since the computed mean of the long-term T_{or} for the full period is 3 for each station over the examined 1961–93 period with relatively little variance (see Table 2), this factor was accepted for use at each station.

The calculated T_{or} value is dependent upon the number of measurements in a day. For stations with a synoptic program (measurements every 6 hours), $T_{or,6}$ was calculated as described above. The other typical observation program is that of the ordinary climate station, where two daily measurements are taken. The resulting trace occurrence ratio was assumed to be proportional to the number of measurements taken. For ordinary climate stations the $T_{or,12}$ value was half (i.e., $T_{or,12} = 1.5$) of the factor applied at synoptic stations.

TABLE 2.   Precipitation trace correction results

Name	Period	Lat	Long	All Tr	Rain	ICR	Snow	ICR%	trace$_1$	trace$_2$	TOR$_{26}$
Big Trout Lake	1961–91	53.8	−89.9	12034	2437	3975	5622	41.4	0.1	0.07	3.14
Kuujjuarapik	1961–93	55.3	−77.8	12765	2957	3850	5958	39.3	0.1	0.07	3.32
Lynn Lake	1961–93	56.9	−101.1	10281	2224	2369	5688	29.4	0.1	0.07	2.83
Kuujjuaq	1961–93	58.1	−68.4	13822	2672	3615	7535	32.4	0.1	0.07	3.28
Inukjuak	1961–92	58.5	−78.1	14055	2335	4833	6887	41.2	0.1	0.07	3.40
Churchill	1961–93	58.8	−94.1	18860	2911	9683	6266	60.7	0.1	0.04	3.40
Fort Smith	1961–93	60.0	−112.0	10405	2083	2283	6039	27.4	0.1	0.07	2.94
Watson Lake	1961–92	60.1	−128.8	9135	1989	850	6296	11.9	0.1	0.07	3.06
Whitehorse	1961–93	60.7	−135.1	9777	2323	1288	6166	17.3	0.1	0.07	2.88
Hay River	1961–93	60.8	−115.8	7892	1451	1324	5117	20.6	0.1	0.07	2.78
Fort Simpson	1964–93	61.8	−121.2	8000	1551	1346	5103	20.9	0.1	0.07	2.79
Yellowknife	**1961–93**	**62.5**	**−114.5**	**10176**	**1822**	**1854**	**6500**	**22.2**	**0.1**	**0.07**	**2.93**
Mayo	1961–93	63.6	−135.9	7171	1487	499	5185	8.8	0.1	0.07	2.61
Iqaluit	1961–93	63.8	−68.5	152964	2187	5244	7865	40.0	0.1	0.07	3.31
Dawson	1962–86	64.1	−139.1	5778	1433	358	3987	8.2	0.1	0.07	2.62
Coral Harbour	1961–92	64.2	−83.4	12423	1815	5254	5354	49.5	0.1	0.06	2.89
Baker Lake	1961–92	64.3	−96.1	14262	1909	6880	5473	55.7	0.1	0.05	2.95
Norman Wells	1961–93	65.3	−126.8	12881	1849	5366	5666	48.6	0.1	0.06	3.42
Cape Dyer	1961–92	66.6	−61.6	12205	1070	4648	6487	41.7	0.1	0.07	3.19
Inuvik	1961–93	68.3	−133.5	13432	1943	4952	6537	43.1	0.1	0.07	3.54
Hall Beach	1961–92	68.8	−81.3	18665	1921	9416	7328	56.2	0.1	0.05	3.23
Cambridge Bay	1961–93	69.1	−105.1	22873	2403	13828	6642	67.6	0.1	0.03	3.35
Cape Parry	1961–92	70.2	−124.7	19659	2593	9579	7487	56.1	0.1	0.05	3.40
Resolute	**1961–92**	**74.7**	**−95.0**	**21789**	**1851**	**12644**	**7294**	**63.4**	**0.1**	**0.04**	**3.28**
Mould Bay	1961–92	76.2	−119.3	21979	1456	11476	9047	55.9	0.1	0.05	3.35
Eureka	1961–92	80.0	−85.9	17492	756	11263	5473	67.3	0.1	0.03	3.26
Alert	1961–90	82.5	−62.3	13845	472	6696	6677	50.1	0.1	0.06	3.32

Rain	# of Rain trace events	trace$_1$	Rainfall trace adjustment factor
ICR	# of Ice Crystal trace events	trace$_2$	Snowfall trace adjustment factor
Snow	# of Snow trace events	$T_{OR,6}$	Trace occurrence ratio, measurements taken 6-hourly
ICR%	Ice Crystal/solid precipitation trace ratio		

The adjustments associated with trace observations for daily rain gauge measurements can be summarized:

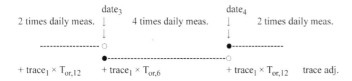

c Adjustments for Daily Snow Ruler Measurements

The depth of freshly fallen snow, measured by ruler, has been the standard climate measurement of snowfall since Canadian climate measurements began. For all stations, prior to the 1960s, and for non-synoptic stations over the entire record, precipitation amount (water equivalent) for snowfall events is determined by assuming a density for fresh snow of 100 kg m^{-3}. At synoptic stations, a Nipher shielded snow gauge was introduced in the 1960s to directly measure snow water equivalent for determination of precipitation amount, but depth of snowfall measurements with the ruler were continued.

The adjustment of ruler measurements to be homogeneous with Nipher gauge data would raise a number of difficulties, including that of the snow density in the past. The limiting factor for the use of Nipher shielded snow gauge data is its restricted availability both in time (Nipher gauges were first installed in the 1960s) and space (there are almost 10 times more ruler measurement stations). As well, inherent to the Nipher measurements are all the problems associated with gauge undercatch due to wind, wetting, etc. Since the process of ruler measurement has undergone fewer changes over time, daily snow ruler data were used for the full period. For computation of snow water equivalent the use of the 100 kg m^{-3} standard density of fresh snow was rejected. Instead, a new fresh snow density adjustment (ρ_{new}) is calculated separately for each station. The adjustment method follows Metcalfe et al. (1994) and is based on the ratio of adjusted Nipher gauge measurements to snowfall ruler depth measurements for events with snow water equivalent ≥ trace during the period of record when snow ruler and gauge measurements were made coincidentally. This procedure attempted to account for gauge losses due to wind, wetting, etc. By applying density adjustments determined in this way, the ruler measurements are adjusted as closely as possible to the true, long-term average, water equivalent of snowfall. Using data from 63 stations and spatial interpolation (kriging), a map of the values of ρ_{new} was produced (Fig. 2) and used to obtain estimates of the average density of new snowfall for each station. This density was applied to all ruler measurements to generate time series of snow water equivalent at each location.

Accumulation of solid precipitation trace measurements is an even more complex problem than that for rainfall trace. No procedure has yet been developed to allow for trace in the ruler measurement but significant work has been done on trace measurements with the Nipher shielded snow gauge. Therefore, procedures for adjustment of trace measurement for the Nipher gauge (Metcalfe et al., 1994) were adapted for use with ruler data. Metcalfe et al. (1994) recommend assignment of 0.07 mm water equivalent for trace of solid precipitation throughout southern Canada. Due to the measurement methodology, snow ruler trace measurements are not exposed to wetting and retention losses, and evaporation losses are minimized by the season. A value less than the rainfall trace allowance (0.1 mm) is quite consistent and 0.07 mm for snowfall trace was deemed appropriate in southern Canada. However, north of 55°N, the value assigned to solid (snow and ice crystal) trace is both less appropriate and more significant. In Canada, observation of ice crystals are considered precipitation. Ice crystals are recorded at very low temperatures (usually below –2°C) and usually contribute extremely little water equivalent to the precipitation totals. Each observation of ice crystals is recorded as a trace of precipitation. But 0.07 mm/trace is too high for a typical ice crystal event. An additional complicating factor is that the number of ice crystal events increases as latitude increases while annual precipitation decreases. Thus, the appropriate value for the trace adjustment varies with climate and location. As well, some locations in the High Arctic have shown dramatic increases in the number of trace observations over the past few decades. It is not known whether this is an artifact of changing observational procedures, a real regional phenomenon or localized in the immediate vicinity of the observation.

The water equivalent of ice crystal events is much less than the water equivalent of snowfall trace events. If the increase in number of trace observations in the Arctic is due to an increase in the number of ice crystal events, then it is very important to properly adjust for the trace observations due

Fig. 2   New fresh snow density adjustments for events > trace.

to ice crystals to avoid introducing an artificial trend in Arctic precipitation. Data differentiating between ice crystal traces and snowfall traces are not available for the entire time series so average adjustments to lessen the accumulation contributed by ice crystals were applied to time series at each station. This was done by reducing the amount assigned to trace observations in proportion to the average number of ice crystal events observed at the site. The ratio of the number of ice crystal events to the number of solid precipitation trace events was computed for 27 stations in Arctic regions (Table 2). The average ratio of ice crystal to solid precipitation trace number is mapped in Fig. 3. The 40% ratio isoline nicely separates regions of different climate behaviour across the country. The ratio rapidly increases above that line. A uniform adjustment of 0.07 mm/trace was applied below the 40% ratio and the trace adjustment was separately computed for each station above that limit. The allowance for trace gradually decreases as the percentage of traces that are ice crystal events increases above 40%. The actual snowfall trace adjustment due to the effect of ice crystal events is given by:

$$trace_2 = 0.07 - [ICR\% - 40\%] \times (0.07 - 0.03)/(70\% - 40\%)$$
$$\text{for } ICR > 40\,\% \text{ events}$$
$$= 0.07 \qquad \text{for } ICR \leq 40\% \text{ events},$$

where the Ice Crystal Ratio (ICR) is the fraction of solid precipitation trace events due to ice crystal events, computed for each station separately (see Table 2).

This methodology generates allowances for trace ranging from 0.07 mm in the south to 0.03 mm at Eureka in the High Arctic. These values correspond to earlier results suggested for Canada and the Arctic Islands by Metcalfe et al. (1994).

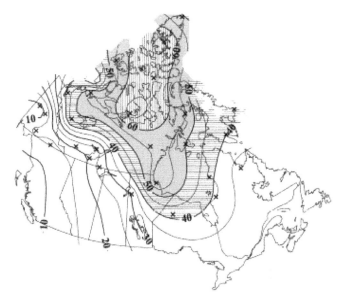

Fig. 3 Ratio of number of ice crystal events/number of solid precipitation trace events [%].

The adjustments associated with daily snowfall ruler data can be summarized by:

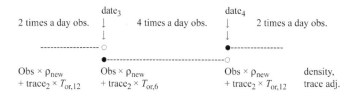

The meaning and computation method of trace occurrence ratio ($T_{or,6}$, $T_{or,12}$) is discussed earlier in the description of rain trace adjustment.

d The Effect of Adjustments on the Time series

To evaluate the effect of adjustments for trace and other changes, a sensitivity study was performed. The results are plotted in Figs 4 and 5. The selected stations represent two different climate regions. For details of the two stations, see Table 2. Resolute is located in the Arctic Islands (75°N). The occurrence of ice crystal trace events is quite frequent (one of the highest in our sample), snowfall density adjustment is high (Fig. 2) and annual precipitation is small, so the amount of adjustment is relatively high compared to the total. Yellowknife is located in the Mackenzie River Basin (62°N). The area is wetter, the density adjustment for fresh snow is nearer to one and the sum of all trace events is approximately half that of Resolute.

For the two stations, step-by-step adjustments were performed and the percentage change along with trends of total annual precipitation amounts were computed. In the first step the actual snow values were multiplied by the new density adjustment. The average change in total precipitation was 28% for Resolute and 3% for Yellowknife. Since ρ_{new} is a multiplicative factor, it has a direct and significant effect on total precipitation amounts but changes in the trend are dependent only upon changes in the fraction of total precipitation falling as snow.

Adjustment for undercatch and wetting loss due to rain gauge change (line 3 in Figs 4 and 5), accounts for 3% and 5% increases in total precipitation for Resolute and Yellowknife respectively. Since the new Type-B gauge has smaller losses and approximates "true" rainfall amount better, the trend is decreased by this step.

Adjustment for rain trace days, which averaged over 20 per year at each location, (line 4 on Figs 4 and 5) adds 7% and 4% accumulation and increases trends slightly at each. In the last step (line 5 on Figs 4 and 5) the solid trace adjustment values were added to the daily snow ruler measurements. In spite of the different solid trace intensity rates, (0.04 mm for Resolute, 0.07 mm for Yellowknife) the effect of this adjustment was much higher on the Northern station (13% vs. 5%) because of the larger percentage of ice crystal events at Resolute. The trend is also increased slightly at both locations because of the recorded increase in number of measurements of trace.

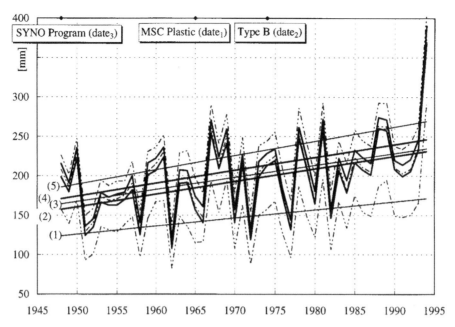

Fig. 4   Sensitivity study of annual total precipitation adjustment at Resolute. Curve (1): archive rain + snow, trend (1) = 1.04 mm/year; curve (2): rain + snow • new density (32% change), trend (2) = 1.61 mm/year; curve (3): (2) + gauge correction (36% change), trend (3) = 1.52 mm/year; curve (4): (3) + rain trace correction (42% change), trend (4) = 1.64 mm/year; curve (5) = (4) + snow trace correction (56% change), trend (5) = 1.81 mm/year.

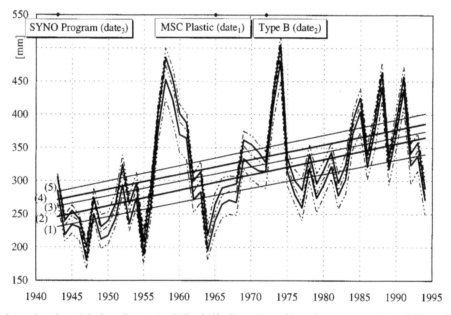

Fig. 5   Sensitivity study of annual total precipitation adjustment at Yellowknife. Curve (1): archive rain + snow, trend (1) = 2.15 mm/year; curve (2): rain + snow • new density (7% change), trend (2) = 2.34 mm/year; curve (3): (2) + gauge correction (12% change), trend (3) = 2.21 mm/year; curve (4): (3) + rain trace correction (16% change), trend (4) = 2.25 mm/year; curve (5) = (4) + snow trace correction (21% change), trend (5) = 2.31 mm/year.

After application of all adjustments to both stations, the trend for Resolute almost doubled, while for Yellowknife the adjustments increased the original relatively high trend only slightly.

The effect of all the adjustments was computed for each station, expressed in percentage change (adjusted/measured) and plotted and mapped for Canada (Fig. 6). As expected, the highest values are located in the North,

decreasing towards the South. The spatial variance is dominated by the variability in snow density (Figs 2 and 6 are quite similar).

e   Station Joining and Missing Value Estimation

Occasionally the selected station was relocated during the observation period. It is important to adjust the time series for

Fig. 6   Map of change due to adjustments.

inhomogeneities caused by relocation. To do this, a simple ratio method (Thom, 1966) was used for combining the records from two stations. The first station's daily values are multiplied by the ratio of total precipitation at the two stations during the overlapping period.

The ratios are computed separately for each month in order to avoid possible seasonal effects. The adjustment used for a specific month was the ratio based on all data for that month in the overlapping period. Adjustment ratios were determined independently for rain and snow. This made it important to introduce a limit for the minimum number of events to be used to compute the adjustment ratio. If fewer than 30 observations common to both sites (less than a month) were available for the given month at the examined stations, the sample was assumed unrepresentative and no adjustment was made.

For trend and variability analysis purposes it is sometimes useful to fill missing data gaps in daily data. In response to this requirement, this project created two adjusted datasets, one of which had missing data gaps filled with values consistent with the probability distribution of the available data. In climatology, the gamma distribution with zero lower bound is commonly used to represent variations in precipitation amounts (Stern and Coe, 1984; Diaz et al., 1989) and was selected for gap filling in this study. The popularity of the gamma distribution for describing precipitation data derives from the fact that it provides a flexible and accurate representation involving only two parameters. It is defined by its frequency or probability density function:

$$f(x) = \frac{1}{\beta^\gamma \Gamma(\gamma)} x^{\gamma-1} e^{-x/\beta} \quad \beta > 0, \gamma > 0$$

where x is the random variable, β scales x and is therefore the scale parameter, γ is the shape parameter, Γ is the usual gamma function, and $f(x) = 0$ for $x < 0$.

To eliminate seasonal effects, the daily precipitation data were placed into 12 subsets, one for each month. Since daily precipitation data records often contain zeros, a mixed distribution function of zeros and continuous precipitation amounts was employed and the dry and wet days were separated from each other. Missing value estimation was applied on rain and snow separately. This created sample size problems for rainfall in some winter months and for snowfall in warm months, so a minimum sample size (#me10) for parameter estimation was introduced. Below this limit the shape and scale parameters are assumed to be zero. The wet-day shape and scale parameters are determined by maximum likelihood estimator (Thom, 1958):

$$\gamma_{est} = 1 + (1 + 4A/3)^{1/2}/4A$$

where

N = sample size
$A = \ln x_{mean} - \Sigma \ln x/N$

and

$\beta_{est} = x_{mean}/\gamma_{est}$.

An additional distribution parameter defining the frequency of zero values (average number of days without precipitation) was computed for the 1961–90 reference period. The maximum observed precipitation amount was also used as an upper limit for final estimation.

Missing precipitation values are filled by generating random numbers from the gamma distribution defined by the station's gamma and beta parameters for the given month. The number of values generated is equal to the average number of days with precipitation and no generated value is allowed to exceed the maximum observed amount of precipitation. The generated values and zero precipitation values are distributed randomly within the missing data interval.

This method makes no attempt to recreate actual data for individual days. If spatially consistent data are required for mapping, etc., the adjusted dataset complete with acknowledged missing data should be used. For the gap-filled dataset, the goal is to create a continuous time series to ease computer processing, without changing the overall statistical properties of the data. The method is inappropriate for replacing extended periods of missing data. Missing data were not estimated for gaps of two consecutive years or more.

4 Regional and national time series computation

After all the adjustments, station joining and missing value estimation have been applied separately on daily rain and snow, time series of precipitation in millimetres are available for each of the selected stations. These time series are analysed for linear trend on a seasonal and annual basis for regions with similar climatological and geographical characteristics. Due to the special interest in and behaviour of stations north of 55°N, these were also combined into one large regional time series. Although the differences within regions were minimized, the derivation of regionally representative annual and seasonal rainfall amounts sometimes combine data from stations that may have greatly differing means and variance. Consistent with Jones and Hulme (1996), to minimize the effects of inclusion of very dry stations (where departures are large compared with the mean rainfall) and to be more comparable with other country's data trends, both the percentage departure and the standardized departure from normal were tabulated. Percentage departure was defined as:

$$P_i' = \frac{P_i - P_{30}}{P_{30}} \times 100\%$$

and standardized departure from normal as:

$$P_i' = \frac{P_i - P_{30}}{\sigma_{30}} \times 100\%$$

where P_{30} = mean for the 1961–90 period,
        σ_{30} = standard deviation for the 1961–90 period.

Time series distributed according to total annual precipitation (based on the calendar year), by season (Winter = DJF, Spring = MAM, Summer = JJA and Autumn = SON) and by the form of precipitation (solid or liquid) were created in this way.

Regional time series are based upon the arithmetic mean of the stations within the region. Since the station periods may differ within the given region (Table 1), the final analysis period of a region is defined by the years where at least half of the stations have useful data. Consequently the periods are not identical for each region within the country. The national time series is the weighted average of the regional values with weights proportional to the area. The computation period of national time series starts when all regional data are available.

In Table 3, results were computed with all three methods. For presentation purposes on Figs 7–14 the actual amount is used. Table 3 contains the summary of the linear trends for all the selected regions, precipitation types and seasons. The occasional empty field means that there was no significant (<5 mm mean) precipitation measured in that season and the trend was not computed. The t-test is used to evaluate whether the magnitudes of the computed linear trends are significantly different from zero at the 5% level. The bold numbers in Table 3 indicate time series with significant trends. The regression/linear trend computation uses the assumption that the residuals are independent of each other. The Durbin-Watson test was applied to total precipitation data to test whether the residuals behave according to this assumption. Underlined values indicate time series where residuals to the linear trend are auto-correlated. In such cases extrapolation beyond the record of the time series or intercomparison of trends covering different periods are inappropriate.

TABLE 3.   Summary of precipitation trends

NAME		Period	Annual				Winter				Spring			
			P	S	R	SD	P	S	R	SD	P	S	R	SD
I.	Atlantic Canada	1895–96	0.5	**2.7**	−0.5	0.6	0.7	**2.2**	−1.8	0.5	0.5	**3.6**	1.0	0.4
II.	Great Lakes/St. Lawrence L	1895–95	<u>**1.1**</u>	−1.8	**2.2**	**1.4**	−0.3	−1.5	2.6	−0.2	1.0	<u>**3.8**</u>	2.6	0.5
III.	Northeastern Forest	1918–96	2.4	<u>**4.3**</u>	1.3	**3.6**	**3.1**	**3.0**	6.2	**2.1**	**2.5**	4.6	0.5	**1.7**
IV.	Northwestern Forest	1938–96	<u>**1.9**</u>	0.8	2.4	**2.0**	−2.7	3.1		−1.5	0.6	<u>**1.2**</u>	0.1	0.3
V.	Prairies	1895–95	0.5	1.5	0.1	0.5	**2.4**	**2.1**		0.8	0.9	1.2	0.8	0.4
VI.	South BC Mountains	1895–95	**1.4**	**2.0**	0.9	**1.3**	**2.4**	**2.7**	1.4	0.9	**1.6**	0.8	**2.5**	0.8
VII.	Pacific Coast	1898–95	**0.9**	<u>**1.9**</u>	0.8	0.9	0.7	2.1	0.6	0.4	0.1	−4.5	0.2	0.0
VIII.	Yukon/North BC Mountains	1939–95	**3.0**	<u>**4.8**</u>	1.6	**3.3**	**5.8**	**5.6**		**2.4**	**4.9**	2.8	7.3	**2.0**
IX.	Mackenzie District	1927–96	<u>**1.2**</u>	<u>**1.3**</u>	1.1	1.0	<u>**1.6**</u>	<u>**1.5**</u>		0.7	0.2	1.0	−1.3	0.1
X.	Arctic Tundra	1948–96	<u>**5.1**</u>	<u>**9.5**</u>	−1.1	**5.2**	**5.6**	**5.9**		1.9	7.7	9.2	−2.7	3.9
XI.	Arctic Mountains & Fjords	1948–96	<u>**2.4**</u>	<u>**3.3**</u>	0.2	**1.8**	0.3	0.4		0.1	6.2	6.1		2.2
>55°N		1948–95	<u>**2.3**</u>	<u>**4.7**</u>	−0.3	**5.8**	2.3	2.2		1.6	3.8	3.9	3.5	3.1
Canada		1948–95	**1.7**	<u>**2.2**</u>	<u>**1.4**</u>	**5.1**	−0.1	0.1	−0.7	−0.1	2.6	1.2	4.0	3.3

			Summer				Autumn			
Bold fields: Trend is significant		I.	0.5		0.4	0.3	0.1	**5.3**	−0.3	0.1
<u>Underline</u>: Correlated residuals		II.	1.1		1.1	0.8	**2.6**	−0.1	**2.9**	**1.5**
(Durbin-Watson test failed)		III.	**1.7**		**1.6**	**0.0**	**2.6**	6.2	0.9	**2.0**
		IV.	**3.0**		**3.0**	**1.6**	<u>**3.5**</u>	5.4	2.2	**1.7**
		V.	−0.4		−0.4	−0.2	0.5	0.3	0.6	0.2
		VI.	1.0		1.0	0.5	0.8	2.0	−0.1	0.4
		VII.	**1.9**		**1.9**	**0.7**	1.2	9.6	0.0	0.7
		VIII.	0.2		0.1	0.1	**3.9**	**5.3**	2.6	**2.2**
		IX.	1.0		1.3	0.4	1.9	2.2	1.6	1.0
		X.	0.5	**13.9**	−1.4	0.4	<u>**7.4**</u>	<u>**11.3**</u>	<u>0.3</u>	**6.3**
		XI.	2.1	<u>**11.5**</u>	−1.2	1.3	<u>**1.5**</u>	<u>0.8</u>	5.2	0.7
		>55°N	−0.2	**13.2**	−1.2	−0.3	<u>**4.2**</u>	<u>**6.5**</u>	0.7	**4.7**
		Canada	0.9	**12.2**	<u>0.5</u>	1.5	**3.4**	<u>**5.3**</u>	2.1	**5.1**

P. Total Precipitation   (mm change/mean) over 10 years [%]
S. Total Snow   (mm change/mean) over 10 years [%]   If mean <5 mm, trend not computed
R. Total Rain   (mm change/mean) over 10 years [%]   If mean <5 mm, trend not computed
SD. Standardized Departures relative to 1961–90   change over 100 years $SD = (P_i - P_{30})/S_{30}$

To date, in-depth analysis of the regional time series based on daily data have not been carried out but preliminary investigations of the seasonal and annual aggregations of these data have proven informative. The series have confirmed knowledge in areas where reliable information is available (Environment Canada, 1995; Skinner and Gullett, 1993; Environment Canada, 1992) and provided important new information in areas of previous ignorance.

Because of sparse station distribution in northern Canada, information on trends for the whole country, as depicted in Table 3, is only available for the period 1948–95. During that time,

annual precipitation has increased by an average of 11 mm per decade (1.7% of mean/decade) for a total increase of 8.2% over the 48 years (Fig. 7) which is significant at the 95% level. This increase is not distributed evenly, either in time or across the country. Most of the snowfall increase (2.2% of mean/decade) occurs in the mid-1960s which corresponds with the introduction of the Nipher shielded snow gauge. Over 50% of the locations used in this study have Nipher shields installed. Even though only ruler measurements were used in the time series, ruler measurements were probably influenced by the Nipher gauge catch where the two measurements were coincident. By

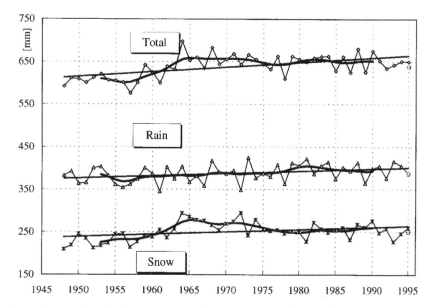

Fig. 7   Annual precipitation for Canada. Trends: total: +1.7; snow: +2.2; rain: +1.4 [% of mean/decade]. The mean value for 1948–95 is depicted by a ▣ to the right of the graph.

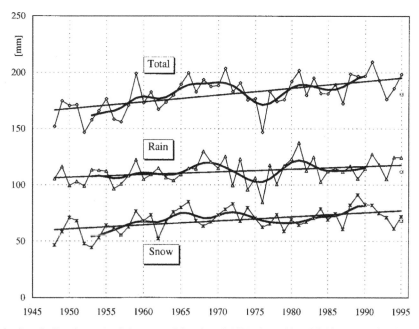

Fig. 8   Autumn precipitation for Canada. Trends: **total: +3.4**; **snow: +5.3**; **rain: +2.1** [% of mean/decade]. The mean value for 1948–95 is depicted by a ▣ to the right of the graph.

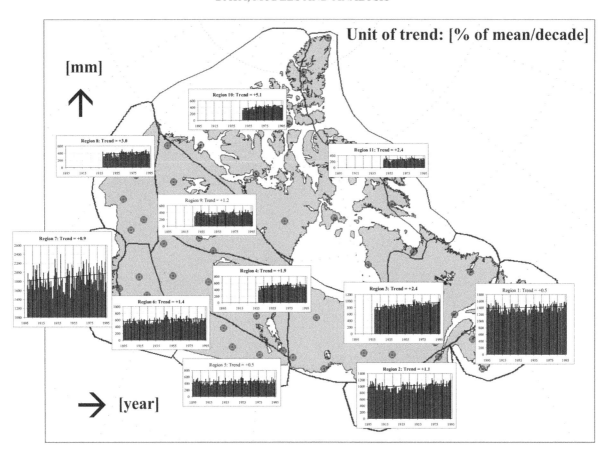

Fig. 9   Adjusted annual total precipitation for each region.

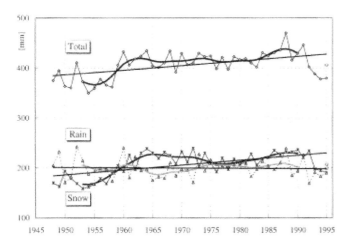

Fig. 10   Annual precipitation for Canada north of 55°N. Trends: **total: +2.3**; **snow: +4.7**; rain: −0.3 [% of mean/decade]. The mean value for 1948–95 is depicted by a ▣ to the right of the graph.

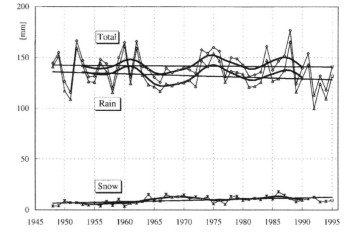

Fig. 11   Summer precipitation for Canada north of 55°N. Trends: total: −0.2; **snow: +13.1**; rain: −1.2 [% of mean/decade]. The mean value for 1948–95 is depicted by a ▣ to the right of the graph.

season, the autumn amounts show the greatest increase (Fig. 8), 3.4% of mean/decade for total precipitation and 5.3% of mean/decade in the form of snow. The snowfall increase is not concentrated in the early 1960s for the autumn.

If a linear trend is assumed, all regions show an increase in precipitation over the period studied. By region the Arctic

Tundra and Yukon/North BC Mountains show the biggest change (Fig. 9), 5.1% and 3.0% of mean/decade respectively. Trends remain positive but decrease to the south and east.

Stations located north of 55°N experienced an average 9.2 mm per decade increase, which translates to 10.8% increase in total annual precipitation over the same 48-year

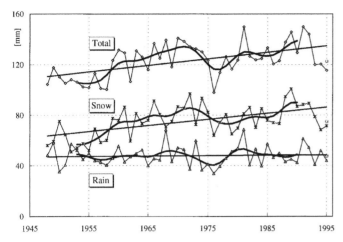

Fig. 12 Autumn precipitation for Canada north of 55°N. Trends: **total: +4.2**; **snow: +6.5**; rain: +0.7 [% of mean/decade]. The mean value for 1948–95 is depicted by a ▧ to the right of the graph.

period (Fig. 10—2.3% of mean per decade). This increase is actually the largest percentage increase in the country but is still much less than the 15–20% suggested by Groisman and Easterling (1994). The authors believe that the current study was able to implement more comprehensive corrections to the rainfall and snowfall data. Figure 11 indicates a slight decrease

in summer precipitation north of 55°N. For the annual and autumn time series, the total precipitation trend is dominated by snowfall trend (4.7% and 6.5% of mean per decade respectively) (Figs 10 and 12). The 4.7% per decade for snowfall trend agrees well with the Groisman and Easterling (1994) value of approximately 20% over 40 years for northern Canada. The summer snowfall trend is less significant, since the average total snow amount in summer is only 9.5 mm water equivalent.

It is also important to study which season contributes the most to the annual increase in the total precipitation trend. For the two regions with highest trends in annual precipitation, spring and autumn contributed most in the Arctic Tundra, while winter and spring contributed most in the Yukon. The graphs of annual, autumn and winter total precipitation (Figs 9, 13 and 14) indicate that both regions' time series actually peak in the 1960s and remain constant or fall over the final 30 years of data. Differing record lengths distort the comparisons somewhat but for spring, the two Arctic regions have the highest trends; for autumn the Arctic Tundra and Yukon (Fig. 13) and for winter the previous regions plus the Northeastern Forest (Fig. 14) have the highest trends. Summer total precipitation time series behave a little differently. Trends are small and even negative in the Arctic and increases are greatest in the Northeastern and Northwestern Forest regions.

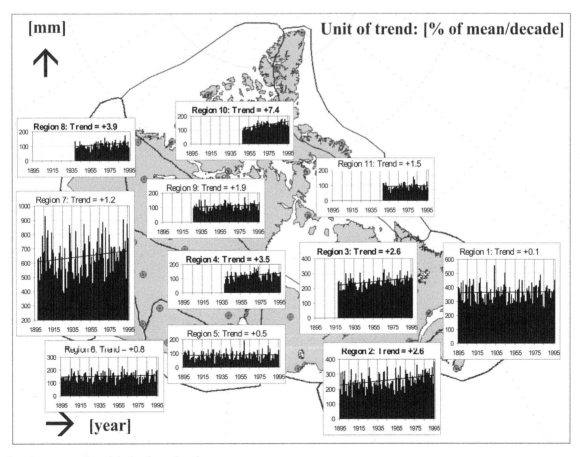

Fig. 13 Adjusted autumn total precipitation for each region.

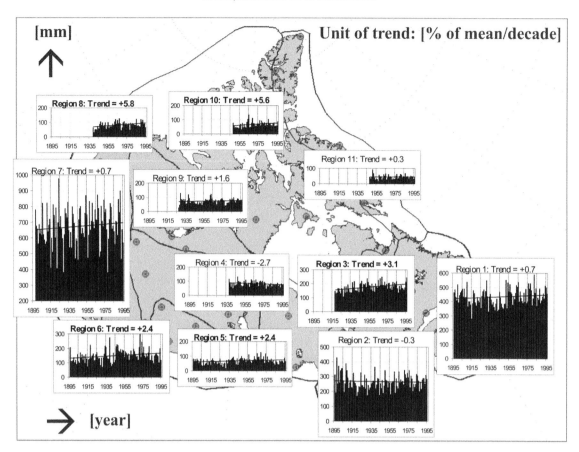

Fig. 14   Adjusted winter total precipitation for each region.

Since the early part of the century, Canada's drought sensitive grain growing area in the Prairies shows a slight increase in winter precipitation and little change in the other seasons. These findings are in agreement with suggestions by IPCC (1995) and Karl et al. (1996) that an increase in precipitation, primarily in the cold season is consistent with climate warming.

It is also evident that almost all annual and seasonal total precipitation trends are higher north of 55°N than the equivalent time series for the whole country. This is especially true for snowfall. IPCC (1995) speculates that the ratio of liquid to solid precipitation would be a useful indicator of climate warming. This ratio is computed and shown for both all Canada and Canada north of 55°N in Fig. 15. Contrary to climate warming expectations, the time series for both show a decreasing trend. The trend for the entire country is very small but for northern Canada, the ratio decreases substantially through the 1950s, before the introduction of the Nipher, with little trend evident afterward. The evidence that the decrease occurred prior to introduction of the Nipher gauge, increases confidence that the trend is real and not an artifact of measurement procedures.

In spite of our efforts to adjust for inhomogeneities, the two activities most likely to induce systematic artificial inhomogeneities in Canadian precipitation data are the introduction of the Nipher shielded snow gauge in the mid-1960s and the switch to a new Type-B rain gauge in the mid-1970s. However, there is

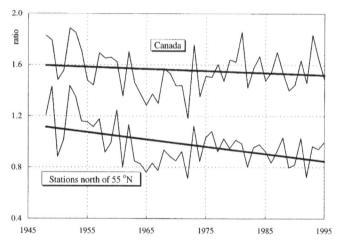

Fig. 15   Annual total rain/snow ratios.

limited visual evidence from any of the regional time series that serious inhomogeneities associated with these events remain in the adjusted data or contribute unduly to the trend computation. While the timing of the peak winter snowfall for the Arctic Tundra occurs close to the period when Nipher shields were introduced, the bulk of the increase occurred in the 1950s, well before introduction of the Nipher, and winter precipitation has remained relatively constant since then.

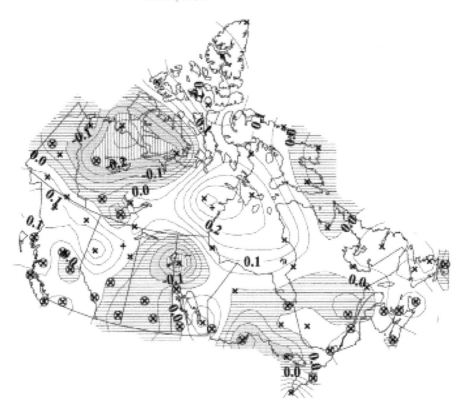

Fig. 16   Trend (1940–95) in fraction of annual precipitation falling in "heavy" (>90th percentile) events. Negative values are shaded.

The scope of the current study does not include an extensive analysis of extreme events but the topic is too important in the context of greenhouse warming to be ignored. For tracking intense daily events, Karl et al. (1996) postulated that the fraction of annual rainfall occurring in extreme events is increasing. He selected a threshold of 2 inches (50.8 mm) and examined the percentage of the United States with greater than normal proportion of precipitation derived from events exceeding this threshold. The examination of the same ratio for the 69 rehabilitated Canadian precipitation stations quickly identified that it is difficult to transport this method across the border. The mean intensity of extreme events decreases rapidly with latitude north of 50°N and for large portions of the country a 50 mm event is extremely rare. A different threshold based upon the statistics of extreme rain events at each station (e.g., the normal 90th percentile) is a better alternative. Sixty-nine stations were insufficient to define areas of Canada with greater than normal values of this ratio, so linear trends of the ratio were determined at each station. These trends for the period 1940–95 are mapped in Fig. 16. The map is dominated by a large area in the northern part of the country, showing an increase in the fraction of precipitation falling in heavy (>90th percentile) events. Figure 17 shows three plots depicting the time series of fraction of annual precipitation falling in heavy events representing averages over three sets of stations. Curve A is the average of all stations with complete data for the period 1940–95 and is representative of the entire country. Linear trend analysis indicates that, averaged over the whole country,

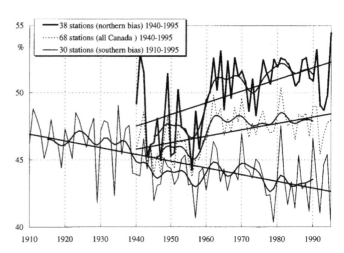

Fig. 17   Time series of fraction of annual precipitation falling in "heavy" events. The 30 stations with full record are circled on Fig. 16.

the fraction of precipitation falling in heavy events, expressed as a percentage, has increased at a rate of about 0.05% per year since 1940, due mainly to a sharp increase in the early 1960s. Curve B is the time series for the 30 stations with complete data from 1910. The linear trend for these data (stations identified by ★ in Fig. 16) which are almost all from southern stations, suggests that the fraction of precipitation in heavy events has been decreasing since 1910 at the rate of about −0.05% per year. Curve C shows the fraction of annual precipitation in heavy events for the remaining 38 stations with data restricted to the period since 1940, and is heavily biased to northern stations. Linear trend for Curve C indicates that the fraction of precipitation in heavy events has increased by about 0.09% per year since 1940 for these northern stations. It is clear from examination of these three curves, and supported by Fig. 16, that the increase in the national average fraction of annual precipitation falling in heavy events is due mainly to an increase in this fraction at northern stations.

5 Conclusion

This project has created a valuable dataset for various forms of Canadian climate analysis. For the first time, daily records were used to create long time series of rainfall and snowfall data. This permitted more detailed adjustments to the precipitation data, including correction for inhomogeneities caused by changes to instrumentation and measurement procedures. Two separate datasets have been compiled, one with missing data simply identified and one with missing data estimated in a manner which preserves the parameters of the probability distribution for each station. The expansion of the adjusted network from 69 to nearly 500 stations is progressing preparatory to replacing regional studies with Canada-wide mapping of results. The precipitation part of the Climate Trends and Variations Bulletin (wwwib.tor.ec.gc.ca/extracurricular) for Canada is planned to be produced based on the adjusted data described in this article. Information on availability of these datasets may be obtained from the authors.

The work reported on here is just one step in the ongoing effort to improve the utility of Canadian precipitation data for use in climate research. Adjustment procedures, especially for snowfall, continue to need refinement. In spite of network density constraints, a technique for assessing station data homogeneity must be developed and applied. Improved joining techniques would also be a priority for future work.

The automation of Canadian climate measurements introduces immense homogeneity problems to precipitation time series. Existing automated procedures do not differentiate between rainfall and snowfall amounts. Not only is the ability to distinguish amounts of rainfall and snowfall lost, but gauge undercatch characteristics are very dependent on precipitation type, so that the ability to accurately estimate total precipitation is diminished. Catch efficiency of gauges suitable for automation is generally lower and much more dependent on wind speed. Identification of intermittent problems associated with things like blowing snow, adhesion of snow to the gauge orifice and freezing precipitation is more difficult in the absence of human observers. On the positive side of automation, increased availability of coincident and continuous measurements of temperature and wind speed increases undercatch adjustment options, while the seemingly random and occasionally large errors associated with human observations are eliminated. As well, automation is often the only method to sustain uninterrupted measurements at long-record or remote locations. However, the dominant consideration is that, in Canada at least, automation of a significant portion of climate observations is inevitable. Almost one third of stations selected for this project have already been automated. New methods to maximize the usefulness and minimize inhomogeneities of precipitation measured at automatic weather stations must be developed and implemented.

Analyses of the annual time series has shown that precipitation has increased significantly during the period of reliable data availability (1948–present), in all regions of the country. Yukon/North BC Mountains (Region 8) and Arctic Tundra (Region 10) exhibited the most rapid increase in precipitation during that period, but the 10.8% increase in total precipitation over the 48-year period for Canada north of 55°N is much less than previously estimated by other authors. Nationally, Table 3 indicates that autumn shows the greatest increase in total precipitation and snowfall, while spring has the largest rain increase. Further study is planned to explore the correlation of temperature change with rainfall and snowfall. Data for the southernmost regions of the country are available back to the turn of the century and all show increases in annual precipitation. Among these southern regions, the largest absolute increase was observed for west coast stations (almost 200 mm) while stations in the mountains of interior British Columbia exhibited the largest percentage increase (1.4%/decade) this century. The fraction of annual precipitation falling in heavy (90th percentile) events has decreased by over 4 percentage points since 1910 in southern portions of the country but has increased by almost 5 percentage points since 1940 at more northern stations. During the nationwide reliable data period (1940–present) the fraction for all Canada of annual precipitation falling in heavy events increased by almost 3 percentage points due to the above noted increase at northern stations.

Much remains to be done with these daily time series. They will be used to examine, among other things, trends in variability and extremes; relationships to temperature change; change of snow cover over time. The adjusted precipitation data have already been used to improve the skill of seasonal climate forecasts (Shabbar and Barnston, 1996), as validation data for the Canadian Regional Climate Model (MacKay et al., 1998), as input data for the water balance computation on the MacKenzie Basin (Hogg et al., 1996), as input data for climate variability computation over the instrumental record (Skinner, 1996) and for optimal climate normal computation (Zhang et al., 1996). The adjusted daily precipitation (rain and snow) dataset will provide a valuable source of information about Canada's changing climate for years to come.

Acknowledgments

The authors wish to acknowledge the extensive contribution of B. Routledge to this project through metadata searches, data preparation and consultation on measurement and adjustment procedures. J. Metcalfe also provided valuable advice on adjustment procedures and P. Louie and S. Ishida computed the density adjustments based upon Nipher gauge measurements. P. Louie also provided useful comments on the draft manuscript, as did D. Gullett.

References

BRADLEY, R.S. and J. ENGLAND. 1978. Recent climatic fluctuations of the Canadian high arctic and their significance for glaciology. *Arctic Alpine Res.* **10(4)**: 715–731.

DIAZ, H.F.; R.S. BRADLEY and J.K. EISCHEID. 1989. Precipitation fluctuations over global land areas since the late 1800's. *J. Geophys. Res.* **94(D1)**: 1195–1210.

ENVIRONMENT CANADA. 1992. The state of Canada's climate: Temperature change in Canada 1895–1991. State of the Environment Reporting, Report No. 92–2, 36 pp. (Available from State of the Environment Directorate, Environment Canada, Ottawa, Ontario, Canada, K1A 0H3)

——. 1995. The state of Canada's climate: Monitoring variability and change. State of the Environment Reporting, Report No. 95–1, 52 pp. (Available from State of the Environment Directorate, Environment Canada, Ottawa, Ontario, Canada, K1A 0H3)

——. 1996. MANOBS – Manual of surface weather observations. Seventh Edition, pp. 3–9, 3.7.2

GOODISON, B.E. and P.Y.T. LOUIE. 1986. Canadian methods for precipitation measurement and correction. *In*: Proc. International Workshop on Correction of Precipitation Measurements, Zurich, Switzerland, 1–3 April 1985, pp. 141–145.

GROISMAN, P.Y. and D.R. EASTERLING. 1994. Variability and trends of total precipitation and snowfall over the United States and Canada. *J. Clim.* **7**: 184–205.

—— and D.R. LEGATES. 1994. The accuracy of United States precipitation data. *Bull. Am. Meteorol. Soc.* **75(2)**: 215–227.

GULLETT, D.W.; W.R. SKINNER and L. VINCENT. 1992. Development of an historical Canadian climate database for temperature and other climate elements. *Climatol. Bull.* **26**: 125–131.

HOGG, W.D.; P.Y.T. LOUIE, A. NIITSOO, E. MILEWSKA and B. ROUTLEDGE. 1996. Gridded water balance for the Mackenzie Basin GEWEX study area. *In*: Proc. Second Int. Scientific Conference on GEWEX, 17–21 June, Washington DC, USA.

IPCC. 1995. Climate Change 1995, *The Science of Climate Change*. Contribution of Working Group I to the Second Assessment Report of the Intergovernmental Panel on Climate Change, J.T. Houghton, L.G. Meira Filho, B.A. Callander, N. Harris, A. Kattenberg and K. Maskell (Eds). Cambridge University Press, 531 pp.

JONES, P.D. and M. HULME. 1996. Calculating regional climatic time-series for temperature and precipitation: Methods and illustrations. *Int. J. Climatol.* **16**: 361–377.

KARL, T.R.; R.W. KNIGHT, D.R. EASTERLING and R.G. QUAYLE. 1996. Indices of climate change for the United States. *Bull. Am. Meteorol. Soc.* **77(2)**: 279–292.

LEGATES, D.R. 1987. A climatology of global precipitation. Publications in Climatology, Mather, J.R. (Ed.), Volume XL, Number 1, 84 pp.

—— and C.J. WILLMOTT. 1990. Mean seasonal and spatial variability in gauge-corrected, global precipitation. *Int. J. Climatol.* **10**: 111–127.

MACKAY, M.D.; R.E. STEWART and G. BERGERON. 1998. Downscaling the hydrological cycle in the Mackenzie Basin with the Canadian Regional Climate Model. *atmosphere-ocean*, **36(3)**: 179–211.

METCALFE, J.R.; S. ISHIDA and B.E. GOODISON. 1994. A corrected precipitation archive for the North-west Territories. Environment Canada – Mackenzie Basin Impact Study, Interim Report #2, pp. 110–117.

——; B. ROUTLEDGE and K. DEVINE. 1997. Rain-fall measurement in Canada: changing observational methods and archive adjustment procedures. *J. Clim.* **10**: 92–101.

ROUTLEDGE, B. 1997. Corrections for Canadian standard rain gauges. Internal Report, Environment Canada, Atmospheric Environment Service, Downsview, 8 pp.

SEVRUK, B. 1982. Methods of correction for systematic error in point precipitation measurement for operational use. Operational Hydrology Report No. 21, World Meteorological Organization, Geneva, WMO No. 589, 91 pp.

SHABBAR, A. and A.G. BARNSTON. 1996. Skill of seasonal climate forecasts in canada using canonical correlation analysis. *Mon. Weather Rev.* **124**: 2370–2385.

SKINNER, W. 1996. Climate variability over the instrumental record with special emphasis on western Canada. *In*: Proc. Inter-American Institute for Global Change Workshop, 6–11 October, Jasper, Alberta, Canada, pp. 21.

—— and B. MAXWELL. 1994. Climate patterns, trends and scenarios in the Arctic. Mackenzie Basin Impact Study (MBIS), Interim Report #2, AES Downsview, pp. 125–137.

—— and D.W. GULLETT. 1993. Trends of daily maximum and minimum temperature in Canada during the past century. *Climatol. Bull.* **27(2)**: 63–77.

STERN, R.D. and R. COE. 1984. A model fitting analysis of daily rainfall data. *J. R. Statis. Soc.* **147(1)**: 1–34.

THOM, H.C.S. 1958. A note on the gamma distribution. *Mon. Weather Rev.* **86(4)**: 117–122.

——. 1966. Some methods of climatological analysis. WMO Technical Note No. 81. pp. 7–9.

WALSH, J.E. 1995. Recent variations of Arctic climate: The observational evidence. *In*: Proc. Sixth Symposium on Global Change Studies, *Am. Meteorol. Soc.*, Dallas, Texas, 15–20 January, pp. 20–25.

WOO, M.K. and P. STEER. 1979. Measurements of trace rainfall at a high arctic site. *Arctic*, **32(1)**: 80–84.

ZHANG, X.; A. SHABBAR and W.D. HOGG. 1996. Seasonal prediction of Canadian surface climate using optimal climate normals. *In*: Proc. 21st Annual Climate Diagnostics and Prediction Workshop, U.S. Dept. Commerce, Huntsville, Alabama, 28 Oct.–1 Nov., pp. 207–210.

Temperature and Precipitation Trends in Canada During the 20th Century

Xuebin Zhang, Lucie A. Vincent, W.D. Hogg and Ain Niitsoo

Climate Research Branch, Meteorological Service of Canada, Downsview, Ontario

ABSTRACT *Trends in Canadian temperature and precipitation during the 20th century are analyzed using recently updated and adjusted station data. Six elements, maximum, minimum and mean temperatures along with diurnal temperature range (DTR), precipitation totals and ratio of snowfall to total precipitation are investigated. Anomalies from the 1961–1990 reference period were first obtained at individual stations, and were then used to generate gridded datasets for subsequent trend analyses. Trends were computed for 1900–1998 for southern Canada (south of 60°N), and separately for 1950–1998 for the entire country, due to insufficient data in the High Arctic prior to the 1950s.*

From 1900–1998, the annual mean temperature has increased between 0.5 and 1.5°C in the south. The warming is greater in minimum temperature than in maximum temperature in the first half of the century, resulting in a decrease of DTR. The greatest warming occurred in the west, with statistically significant increases mostly seen during spring and summer periods. Annual precipitation has also increased from 5% to 35% in southern Canada over the same period. In general, the ratio of snowfall to total precipitation has been increasing due mostly to the increase in winter precipitation which generally falls as snow and an increase of ratio in autumn. Negative trends were identified in some southern regions during spring. From 1950–1998, the pattern of temperature change is distinct: warming in the south and west and cooling in the northeast, with similar magnitudes in both maximum and minimum temperatures. This pattern is mostly evident in winter and spring. Across Canada, precipitation has increased by 5% to 35%, with significant negative trends found in southern regions during winter. Overall, the ratio of snowfall to total precipitation has increased, with significant negative trends occurring mostly in southern Canada during spring.

Indices of abnormal climate conditions are also examined. These indices were defined as areas of Canada for 1950–1998, or southern Canada for 1900–1998, with temperature or precipitation anomalies above the 66th or below the 34th percentiles in their relevant time series. These confirmed the above findings and showed that climate has been becoming gradually wetter and warmer in southern Canada throughout the entire century, and in all of Canada during the latter half of the century.

RÉSUMÉ *On a analysé les tendances de la température et des précipitations au Canada durant le 20ème siècle en utilisant les données de stations récemment mises à jour et ajustées. On examine six éléments: les températures maximales, minimales et moyennes ainsi que l'écart diurne de température (EDT), les précipitations totales et le rapport des chutes de neige sur les précipitations totales. On a d'abord obtenu des anomalies aux stations individuelles pour la période de référence 1961–1990, dont on a ensuite engendré des ensembles de données maillées utilisées pour les analyses de tendances. Les tendances ont été calculées pour 1900–1998 sur le sud du Canada (au sud de 60° de latitude nord) et séparément pour 1950–1998 sur tout le pays, à cause de l'insuffisance des données sur l'Arctique septentrional pour les années précédant 1950.*

De 1900 à 1998, la température annuelle moyenne a augmenté de 0,5 à 1,5°C dans le sud. Le réchauffement est plus marqué pour la température minimale que pour la température maximale durant la première partie du siècle, ce qui diminue aussi le EDT. Le réchauffement le plus important s'est produit sur l'ouest, et on y voit des augmentations statistiquement significatives surtout durant les périodes de printemps et d'été. La précipitation annuelle a aussi augmenté de 5 à 35% sur le sud du Canada durant la même période. En général, le rapport des chutes de neige sur les précipitations totales a augmenté surtout à cause des précipitations hivernales, qui tombent principalement sous forme de neige, et aussi à une augmentation du rapport en automne. On a identifié des tendances négatives sur certaines régions du sud au printemps. De 1950 à 1998, le genre de changement de température est distinct: réchauffement sur le sud et l'ouest et refroidissement au nord-est, avec des amplitudes semblables pour les températures maximales et minimales. Ces caractéristiques se manifestent surtout en hiver et au printemps. Sur l'étendue du Canada, les précipitations se sont accrues de 5 à 35%, mais on trouve des tendances négatives sur les régions du sud en hiver. Dans l'ensemble, le rapport des précipitations nivales sur les précipitations totales a augmenté, mais avec des tendances négatives statistiquement significatives surtout sur le sud du Canada au printemps.

On examine aussi des indices de conditions climatiques anormales. On a défini ces indices comme les régions du Canada pour la période 1950–1998, ou du sud du Canada pour la période 1900–1998, démontrant des anomalies au-dessus du 66ème ou sous le 34ème percentile de leur série temporelle propre. Ces indices ont confirmé les résultats ci-dessus et montré que le climat est graduellement devenu plus chaud et qu'il y a plus de précipitations au sud du Canada tout au long du siècle, ainsi que partout au Canada durant la deuxième moitié du siècle.

1 Introduction

Recent analyses of climate trends indicate that the global mean surface temperature has increased by about 0.3 to 0.6°C since the late 19th century, and by about 0.2 to 0.3°C over the last 40 years (e.g., Nicholls et al., 1996). These studies have also shown that daily minimum temperatures have often increased at a greater rate than maximum temperatures, resulting in a decrease in the diurnal temperature range (DTR) for several regions of the world. There has been a small positive trend in global precipitation, of about 1% during the 20th century over land, with a greater increase in the high latitudes of the Northern Hemisphere, especially during the cold season. Considerable spatial and temporal variations have occurred over the past 100 years, and these tendencies of warming, increased precipitation and reduction of DTR have not been globally uniform. For example, Nicholls et al. (1996) showed warming in the mid-latitude Northern Hemisphere continents in winter and spring, and year-round cooling in the northwest North Atlantic and mid-latitudes over the North Pacific in the past four decades. Understanding the observed climate trends for Canada in terms of regional characteristics and hemispheric perspectives is important to assess regional changes and to understand further the global climate system. As well, the anthropogenic climate change signal is projected to be stronger in the high-latitudes (Nicholls et al., 1996). This suggests that it might be easier to detect climate change in a country like Canada.

The detection of climatic trends, including those predicted to occur from rising concentrations of atmospheric greenhouse gases (Wigley and Barnett, 1990), may be sought in historical climate records providing that such records are representative and cover a long enough period of time (typically more than 100 years). A number of data-related difficulties arise when attempting to analyze Canadian climate trends. Firstly, because of its vast land mass and location in high latitudes, the country experiences many different types of climate and relatively large spatial variability (Phillips, 1990). This requires a proper identification of the climate change signal from very noisy fields. Secondly, the observational network was not established in the north until the late 1940s and is very sparse. Thirdly, there have been changes in station location, in instrumentation and in observing practices which have caused inhomogeneities in the climatological records. Reliable trend estimates cannot be made before these inhomogeneity issues are adequately resolved. Many of these data-related issues have been addressed in Vincent (1990) and Gullett et al. (1992) for temperature records. Preliminary trend analysis was performed based on these datasets (Boden et al., 1994; Environment Canada, 1995; Skinner and Gullett, 1993; Gullett and Skinner, 1992). In these studies, seasonal and annual trends computed for maximum, minimum and mean temperatures for 11 climate regions established that significant warming has occurred in western Canada during the past 100 years.

Precipitation trend analyses were also reported in Boden et al. (1994) and by Environment Canada (1995). The data used in these studies were raw archive records, and hence were not rigorously assessed for temporal homogeneity. Results showed small increasing trends in the eastern part of the country over the period 1948–1992, and they also indicated that precipitation in many regions has been below average since the mid-1980s. Groisman and Easterling (1994) identified some problems in Canadian precipitation data. Prior to trend computation, they adjusted monthly rainfall data before 1975 by a factor of 1.025 to account for temporal inhomogeneity caused by a gauge change that occurred in the mid-1970s. It was found that over the 1950–1990 period, both annual snowfall and total precipitation have increased by about 20% in northern Canada (north of 55°N), and that total precipitation has also increased in the southern part of the country. Precipitation data are more problematic compared with temperature data; Mekis and Hogg (1999) have shown that, due to cumulative daily problems such as wetting loss, evaporation and trace measurements in the Canadian data, it is difficult, if not impossible, to correct Canadian precipitation data on a monthly basis and that the Groisman and Easterling (1994) adjustment needed to be improved.

Since the early computations of trends in Canadian temperature and precipitation, data have been substantially improved and updated. A recently developed technique has made it possible to address better the homogeneity problems caused by changes in the station location and/or measurement programs in temperature records (Vincent, 1998). As well, a new temperature database which includes monthly mean maximum and minimum temperatures at 210 stations has been created (Vincent and Gullett, 1999). Comprehensive adjustments of daily rainfall and snowfall have been performed for known changes of instruments, as well as for gauge undercatch, wetting loss and trace measurements (Mekis and Hogg, 1999) for about 500 stations. The computation of trends has also been improved. It is recognized that autocorrelation is quite common in temperature and precipitation time series, and may affect the linear trend computation (e.g. von Storch, 1995). Autocorrelation, a factor that was not taken into account in any of the previous studies, has been considered in this analysis.

Different components of the climate system interact and changes in one element can affect others. For example, warming in minimum temperature over global land surfaces is almost three times the magnitude of that for maximum temperature during 1951–1990 (Karl et al., 1993a). This was likely caused by changes in global cloud cover, and resulted in a decrease in the DTR. The outputs from simulations by three General Circulation Models (GCMs) for double CO_2 showed that a significant increase in snow accumulation in the arctic region is expected as a result of global warming (Ye and Mather, 1997). These examples suggest trends be analyzed on multiple variables. Providing trends in the most important climate variables for Canada is therefore a major objective of this study. We will present spatial distributions of trends along with their statistical significance for the following six elements: maximum, minimum and mean temperatures;

Fig. 1 Major Canadian political boundaries with provincial and territorial names. The Canadian Prairies include the provinces of Alberta, Saskatchewan and Manitoba. The Mackenzie Basin is shaded. Southern Canada is the region south of the 60°N.

DTR; precipitation totals and the ratio of snowfall to total precipitation.

In an effort to provide a set of indices which represents complicated multivariate, multidimensional climatic changes that can be readily understood and used by policy makers, Karl et al. (1996) developed and analyzed a climate extremes index (CEI) and a greenhouse climate response index (GCRI) for the United States. They found a positive trend in the U.S. GCRI during the 20th century that is consistent with projections resulting from increased emissions of greenhouse gases. The computation of similar indices such as indices of abnormal climate for Canada (to be defined in the next section) will provide a comparison with those computed for the U.S., and may also be used to verify the results of trend analysis. A preliminary examination of time variations of indices of abnormal climate is another objective of this study.

Due to the limited data availability in northern Canada prior to the 1950s, two periods have been analyzed in this study: 1900–1998 for southern Canada (south of about 60°N), and 1950–1998 for Canada as a whole (Fig. 1). After the description of data and methodologies used in Section 2, we will present temperature and precipitation trends during 1900–1998 for southern Canada and during 1950–1998 for the whole of Canada in Sections 3 and 4, respectively. The indices of abnormal climate which are similar to the GCRI of Karl et al. (1996) are examined in Section 5. Conclusions and discussion follow in Section 6.

2 Data and methods

a Data

The basic data used in this study are from the recently created database of monthly temperature at 210 stations (Vincent and Gullett, 1999), and daily rainfall and snowfall at 489

stations (Mekis and Hogg, 1999). To facilitate analysis and interpretation, these datasets were first gridded with a spatial resolution of 50 km on a polar stereographic projection using the Gandin optimal interpolation technique (Milewska and Hogg, unpublished manuscript).

The temperature data are monthly means of daily maximum and minimum temperatures. The 210 stations are relatively evenly distributed across Canada (Fig. 2). Records from different sites were sometimes joined to extend temporally the series backward to maximize coverage during the 20th century. However, record lengths still differ from one station to another. These data have undergone rigorous quality control, and have been adjusted for identified inhomogeneities caused by station

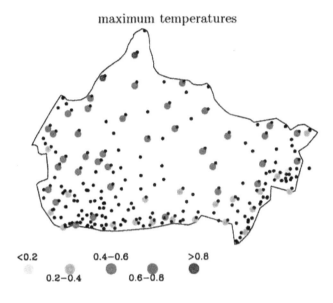

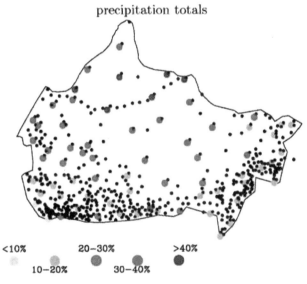

Fig. 2 Locations (black dots) of stations used for gridding (210 for temperature, 489 for precipitation). Coloured dots represent root mean square errors between anomalies of the station data and the nearest grid point at 59 grids for maximum temperature (upper panel, in °C) and at 68 grids for the precipitation totals (lower panel) during 1961–1990.

relocation, changes in instrumentation and in observing practices. A significant improvement over previous versions of the temperature database is the adjustment in the minimum temperature to account for changes in observing procedure at principal stations in 1961 (Vincent and Gullet, 1999). Adjustment for this bias results in a less pronounced cooling in minimum temperature in terms of magnitude (by as much as 0.5°C) and spatial extent in eastern Canada (Ontario, Quebec and the Atlantic provinces) during 1946–1998. The monthly mean temperature is derived from computing the average of the monthly maximum and minimum temperatures, while the monthly mean range was obtained by subtracting the monthly mean minimum temperature from the monthly mean maximum temperature. Seasonal and annual time series were calculated from the monthly values. The standard climatological seasons (i.e., December to February for winter, March to May for spring, June to August for summer, September to November for fall) were used in computing seasonal values.

Data collected from the 489 stations recording daily precipitation were also subjected to rigorous quality control. In addition to being strategically selected to cover as much of the country as evenly as possible, the data covered the period 1900–present. Based upon station history information, all known inhomogeneities in the precipitation time series resulting from changes in instrumentation or observing practices were carefully minimized. Wind undercatch and wetting loss were corrected according to different types of rain gauges. A 4% or 2% increase was added to observations to account for wind undercatch of the Meteorological Service of Canada gauge and the Type-B gauge, respectively. Trace events and spatially varying snow densities were also considered when adjusting the data. It is probable that trace events were not systematically reported/recorded throughout the century at all stations, but it was found that adjustment for trace events had little impact on trend computation. As a result, this dataset is the most homogeneous available and represents a significant improvement over other Canadian precipitation data. Mekis and Hogg (1999) detail the procedure and effects of the adjustment of precipitation data. Monthly, seasonal and annual totals of rainfall and snowfall were computed from daily amounts. The total precipitation was computed by summing the amounts of rainfall and snowfall, and the ratio of snowfall to total precipitation was also obtained.

The station data were gridded using a procedure developed by Hogg et al. (1997). The procedure uses statistical optimal interpolation and employs climatology as a "first guess" field and interpolating only the departures from the climatology of the relevant field. Hogg et al. (1997) demonstrated that this procedure is effective when interpolating monthly precipitation in data-sparse, strong-relief regions. Compared with the original field, the anomaly field is more homogeneous and isotropic and therefore better meets the essential requirements for proper statistical optimal interpolation. The interpolated anomaly grids were the basis of most of the analysis described here. Trends or abnormal climate indices were directly computed from gridded anomaly fields. Departures from the

reference period 1961–1990 were first calculated at individual stations. For precipitation, the anomalies were further normalized by dividing by the 1961–1990 period means. These normalized anomalies were then interpolated to the grid using the Gandin Optimal Interpolation technique (Alaka and Elvander, 1971; Milewska and Hogg, unpublished manuscript). The anomalies of ratio of snowfall to total precipitation were also gridded.

The error of interpolation was assessed in a cross-validation framework. One station at a time was withheld and the remaining stations were used to generate gridded values. The station data were then compared with the values at the nearest grid. To make the cross-validation computationally manageable while providing reliable information on the quality of the gridded datasets, the cross-validation was performed on selected parameters and stations. It was carried out on seasonal time series for the period 1961–1990 on maximum temperature at 59 stations, precipitation and the ratio of snowfall to total precipitation at 68 stations. The selected stations were evenly distributed across the country, including several in the datasparse north. It is reasonable to assume that the other three temperature related elements – anomalies of minimum and mean temperature, and DTR – are as spatially homogeneous as the maximum temperature anomalies, and their gridded datasets should be of the same quality as the gridded maximum temperature anomalies. Figure 2 shows the root mean square errors (RMSE) between the anomalies of station data and the nearest grid point. The average RMSE seasonal mean maximum temperature is 0.47°C with smaller values (typically between 0.2–0.4°C) in southern Canada where the observational network is denser. On average, this RMSE is about 0.26 standard deviations of the seasonal mean maximum temperature. The correlation coefficients for temperatures at grid and station are generally larger than 0.95 in southern Canada, and larger than 0.85 in the northern part of the country. These coefficients indicate that the optimal statistical interpolation technique performed reasonably well for the temperature data even in the very remote and data-sparse area. It may be concluded that the anomalies of maximum temperature provide a reasonably homogeneous field.

The RMSE for normalized seasonal precipitation anomalies is also shown in Fig. 2. The average RMSE is 0.23 (or 23% of the mean) with errors less than 0.2 (20%) in the southern part of the country. This translates to about 0.5 standard deviations. The grid to station correlation coefficients are generally not as high as those for temperature; nevertheless, given the spatial variability of the element, the interpolation was considered to be satisfactory. The correlation coefficients for the ratio of snowfall to total precipitation were much higher than those for precipitation anomalies, which indicates that this ratio is more homogeneous than the precipitation anomaly field.

The interpolation procedure plays the role of a spatial filter and thus removes local noise to some extent. The gridded fields are smoother than the station data and are more suitable for describing large-scale features such as long-term trends. To reduce the amount of calculation, trend analysis was performed on the time series of a coarse 200 × 200 km grid obtained by averaging the values of 16 grid points from the 50 × 50 km grid. This procedure further smoothes the fields. Since the climate observation network was not established in northern Canada until the late 1940s, there were large portions of the country with no data during the first half of the century. Based loosely on error analyses reported in Milewska and Hogg (unpublished manuscript) and those reported here, interpolation limits were established. No more than five stations with the shortest distance (less than 750 km for precipitation and 1000 km for temperature) to a grid point are used to interpolate the value for the grid. The distance limits are seldom reached. For example, more than 80% of grid values for precipitation have stations within 200 km. No grid values were generated for northern Canada (north of about 60°N) prior to the 1950s due to insufficient data. Trend analysis was performed on datasets for 1900–1998 for the southern part of the country (south of about 60°N) where grid values were complete for both temperature and precipitation, and for 1950–1998 for the nation as a whole.

b Extreme and Abnormal Climate Indices

Karl et al. (1996) used a set of indicators which represent a projected response of U.S. climate change caused by the increase of greenhouse gases in the atmosphere (e.g., Nicholls et al., 1996). These indicators include the percentages of the U.S. with much higher than normal (larger than 90th percentile) mean temperature, with much higher than normal precipitation during the cold season. Jones et al. (1999) have used similar indicators to determine the areas of the world affected by extreme temperature. A detailed investigation of all those indicators for Canadian climate merits a separate study. Here, we present some extreme climate indices such as extreme cold (dry) and warm (wet) indices. These are defined as percentages of the nation affected by extremely cold (dry) and warm (wet) conditions, respectively, and represent temperature (precipitation) below the 10th and above the 90th percentiles in the relevant time series. For example, for a particular year (season), the extreme dry index is obtained by counting the number of grids in that year (season) when the precipitation is below the 10th percentile and then dividing by the total number of grids. In many regions of Canada, climate does not need to be at an extreme to have severe impacts on environmental processes and economic activities like agriculture. For example, there is normally just enough precipitation to sustain agriculture in the semi-arid regions of southwestern Saskatchewan and southern Alberta. Any departure from normal may have severe effects (Bonsal et al., 1999). Moreover, a combination of anomalies in different parameters, such as warm temperature with lack of precipitation, may have even more severe repercussions. It should be noted that extreme conditions occurring in several climate variables at the same time, such as precipitation and temperature, are rare; therefore, we also

computed "abnormal" climate indices, defined similarly to those of extreme climate indices, with thresholds being the 34th and the 67th percentiles. Indices were then computed to represent the joint abnormal conditions in both precipitation and temperature.

c Methods

In most climate trend studies, it is generally assumed that the climate time series consists of a long-term trend component and a white noise residual component (Wigley and Jones, 1981). This assumption was used in previous studies of Canadian climate trends (e.g., Skinner and Gullett, 1993; Boden et al., 1994; Environment Canada, 1995). Most climate series, however, actually contain red noise and are serially correlated due to the multi-year nature of natural climate variability. Assuming white noise residuals will result in an overestimation of the effective sample size of the residuals (Leith, 1973; Trenberth, 1984), and hence overestimation of the significance of a trend (von Storch, 1995). To overcome such problems, recent approaches have introduced regression models with serially correlated residuals (Bloomfield, 1992; Bloomfield and Nychka, 1992; Woodward and Gray, 1993). Some studies have also included so called "explanatory variables" such as the southern oscillation index (Zheng et al., 1997; Zheng and Basher, 1998). Including explanatory variables in the equation will help to increase the fit of the model, but this may also bias the magnitude of the trend since the explanatory variables themselves generally are not time independent and contain the signal of trend to be detected.

Climate trend at the regional scale is by no means easy to detect; regional climate has relatively greater variability associated with natural climatic processes. This, together with the relatively short periods of time of observations in Canada, interferes with the proper identification of statistically significant regional trends. In this study, we have attempted to produce more reliable estimates of the magnitude and the statistical significance of the trend. Since serial correlation in the climate time series can influence the estimate of trend, and that trend can in turn have an impact on computation of the serial correlation, we have investigated a new approach to obtain the linear change that takes the serial correlation into account. To accomplish this, we have used the following statistical model:

$$Y_t = \mu + T_t + v_t \qquad (1)$$

where

Y_t is a climate variable at time t,
μ is the constant term,
T_t is the trend, and
v_t is the noise at time t.

To simplify computation, the trend is assumed to be linear, i.e.,

$$T_t = \beta_t \qquad (2)$$

where φ is the slope of the linear regression between the climate variable and time. The noise is represented by a *p*th order autoregression, AR(*p*),

$$v_t = \sum_{r=1}^{p} \phi_r \, v_{t-r} + \varepsilon_t, \qquad (3)$$

where ϕ_r are the parameters of the autoregressive process and $\{\varepsilon_t\}$ is white noise.

The order of the autoregression *p* was determined by computing the autocorrelation and the partial autocorrelation functions of each climate element at individual grid points. The partial autocorrelation for lags larger than one is, in general, not significantly different from zero. Therefore, we used an AR(1) process to model the noise. Thus, equation 1 becomes:

$$Y_t = \mu + \beta_t + \phi Y_{t-1} + \varepsilon_t \qquad (4)$$

The following iteration procedure was used to estimate the parameters in equation 4. A first guess of φ is computed directly from the dataset, and is used to remove the autocorrelation from the time series. Then β and its significance level are obtained from the de-autocorrelated time series by computing a simple and robust estimator of β based on the non-parametric Kendall's rank correlation tau (Sen, 1968), or the median of the slopes obtained from all possible combinations of two points in the series. We have used the Kendall estimate instead of the least squares estimate because it is less sensitive to the non-normality of the distribution and less affected by extreme values or outliers in the series. Both conditions can often be found in climatological time series, and in particular, in precipitation datasets. The estimated β is then used to remove the trend from the original series and the resulting residuals are used to obtain a second and more accurate estimation of the lag-1 autocorrelation coefficient. Once again, the newly estimated φ is used to pre-whiten the original series and a second estimate of β is obtained. This procedure continues until the differences in the estimates of both β and φ in two consecutive iterations are small enough. The procedure usually converges within 3–12 iterations. Monte-Carlo tests involving 1000 simulations each for a variety of autocorrelation and trend combinations confirmed that our procedure does produce satisfactory results.

To determine the statistical significance of the trend, the confidence interval for the slope β is obtained in terms of lower and upper bounds based on the order of all possible slopes. This confidence interval is not sensitive to non-normality of the datasets, nor to outliers in the series. Using a limited number of series, we have compared the magnitude and statistical significance of the trend when the traditional linear model was fitted and when our approach was used. Generally, we have found that our approach produced a slightly smaller magnitude than that obtained by the traditional linear model, and, in some cases, trends identified as statistically significant using linear regression are not significant using our procedure due to positive autocorrelations in the time series. Throughout this paper, we use the 5% level to define statistically significant trend.

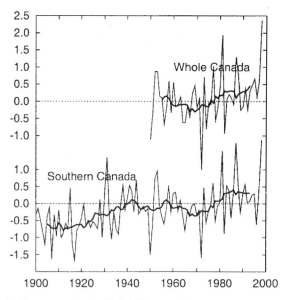

Fig. 3   Departures from the 1961–1990 mean of area average mean temperature (°C). Bold curves are 11-year moving averages.

3 Trends for 1900–1998 (southern Canada)

a Temperatures

Based upon the gridded datasets, the time series of annual mean temperature anomalies relative to the 1961–1990 mean, are computed and shown for southern Canada (1900–1998) in Fig. 3. There is a statistically significant positive trend, which accounts for an increase of 0.9°C, for the region during the period. The linear trend is not exactly monotonic. The rises of temperature prior to the 1940s and after the 1970s account for the significant trend. There is a modest decrease during 1940–1970. Trends differ for different regions, and for different seasons, as well as for daily maximum and minimum temperatures.

Mapped trends in annual and seasonal mean daily maximum temperature are depicted in Fig. 4 for 1900–1998. The annual mean daily maximum temperature has increased since the beginning of this century in the southern part of the country. Statistically significant trends are observed in Alberta and Saskatchewan, as well as in eastern Quebec. The greatest warming, which is in the Prairies, is about 1.5°C over the 99-year period. The spatial patterns of the trends differ from season to season. The mean daily maximum temperature has increased over all of southern Canada in both winter and spring. However, it has increased in some areas but decreased in other areas during summer and fall. Among the four seasons, spring shows the greatest warming. The spatial pattern in this season is similar to the annual one, except that spring warming is stronger and the area with significant upward trend has expanded from the Prairies to include northern B.C. and Manitoba. The greatest warming during spring is well over 2°C for the 1900–1998 period in the Prairies. Warming during winter is more than 1.5°C during 1900–1998 in western Canada, however, the trends are significant only in southwestern B.C. Summer maximum temperature shows significant positive trends in Quebec and

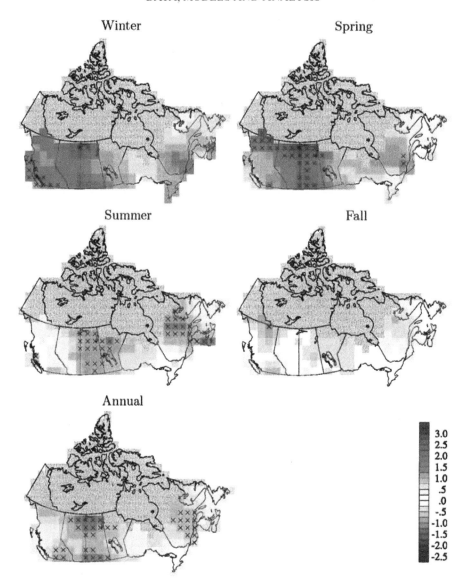

Fig. 4   Trends in daily maximum temperature from 1900–1998. Units are °C per 99-year period. Grid squares with trends statistically significant at 5% are marked by crosses. Grey areas indicate insufficient data.

the Prairies, and significant negative trends in southwestern B.C. Fall shows no significant warming or cooling trends. It is apparent that warming in spring maximum temperature contributed the most to the positive trend in the annual mean of daily maximum temperature.

Strong warming is the sole characteristic of the minimum temperature. This is clearly shown in Fig. 5, where no negative trends can be found. Annual mean minimum temperature has increased from about 1 to 2.5°C during the last 99 years, with strongest warming in the Prairies and southern Quebec. The trends are statistically significant over all of southern Canada. Spring minimum temperatures have warmed the most, with a rate over 3°C during 1900–1998 in the northern Prairies. Winter has the second highest warming rate with some areas in the Prairies and B.C. reaching as high as 3°C, however, some of the warming in Manitoba, Ontario and along the east coast is not significant. In summer, the spatial

pattern is quite uniform with significant trends from 1.3 to 2.0°C. During the fall, the minimum temperatures have warmed with significant increase in eastern Canada and along the west coast. Overall, these results clearly show that daily minimum temperatures, indicators of nighttime temperatures, have significantly increased throughout southern Canada over the past century.

Significant positive trends were also found in the annual and seasonal daily mean temperatures. The spatial patterns (not shown) are similar to those of minimum temperatures but the trends are of a lesser magnitude. This suggests that it is the strong and positive trends of the minimum temperatures, especially during spring and summer, that contributed the most to the trend in the mean temperatures. The annual mean temperatures have increased by about 0.5 to 1.5°C during the last 99 years over southern Canada, with the highest magnitude being a 1.5°C increase in the Prairies. The spatial

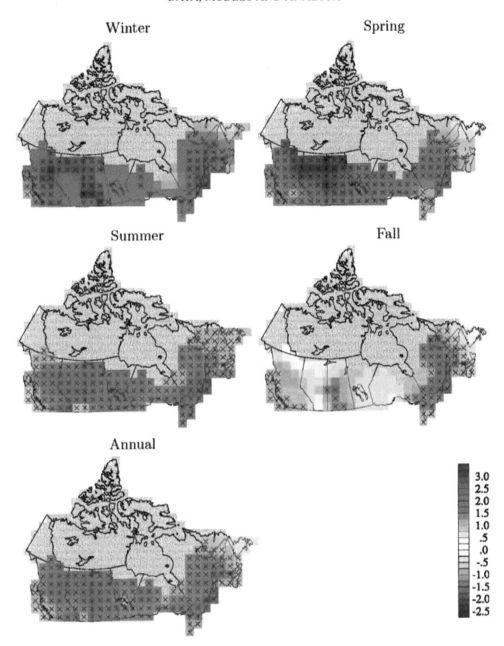

Fig. 5   As in Fig. 4, but for daily minimum temperature.

distribution of trend in the annual mean temperature agrees well with those identified for the U.S. (Karl et al., 1996) across the Canada–U.S. border. In fact, except in the southeast U.S., which showed mostly cooling, and the Canadian High Arctic where data were insufficient, North America has exhibited warming in annual mean temperature during the 20th century. The greatest warming has been in the Canadian Prairies extending to the central U.S. It appears that warming in southern Canada is a part of the larger mid-high latitude continental warming of North America.

The trends in annual and seasonal, maximum, minimum and mean daily temperatures confirm the results for climate regions in southern Canada reported in Environment Canada (1995) for the period 1895–1991. Although the time periods

analyzed are not identical, both studies showed similar results.

Greater warming in the minimum than in the maximum temperatures results in significant decrease in the daily temperature range since the beginning of this century (Fig. 6). Significant decrease in the DTR is observed from coast to coast and for all seasons, with a trend of 0.5 to –2.0°C during 1900–1998. We can conclude that night-time temperature has increased more than daytime temperature in all seasons in the southern part of Canada during the last century. Most of the decrease in the DTR occurred prior to the 1950s, especially late in the first half of the 20th century, coinciding with an increase in total cloud amount in Canadian mid-latitudes during the first half of the 20th century (Henderson-Sellers,

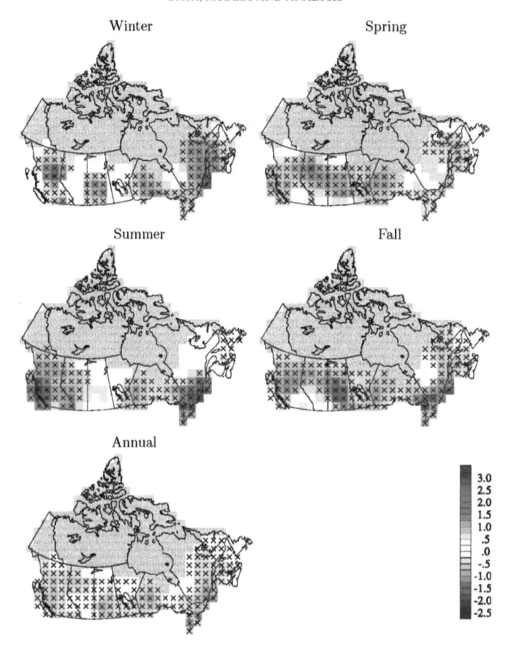

Fig. 6   As in Fig. 4, but for daily temperature range.

1989; McGuffie and Henderson-Sellers, 1988). Henderson-Sellers (1989) did not propose any specific reasons for the cloud increase. Significant decrease in the DTR did not occur in the second half of the century when the greatest increase in greenhouse gases took place. This suggests that trends in the DTR are closely related to changes in total cloud amount. The trends in both DTR and total cloud cover differ from one season to the other. Future investigation into the relationships between the changes in DTR and cloud cover is needed.

b Precipitation
Time series of precipitation anomalies (in percentages of 1961–1990 mean) are displayed in Fig. 7. Annual

precipitation increased by 12% in southern Canada during 1900–1998. The increase in total precipitation resulted from a steady increase during the 1920s to 1970s. There was substantial spatial variability in the precipitation trends which will be detailed in the following.

The trends in annual and seasonal normalized precipitation are provided in Fig. 8. Annual precipitation has increased by 5% to 30% during 1900–1998 in different areas of southern Canada. This annual positive trend is significant from coast to coast, with the exception of southern Alberta and Saskatchewan. The trend is also spatially consistent with the precipitation trend in the U.S. (Karl et al., 1996). While the significant increase in the annual total precipitation

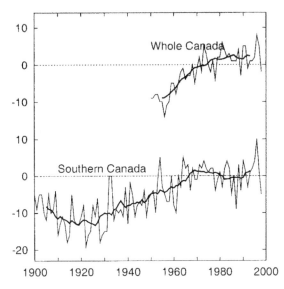

Fig. 7   Departures from the 1961–1990 mean of area average annual precipitation. The departures are the relative changes (in %) to the 1961–1990 mean. Bold curves are 11-year moving averages.

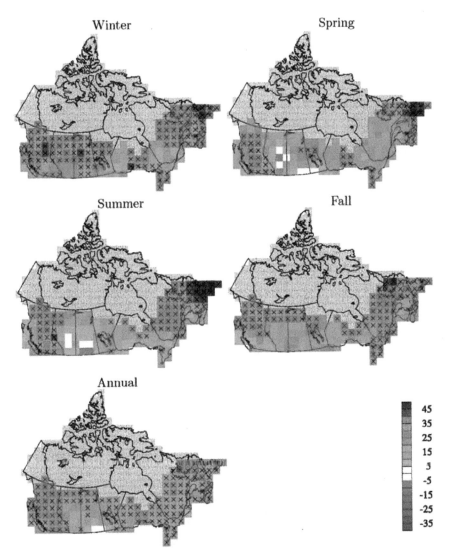

Fig. 8   Trends in precipitation totals from 1900–1998. Units are percent change over the 99-year period. Grid squares with trends statistically significant at 5% are marked by crosses. Grey areas indicate insufficient data.

49

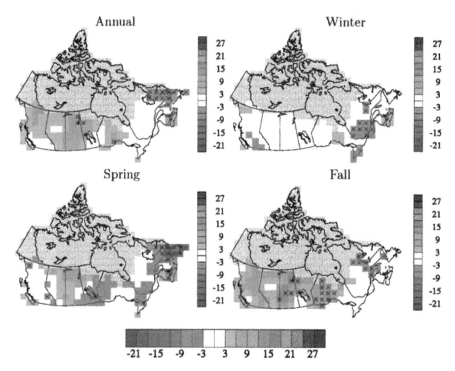

Fig. 9   As in Fig. 8, but for the ratio of snow to precipitation totals.

in B.C. and southeastern Canada extended to the U.S., the insignificant, positive trend in the Canadian Prairies gradually changed to a negative trend south of the Canada-U.S. border with the largest negative trend in Montana and Wyoming. Precipitation increase was greatest, in terms of percentages, in eastern Canada throughout the four seasons, although the trends are not statistically significant in southern Quebec during winter and spring. Precipitation has increased the least in the Prairies with significant trends during the winter, and in B.C. the increase in precipitation is significant in all four seasons. These findings also generally agree with the results of Groisman and Easterling (1994). Although results were based on a much less rigorous adjustment procedure and a smaller dataset, these researchers showed an increase of more than 10% in the annual precipitation over the last century in southern Canada. Our investigation provided spatial distribution of the trends and indicated that the major source of this increase was due not only to changes in eastern Canada (as suggested by Groisman and Easterling), but also due to the increase in the west in B.C.

Overall, the annual ratio of snowfall to precipitation has increased during the period 1900–1998 in southern Canada with a few regions with negative trends (Fig. 9). These increasing trends are due mostly to the increase in winter precipitation (Fig. 8) which generally falls as snow as well as to an increase in the ratio during autumn. Strong and significant negative trends are also observed in eastern Canada, especially in spring.

4 Trends for 1950–1998 (whole of Canada)

a Temperatures

The annual mean temperature anomalies relative to the 1961–1990 mean for the whole of Canada (1950–1998) are also shown in Fig. 3. No significant linear trend is detected for the entire country during the period.

The most striking feature in the pattern of the trends in maximum temperature during this period is the contrast between the west, where strong positive trends are observed, and the northeast, where strong negative trends prevail (Fig. 10). The annual maximum temperature has significantly increased by 1.5 to 2.0°C over the 49-year period in northern B.C. and the Mackenzie basin, and significantly decreased in the northeast part of the country with largest cooling trends of about 1.5°C over the same time period. The spatial patterns of trends are similar for winter and spring. Mean daily maximum temperatures have warmed by more than 3.0°C during the last 49 years in some regions of western Canada in both winter and spring, whereas they have cooled by more than 2°C in some northeast areas. It should be noted that the increase in winter maximum temperature is statistically significant only in some parts of the Mackenzie basin while the area with significant warming is much larger in spring. There are few significant trends during both summer and fall. Mean daily maximum temperatures have increased over most of the country during the summer. In fall, increases have occurred only in the High Arctic and B.C.; decreasing values have occurred in the rest of Canada.

50

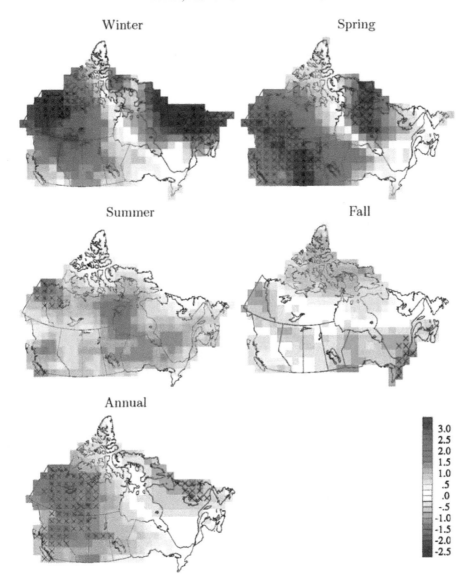

Fig. 10   Trends in daily maximum temperature from 1950–1998. Units are °C per 49-year period. Grid squares with trends statistically significant at 5% are marked by crosses.

Figure 11 shows the spatial distribution of the trends in annual and seasonal minimum temperatures. These have only small differences with respect to their maximum temperature counterparts. The areas with significant warming have shifted from northern Alberta to southern B.C. in the annual temperature. Contrary to findings for the 1900–1998 period during the winter and spring, the magnitude of warming is slightly less in the minimum than in the maximum temperature. The significant summer warming extends over a larger area, while the warming/cooling pattern during the fall remains almost the same as that for maximum temperatures.

The analysis of trends in mean temperature for the period 1950–1998 (not shown) displays similar results to those obtained from maximum and minimum temperatures. The annual mean temperature for Canada has increased by 0.3°C over the last 49 years, but this increasing trend was not statistically significant. It should be noted that the strong warming in winter mean temperature observed in the Mackenzie basin was generally not significant at the 5% level, due to the high variance of winter temperatures in this region.

The trends in DTR calculated for the period 1950–1998 (Fig. 12) are quite different from those obtained for 1900–1998 (Fig. 6). Areas with positive trends are much larger than those with negative trends; but the statistical significance of trends is weak. Some small regions in the High Arctic and across the southern part of the nation showed significant negative trends. Areas with significant increasing trends are located in northeast Canada in the annual, winter and spring time series and correspond to areas of regional cooling. Significant negative trends are found in the High Arctic and populous southeastern regions during the summer and fall. In general, these results agree with the findings of Karl et al. (1993a) who

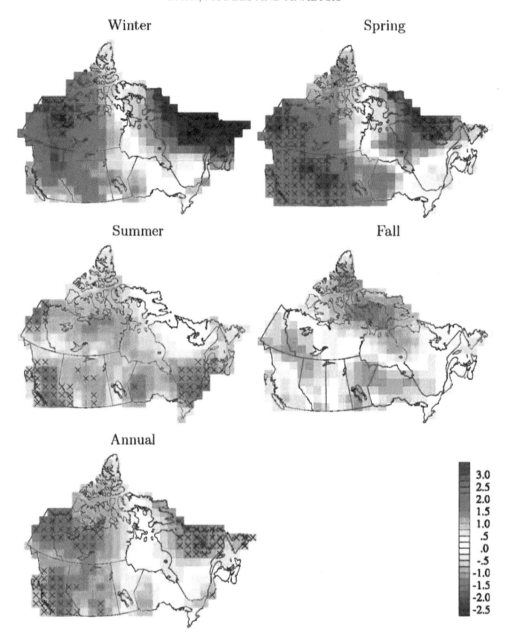

Fig. 11   As in Fig. 10, but for daily minimum temperature.

noted that the DTRs have not been increasing over the interval 1951–1990 in central Canada, and that the country experienced only a moderate decrease during the summer and fall.

The trend in Canadian temperature during the second half of the 20th century has a large-scale, hemispheric background, consistent with the mid-latitude Northern Hemisphere continental warming (during summer and spring) and oceanic cooling seen by Nicholls et al. (1996). It may also reflect changes over the global ocean-atmosphere circulation in the time period. In fact, cooling in northeastern Canada in the last five decades is part of a general decline in northern North Atlantic temperature (Morgan et al., 1993). Winter temperature variability in northeastern Canada at decadal and longer timescales is closely connected with variability of the North Atlantic Oscillation at the same timescales (Shabbar et al., 1997). The warming trend in western Canada is also closely matched by strong warming in sea surface temperature (SST) of the eastern Northern Pacific. This is a part of the El Niño Southern Oscillation (ENSO) like interdecadal variability of Pacific SST, i.e., warming in the tropical and eastern North Pacific and cooling over the central North Pacific (Zhang et al., 1997; Zhang et al., 1998).

Comparison of our findings for southern Canada for the periods 1900–1998 and 1950–1998 reveals a striking difference in DTR; significant negative trends for 1900–1998 in southern Canada are replaced by spatially non-coherent and generally positive trends in the same region. This discrepancy

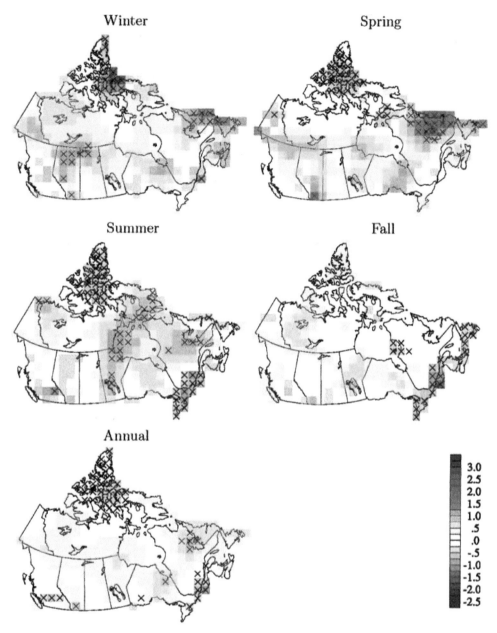

Winter   Spring

Summer   Fall

Annual

3.0
2.5
2.0
1.5
1.0
.5
.0
-.5
-1.0
-1.5
-2.0
-2.5

Fig. 12   As in Fig. 10, but for daily temperature range.

in the trends between the time periods may reflect changes in the total cloud cover. The increase of cloud amount during the first half of the century (Henderson-Sellers, 1989) probably contributed to a faster increase in minimum than in maximum temperatures. In the second half of the century, cloud amount was virtually unchanged and maximum and minimum temperatures changed at more or less the same rate. Thus no spatially consistent trend in DTR, or difference in the trends for minimum and maximum temperatures, was found during 1950–1998.

b *Precipitation*
Time series of precipitation anomalies (in percentages of 1961–1990 mean) for all of Canada are also displayed in Fig. 7.

Annual precipitation increased by 5% during the 1950–1998 period. There was also substantial spatial variability in the precipitation trends.

As depicted in Fig. 13, annual precipitation totals have generally increased by 5% to 35% across the nation during 1950–1998. Significant increases occurred mostly in the Arctic (north of 60°N). Overall, precipitation totals have also shown significant increases in all seasons, and some areas of decrease mostly during the winter time. As noted by Mekis and Hogg (1999), these increases in precipitation are not likely caused by the introduction of the Nipher gauge, the most likely source of artificial increase in the mid-1970s, since most of the changes occurred prior to its introduction.

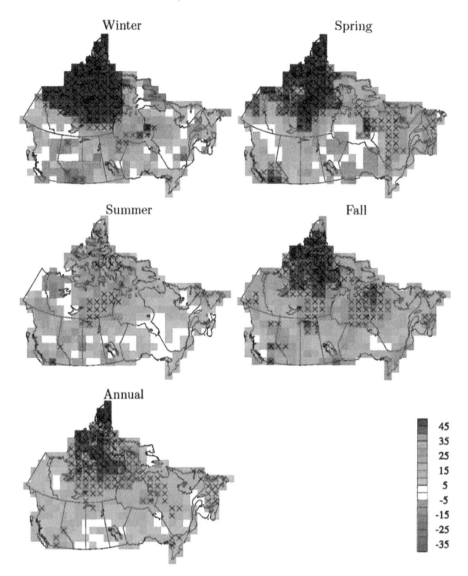

Fig. 13   Trends in precipitation totals from 1950–1998. Units are percent change over the 49-year period. Grid squares with trends statistically significant at 5% are marked by crosses.

The trends in the ratio of snow to total precipitation essentially reflect the combined effects of both precipitation and temperature (Fig. 14). In the annual series, the trends are negative in the south and significantly positive in the north, which fits well with findings for southern Canada by Karl et al. (1993b). The decreasing trend in the annual ratio for southern Canada has resulted primarily from the decrease in winter precipitation, which usually falls as snow (Fig. 13), and from the decrease in spring snowfall. As well, snowfall amounts have decreased in both winter and spring. Brown and Goodison (1996) reported similar trends in the depth of snow on ground measurements. For winter, there is not much change found in the ratio. It is assumed that the increase of spring temperature is so high that it effectively decreased (increased) the proportion of snow (rain) in the total precipitation for the season, although total precipitation was essentially unchanged (see Fig. 13). Such linkage between spring temperature and

precipitation was also reported by Brown and Goodison (1996) and Brown and Braaten (1998).

Winter precipitation during 1900–1998 has increased in southern Canada while it has decreased over large areas during 1950–1998. As well, there are more areas with a decreasing ratio of snow to total precipitation during the spring for the period 1950–1998 than for the period 1900–1998. This emphasizes the need to study decadal and multi-decadal variability of different climate variables in order to gain a better understanding of climate trends and variability.

Trends identified in Canadian precipitation are most likely a regional manifestation of the general tendency of precipitation to increase for high and mid-latitude continents of the Northern Hemisphere in the 20th century. Similar trends were identified in Russia. Wang and Cho (1997) showed upward trends in Russian precipitation (50°N to 70°N) during

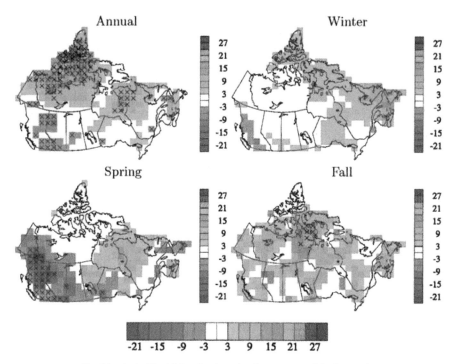

Fig. 14   As in Fig. 13, but for the ratio of snow to precipitation totals.

1881–1989. Ye et al. (1998) reported increases in winter snow depth over most of northern Russia, but decreases over most of southern Russia during 1936–1983.

5 Indices of abnormal and extreme climate

a 1900 to 1998

Annual time series of indices of abnormal climate for southern regions for the period 1900–1998, are consistent with the results of the above trend analysis and are displayed in Fig. 15. Areas affected by abnormally dry conditions (precipitation is less than the 34th percentile) have been gradually decreasing. At the same time, areas affected by abnormally wet conditions (precipitation is more than the 66th percentile) increased. These trends agree with an overall increase in annual precipitation. The areas affected by abnormally low maximum temperatures (temperature is less than the 34th percentile) have decreased, particularly since the 1960s. Areas experiencing abnormally high maximum temperatures, however, showed little trend. The portion of southern Canada experiencing abnormally low minimum temperature was drastically decreased. The reverse was the case for the minimum temperature in its higher quantile. More areas with temperatures in the higher quantile and fewer areas affected by lower quantile temperature vividly displayed the warming picture in southern Canada. Like other parts of the world, Canada has not become hotter (no increase in higher quantiles of maximum temperature), but has become less cold. Because of gradual changes in these areal indices over time, it appears that the warming

TABLE 1.   Changes in the area of southern Canada affected by abnormal climate conditions computed as the differences between the 1950–1998 mean and the 1900–1949 means. Units are percent.

	Precipitation		Max. Temperature		Min. Temperature	
	Dry	Wet	Cold	Warm	Cold	Warm
Annual	−23.4	27.3	−12.8	5.2	−32.2	20.1
Winter	−25.6	21.4	−10.8	7.1	−15.7	9.8
Spring	−12.1	12.3	−10.2	11.8	−21.1	20.3
Summer	−13.9	14.3	−8.5	6.7	−30.6	30.3
Fall	−14.4	17.3	−5.9	−5.1	−20.5	11.6

trend, especially that, in minimum temperature, is not likely caused by sudden step jumps in the data.

On seasonal timescales, the time variations of the areal extent of southern Canada affected by abnormal climate conditions also agreed well with the results of trend analysis. Overall, there was a decrease (an increase) in the areas affected by dry (wet) conditions for all seasons, but the amplitudes of the changes differ from one season to the other. Areas, affected by cold (warm) conditions also decreased (increased). The extent of decrease in the area affected by colder, and of increase by warmer, maximum temperature are about half of their counterparts in the minimum temperature.

Areas affected by abnormal climate conditions were quite different between 1950–1998 and 1900–1949. Table 1 displays the changes in indices of abnormal climate computed as differences between the 1950–1998 mean and the 1900–1949 mean. In winter, areas affected by dry conditions were reduced

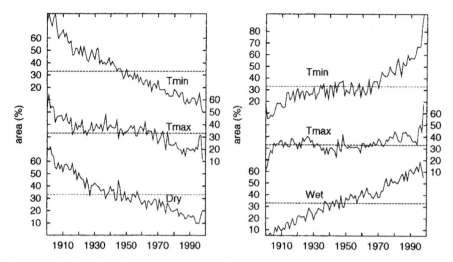

Fig. 15   Percentages of southern Canada affected by abnormally dry, cold conditions (left panel) and wet, warm conditions (right panel). Tmin and Tmax correspond to minimum and maximum temperatures, respectively.

by 25.6% and those affected by wet conditions were increased by 21.4% from the first half to the second half of the century. The largest reduction in the area experiencing colder climate appeared in minimum temperature in summer. Accompanying this reduction is a sharp increase (30.3%) in the areas affected by warm conditions. Since the long-term averaged values of the abnormal climate indices must be 33.3%, such an increase in the index indicates that 90% of abnormally warm daily minimum temperatures in summer occurred in the second half of the century.

Changes in the area affected by extreme climate conditions (precipitation or temperature below the 10th or above the 90th percentiles of their relevant time series) between the two halves of the century generally follow the changes in the abnormal climate conditions; but there are differences, especially in summer for precipitation (Table 2). More specifically, areas affected by both extreme dry and extreme wet conditions during summer increased. That is, while it was generally wetter in the second half of the century, areas affected by extreme dry conditions actually increased (17% of century-long average) in summer. The fact that summer temperature increased, and that areas affected by both extreme dry and extreme wet conditions increased as well during the season, may be an indication of an enhanced hydrological cycle. This is consistent with GCM simulations that hydrological cycles will be enhanced in a warmer world caused by the increase of greenhouse gases in the atmosphere (Kattenberg et al., 1996). All three indices which represent areas of Canada affected by extreme dry conditions during summer, by extreme warm temperatures, and by extreme wet conditions during cool seasons have increased. Karl et al. (1996) also showed an increase in the GCRI which uses those three (with two other) indicators for the U.S. It appears that the changes in the extreme climate indices for Canada and the GCRI for the U.S. point in the same direction: a possible response of regional climate in North America to the increase of greenhouse gases

TABLE 2.   Changes in the area of southern Canada affected by extreme climate conditions computed as the differences between the 1950–1998 mean and 1900–1949 means. Units are percent.

	Precipitation		Max. Temperature		Min. Temperature	
	Dry	Wet	Cold	Warm	Cold	Warm
Annual	−3.6	10.1	−3.1	8.3	−11.4	11.5
Winter	−7.3	5.1	−1.1	5.0	−3.0	4.6
Spring	−0.0	5.1	−3.5	6.4	−7.5	8.5
Summer	1.4	5.2	−7.4	9.0	−10.7	16.4
Fall	−5.5	7.8	0.4	−1.3	−6.2	4.8

in the atmosphere. More studies are needed before we can conclude that such changes are the manifestations of anthropogenic climate change.

Precipitation and temperature are the two most important climate variables affecting society. Coincident changes in both variables, such as abnormally dry when warm, may have more severe impacts on society than changes to only one variable. It is both interesting and important to examine the areas affected by joint abnormal conditions (defined as below 34th or above 66th percentiles of their relevant time series) of these two variables, i.e., areas affected by four different combinations of wet/dry and warm/cold conditions. Figure 16 presents the time series of areas affected by abnormal values of both precipitation and mean temperature (average of maximum and minimum temperature) combined. Areas affected by combined low precipitation and low temperature decreased dramatically throughout the 20th century in annual and seasonal time series. The area of southern Canada experiencing wet and warm conditions was, meanwhile, getting larger. This is a clear reflection of the fact that both temperature and precipitation in the region have generally been increasing. Changes in the areas affected by abnormally dry and warm conditions, and by abnormally wet and cold conditions between the two halves of the century, are much smaller and show some seasonal variability. The seasonal distribution of trends in abnormally dry

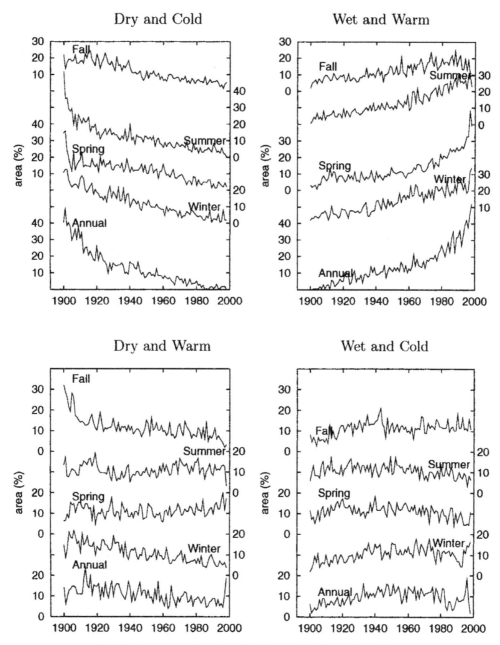

Fig. 16   Percentages of southern Canada covered by abnormal conditions.

and warm conditions, which can be considered to be a proxy of drought, are very interesting. The fall and winter time series of areas affected by abnormally dry and warm conditions have systematically decreased during the century, but in a separate study we found that the same index for the growing seasons, spring and summer, has respectively increased and remained constant over the same period.

b 1950 to 1998
The areas of Canada as a whole affected by abnormally dry conditions gradually decreased in the last five decades of the century, whereas areas affected by abnormally wet conditions increased decade by decade (not shown). Because it has

warmed in the west and cooled in the northeast during this period, no consistent trends were found in the abnormal temperature indices (not shown). It is worth noting that the last two decades show more areas affected by abnormally high temperature (in both maximum and minimum temperatures). Indices of extreme climate for the past five decades displayed similar patterns to those for indices of abnormal climate. Areas affected by combined dry and cold conditions decreased with an increase in areas affected by wet and warm conditions throughout the period. However, these may reflect dominance of these indices by precipitation trends because areas affected by dry and warm conditions also decreased.

6 Conclusions and discussion

Statistical optimal interpolation was employed to generate a high quality Canadian monthly gridded dataset of temperature and precipitation anomalies for the 20th century with the best available station time series as input. Using the gridded dataset, spatial distribution of trends obtained from annual and seasonal time series of six climate variables, maximum, minimum and mean temperatures, diurnal temperature ranges, precipitation totals and ratio of snowfall over total precipitation, have been presented for southern Canada (south of 60°N) for the period 1900–1998, and for the whole country for the period 1950–1998. The statistical significance of the trends has also been assessed.

Annual mean temperature has warmed an average of 0.9°C in southern Canada over the last century. Associated with this increase in mean temperature is a relatively smaller increase in daily maximum temperature and a larger increase in daily minimum temperature. In this century, the increases have resulted in a decrease in diurnal temperature range by 0.5 to 2.0°C. The bulk of decline in DTR occurred during the first half of the century, coinciding with an increase in cloud cover during that period. Both of these results are broadly consistent with greenhouse gas induced climate change, but the timing of the changes, coming prior to the most significant increase in greenhouse gases, suggests that other mechanisms may be responsible. Examining the areas affected by abnormal and extreme temperature confirmed the above analysis. It also suggested that the probability distribution of minimum temperature has shifted with a higher mean but only the left-hand side of maximum temperature distribution has been shifted upward. This indicates that southern Canada has not become hotter but less cold.

Total precipitation has also increased over the last 99 years by 12% in southern Canada. It should be mentioned that the increases in annual precipitation totals do not directly relate to the period of increased cloud cover. Precipitation has a steady increasing trend from the 1920s to 1970, while the major increase in cloud cover occurred during 1936–1950 in mid-latitude Canada (Henderson-Sellers, 1989). The precipitation trend appears to have stopped in about 1970 for the annual time series but not for seasonal time series. There were increasing trends in winter and autumn, decreasing trends in spring and no trend in summer (not shown).

The ratio of solid to total precipitation has also increased, but the trend is not significant. Decreasing trends were observed mostly in southeast Canada in spring. These may be related to changes in both precipitation and temperature and will be discussed later. The time series of the areas affected by abnormal/extreme precipitation and temperature show gradual changes, suggesting the trend detected in precipitation and temperature was not caused by climate jumps.

For the period of complete data coverage, 1950–1998, Canada as a whole has warmed by 0.3°C. There is a strong pattern of warming in the southwest and cooling in the northeast. The maximum and minimum temperatures have increased at a similar rate which has resulted in a slight overall increase in the DTR. Areas affected by abnormal temperatures did not show increasing or decreasing trends (not shown). Overall, there has been an increase of 5% to 35% in annual precipitation totals in Canada during 1950–1998. Although strongest increases were in the north, winter precipitation exhibited decreasing trends in some areas over southern Canada. The increasing trend in precipitation is reflected in the abnormal precipitation indices which showed more areas being affected by abnormally wet conditions, and fewer areas being affected by abnormally dry conditions.

The most interesting finding based on the analysis of the joint occurrence of abnormal temperature and precipitation involved the drought-like conjuncture of warm and dry conditions. In spite of the general increase in precipitation throughout the century, the area affected by abnormally warm and dry conditions in summer remained constant, and has increased in the important spring growing season.

The trend of the ratio of snowfall to total precipitation is complex. It is related to the trends in both precipitation and temperature, as well as the temperature itself. Based on 50 years of monthly snowfall water equivalent and mean temperature data, Davis et al. (1999) analysed the snowfall temperature relationship in Canada. They found that the relationship is positive in high latitudes but is negative in southern Canada, along both coasts, and east of the Rockies. The transition zone, north of which warmer months receive more snowfall than colder months, migrates south-ward from autumn to winter and northward from winter to spring. In southern Canada, spring temperatures have increased, greatly shortening the period of freezing temperatures suitable for snowfall and resulting in a decrease in the ratio. In addition, the decreased snowfall has also helped to enhance the spring temperature response through the snow cover feedback (Brown and Goodison, 1996; Brown and Braaten, 1998). In northern Canada, most temperature increases have occurred in seasons with below freezing temperatures. Warmer conditions have made more moisture available to precipitation events, and these have been in the form of snow so that the ratio of snowfall to total precipitation has increased. These findings are consistent with GCM simulations which have attributed snow accumulation in the arctic region to global warming (Ye and Mather, 1997). The increases in total snowfall amount have also resulted in an increase in the number of heavy snow events (Zhang et al., unpublished manuscript) in northern Canada.

The fact that many previous findings on Canadian climate trends (e.g. Skinner and Gullett, 1993; Environment Canada, 1995) are supported by our analysis of the best available adjusted data indicates that these trends are not artificial (e.g., caused by changes in location of observation sites or programs). Using the most up-to-date and comprehensive dataset confirms the trends to be reliable and robust. Comparison with trends identified in 20th century annual total precipitation and mean temperature for the U.S. (Karl et al., 1996), shows that trends computed from Canadian data for the same period and the same variables are spatially consistent across the Canada–U.S. border.

The results of trend analyses could be influenced by the time period used for the analyses. Quantitative analyses of the influence of decadal and multidecadal variations on the trends are difficult. We are investigating climate variability at both interannual and interdecadal timescales. The results will be reported in the near future. Preliminary findings indicate that spatial distribution of trends similar to those reported in this paper can still be identified after the removal of variability at interannual and interdecadal timescales.

We offered several improvements in this study. Firstly, it is based on the most reliable and recently adjusted temperature and precipitation data available. Secondly, the model used for the trend estimation takes into account the autocorrelation observed in the climate time series, thus the accuracy of estimation and statistical significance is improved. Employing Kendall's tau instead of the least squares method strengthens the computation since Kendall's tau method is less affected by outliers and the non-normality of datasets. We computed trends in annual mean temperature over southern Canada for the last century using both least squares method and our method. The difference in the resultant trends by including/excluding the year 1998, the hottest year on record, in the computations was 10% with the least squares method, but was only 1% with our method. Our method ensures that trend calculation is accurate and robust. Thirdly, rather than using area averages for a number of broad climate regions computed from station data, the station data have been gridded prior to trend analysis. The effect of various record sizes on the computation of trend is reduced. This provided an additional reduction in the local noise and enhanced the signal to noise ratio for trend detection. Finally, we took a new look at the trends by examining the areas of the region affected by abnormal/ extreme climate conditions.

The causes of the different spatial and temporal trends, such as increasing atmospheric greenhouse gases or natural climate variability, cannot be addressed by a study of this nature. There is evidence, nevertheless, suggesting that a certain degree of agreement exists between observed trends in Canadian climate and those predicted by GCMs incorporating an increase in atmospheric greenhouse gases. With new developments in GCMs, and better understanding of global and regional climate, new light will be shed on the nature of trends in Canadian climate.

Acknowledgments

The authors would like to thank Ewa Milewska for her technical support on the Gandin Optimal Interpolation technique and Éva Mekis for the provision of the precipitation datasets. We would also like to thank Walter Skinner and Ross Brown and three anonymous reviewers for their critical and constructive comments which helped to improve this manuscript.

References

ALAKA, M.A. and R.C. ELVANDER. 1971. Optimum interpolation from observations of mixed quality. *Mon. Weather Rev.* **100**: 612–624.

BLOOMFIELD, P. 1992. Trends in global temperature. *Clim. Change*, **21**: 1–16.

—— and D.W. NYCHKA. 1992. Climate spectra and detecting climate change. *Clim. Change*, **21**: 275–287.

BODEN, T.A.; D.P. KAISER, R.J. SEPANKSI and F.W. STOSS (Eds). 1994. Trends'93: A compendium of data on global change. ORNL/CDIAC-65, pp. 800–828.

BONSAL, B.R.; X. ZHANG and W.D. HOGG. 1999. Canadian Prairie growing season precipitation variability and associated atmospheric circulation. *Clim. Res.* **11**: 191–208.

BROWN, R.D. and B.E. GOODISON. 1996. Interannual variability in reconstructed Canadian snow cover, 1915–1992. *J. Clim.* **9**: 1299–1318.

—— and R.O. BRAATEN. 1998. Spatial and temporal variability of Canadian monthly snow depths, 1946–1995. *atmosphere-ocean*, **36**: 37–54.

DAVIS, R.D.; M.B. LOWIT, P.C. KNAPPENBERGER and D.R. LEGATES. 1999. A climatology of snowfall-temperature relationships in Canada. *J. Geophys. Res.* **104(D)**: 11985–11994.

ENVIRONMENT CANADA. 1995. The states of Canada's climate: Monitoring variability and change. State of the Environment Report No. 95-1. Minister of Public Works and Government Services Canada, 52 pp.

GROISMAN, P. YA. and D.R. EASTERLING. 1994. Variability and trends of total precipitation and snowfall over the United States and Canada. *J. Clim.* **7**: 184–205.

GULLETT, D.W.; W.R. SKINNER and L. VINCENT. 1992. Development of a historical Canadian climate database for temperature and other climate elements. *Climatol. Bull.* **26**: 125–131.

—— and ——. 1992. The state of Canada's climate: Temperature change in Canada, 1895–1991. State of the Environment Report No. 92-2. Minister of Supply and Services Canada, 36 pp.

HENDERSON-SELLERS, A. 1989. North American total cloud amount variations this century. *Palaeogeogr. Palaeoclimatol. Palaeoecol. (Global Planet. Change Sect.)* **75**: 175–194.

HOGG, W.D.; P.Y.T. LOUIE, A. NIITSOO, E. MILEWSKA and B. ROUTLEDGE. 1997. Gridded water balance climatology for the Canadian Mackenzie Basin GEWEX study area. *In*: Proc. Workshop on the Implementation of the Arctic Precipitation Data Archive at the Global Precipitation Climatology Centre. 10–12 July, Offenbach, Germany. WMO/TD No. 804, pp. 47–50.

JONES, P.D.; E.B. HORTON, C.K. FOLLAND, M. HULME, D.E. PARKER and T.A. BASNETT. 1999. The use of indices to identify changes in climate extremes. *Clim. Change*, **42**: 131–149.

KARL, T.R.; P.D. JONES, R.W. KNIGHT, G. KULKA, N. PLUMMER, V. RAZUVAYEV, K.P. GALLO, J. LINDSEAY, R.J. CHARLSON and T.C. PETERSON. 1993a. A new perspective on recent global warming: Asymmetric trends of daily maximum and minimum temperature. *Bull. Am. Meteorol. Soc.* **74**: 1007–1023.

——; P. YA. GROISMAN, R.W. KNIGHT and R.R. HEIM JR. 1993b. Recent variations of snow cover and snowfall in North America and their relation to precipitation and temperature variations. *J. Clim.* **6**: 1327–1344.

——; R.W. KNIGHT, D.R. EASTERLING and R.G. QUAYLE. 1996. Indices of climate change for the United States. *Bull. Am. Meteorol. Soc.* **77**: 279–292.

KATTENBERG, A.; F. GIORGI, H. GRASSL, G.A. MEEHL, J.F.B. MITCHELL, R.J. STOUFER, T. TOKIOKA, A.J. WEAVER and T.M.L. WIGLEY. 1996. Climate Models-Projection of future climate. In: *Climate Change 1995, The Science of Climate Change*. HOUGHTON, J.T.; L.G. MEIRA FILHO, B.A. CALLANDER, N. HARRIS, A. KATTENBERG and K. MASKELL (Eds). Cambridge University Press, Cambridge, UK, pp. 285–357.

LEITH, C.E. 1973. The standard error of time-average estimates of climatic means. *J. Appl. Meteorol.* **12**: 1066–1069.

MCGUFFIE, K. and A. HENDERSON-SELLERS. 1988. Is Canadian cloudiness increasing? *Atmosphere-Ocean*, **26**: 608–633.

MEKIS, E. and W.D. HOGG. 1999. Rehabilitation and analysis of Canadian daily precipitation time series. *Atmosphere-Ocean*, **37**: 53–85.

MORGAN, M.R.; K.F. DRINKWATER and R. POCKLINGTON. 1993. Temperature trends at coastal stations in eastern Canada. *Climatol. Bull.* **27**: 135–152.

NICHOLLS, N.; G.V. GRUZA, J. JOUZEL, T.R. KARL, L.A. OGALLO and D.E. PARKER. 1996. Observed climate variability and change. In: *Climate Change 1995, The Science of Climate Change.* HOUGHTON, J.T.; L.G. MEIRA FILHO, B.A. CALLANDER, N. HARRIS, A. KATTENBERG and K. MASKELL (Eds). Cambridge University Press, Cambridge, UK, pp. 132–192.

PHILLIPS, D. 1990. *The Climates of Canada.* (Available from the Canadian Government Publishing Centre, Supply and Services Canada, Catalogue No. EN 56-1/1990E). 176 pp.

SEN, P.K. 1968. Estimates of the regression coefficient based on Kendall's tau. *J. Am. Stat. Assoc.* **63**: 1379–1089.

SHABBAR, A.; K. HIGUCHI, W. SKINNER and J.L. KNOX. 1997. The association between the BWA index and winter surface temperature variability over eastern Canada and west Greenland. *Int. J. Climatol.* **17**: 1195–1210.

SKINNER, W.R. and D.W. GULLETT. 1993. Trends of daily maximum and minimum temperature in Canada during the past century. *Climatol. Bull.* **27**: 63–77.

VON STORCH, H. 1995. Misuses of statistical analysis in climate research. In: *Analysis of Climate Variability: Applications of Statistical Techniques.* H. VON STORCH and A. NAVARRA (Eds). Springer-Verlag Berlin, pp. 11–26.

TRENBERTH, K.E. 1984. Some effects of finite sample size and persistence on meteorological statistics. Part I: Autocorrelation. *Mon. Weather Rev.* **112**: 2359–2368.

VINCENT, L. 1990. Time series analysis: Testing the homogeneity of monthly temperature series. Survey Paper No. 90-05, (Available from York University, 4700 Keele Street, North York, Ontario, Canada, M3J 1P3). 50 pp.

——. 1998. A technique for the identification of inhomogeneities in Canadian temperature series. *J. Clim.* **11**: 1094–1104.

—— and D.W. GULLET. 1999. Canadian historical and homogeneous temperature datasets for climate change analyses. *Int. J. Climatol.* **19**: 1375–1388.

WANG, X.L. and H.-R. CHO. 1997. Spatial-temporal structures of trend and oscillatory variabilities of precipitation over northern Eurasia. *J. Clim.* **10**: 2285–2298.

WIGLEY, T.M.L. and P.D. JONES. 1981. Detecting CO_2-induced climate change. *Nature,* **292**: 205–208.

—— and T.B. BARNETT. 1990. Detection of the greenhouse effect in the observations. In: *Climate Change: The IPCC Scientific Assessment* J.T. HOUGHTON; G.J. JENKINS and J.J. EPHRAUMS (Eds). Cambridge University Press, pp. 195–238.

WOODWARD, W.A. and H.L. GRAY. 1993. Global warming and the problem of testing for trend in time series data. *J. Clim.* **6**: 953–962.

YE, H. and J.R. MATHER. 1997. Polar snow cover changes and global warming. *Int. J. Climatol.* **17**: 155–162.

——; H.-R. CHO and P.E. GUSTAFSON. 1998. The changes in Russian winter snow accumulation during 1936–83 and its spatial patterns. *J. Clim.* **11**: 856–863.

ZHANG, X.; J. SHENG and A. SHABBAR. 1998. Modes of interannual and interdecadal variability of Pacific SST. *J. Clim.* **11**: 2556–2569.

ZHANG, Y.; J.M. WALLACE and D.S. BATTISTI. 1997. ENSO-like interdecadal variability: 1900–93. *J. Clim.* **10**: 1004–1020.

ZHENG, X.; R.E. BASHER and C.S. THOMPSON. 1997. Trend detection in regional-mean temperature series: maximum, minimum, mean, diurnal range, and SST. *J. Clim.* **10**: 317–326.

—— and ——. 1998. Structural time series models and trend location in global and regional temperature series. *J. Clim.* **12**: 2347–2358.

Changes in Daily and Extreme Temperature and Precipitation Indices for Canada over the Twentieth Century

Lucie A. Vincent and Éva Mekis

Climate Research Division, Science and Technology Branch, Environment Canada, Toronto, Ontario

ABSTRACT *This study examines the trends and variations in several indices of daily and extreme temperature and precipitation in Canada for the periods 1950–2003 and 1900–2003 respectively. The indices are based on homogenized daily temperature and adjusted daily precipitation measurements which are special datasets that include adjustments for site relocation, changes in observing programs and corrections for known instrument changes or measurement program deficiencies. For 1950–2003, the analysis of the temperature indices indicates the occurrence of fewer cold nights, cold days and frost days, and conversely more warm nights, warm days and summer days across the country. The results are generally similar for 1900–2003 but they also include a decrease in the diurnal temperature range in southern Canada and a decrease in the standard deviation of the daily mean temperatures for many stations in western Canada.*

The analysis of the precipitation indices for 1950–2003 reveals more days with precipitation, a decrease in daily intensity and a decrease in the maximum number of consecutive dry days. The annual total snowfall significantly decreased in the south and increased in the north and north-east during the second half of the twentieth century. The results are generally similar for 1900–2003. The national series for the century shows an increase in annual snowfall from 1900 to the 1970s followed by a considerable decrease until the 1980s which also corresponds to a pronounced downward trend in the frequency of frost days. No consistent changes were found in most of the indices of extreme precipitation for both periods.

RÉSUMÉ *[Traduit par la rédaction] Cette étude examine les tendances et les variations dans plusieurs indices de température et de précipitation quotidiennes et extrêmes au Canada durant les périodes 1950–2003 et 1900–2003, respectivement. Ces indices sont basés sur des mesures de températures quotidiennes homogénéisées et de précipitations quotidiennes ajustées, qui sont des données spéciales comportant des ajustements pour tenir compte de la relocalisation de sites, de changements dans les programmes d'observation et de corrections relatives à des changements d'instruments ou à des lacunes des programmes de mesure connues. Pour la période 1950–2003, l'analyse des indices de température indique une fréquence plus faible de nuits froides, de jours froids et de jours de gel et, réciproquement, plus de nuits chaudes, de jours chauds et de jours d'été dans le pays. Les résultats sont à peu près semblables pour la période 1900–2003 mais ils incluent également une diminution dans l'amplitude quotidienne de la température dans le sud du Canada et une diminution dans l'écart type des températures quotidiennes moyennes pour plusieurs stations de l'ouest du Canada.*

L'analyse des indices de précipitations pour la période 1950–2003 révèle un plus grand nombre de jours avec précipitation, une diminution dans l'intensité quotidienne et une diminution dans le nombre maximum de jours secs consécutifs. La chute de neige totale annuelle a diminué dans le sud et augmenté dans le nord et le nord-est de façon marquée au cours de la deuxième moitié du vingtième siècle. Les résultats sont à peu près semblables pour la période 1900–2003. La série nationale pour le siècle montre une augmentation dans la chute de neige annuelle de 1900 jusqu'aux années 1970, suivie d'une diminution considérable jusque dans les années 1980, qui correspond aussi à une tendance à la baisse prononcée dans la fréquence des jours de gel. On n'a trouvé aucun changement persistant dans la plupart des indices de précipitations extrêmes pour les deux périodes.

1 Introduction

Recent studies of Canada's climate have shown that significant changes in temperature and precipitation occurred during the twentieth century. The annual mean temperature increased by about 0.9°C in the southern part of the country (south of 60°N) from 1900–1998 (Zhang et al., 2000). This warming is associated with a stronger increase in the night-time temperature as opposed to the daytime temperature, leading to a significant reduction in the diurnal temperature range. The same study also indicated that the annual precipitation has increased by about 12% in southern Canada. Although the ratio of snowfall to total precipitation has increased in most parts of the country, negative trends were also found in eastern Canada, especially during the spring. Considering these important changes in the Canadian climate, it is pertinent to investigate if the past warming was accompanied by detectable changes in temperature and precipitation extremes. Climate modelling studies involving enhanced greenhouse gases have also suggested an increase in the frequency and intensity of climatic extremes in a warmer world (Cubasch et al., 2001). Although changes in long-term climatic means are important, extremes usually have the greatest and most direct impact on our everyday lives, community and environment. For these reasons, detection of changes in extremes has become important in current climatological research.

It is difficult to analyse the changes in specific climatic extreme events such as hurricanes, floods and droughts because these events do not necessarily occur very often and do not happen at the same location. Climate change indices based on daily temperature and precipitation observations have been developed to provide some insights into changes in these extremes (Peterson et al., 2001). They can be obtained from simple climate statistics to describe extremes such as very warm daily temperatures or heavy rainfall amounts. These indices are valuable for studying the impact of climate changes on regional activities, agriculture and economy. They are also helpful for monitoring climate change itself and can be used as benchmarks for evaluating climate change scenarios (Gachon et al., 2005).

During the past few years, there have been numerous studies of trends in extreme temperature and precipitation indices for various regions of the globe. Overall, the findings have suggested a significant decrease in the number of days with extreme cold temperatures, an increase in the number of days with extreme warm temperatures and some detectable increase in the number of extreme wet days in many parts of the world. A recent analysis over Europe has revealed the occurrence of fewer cold nights, more warm days and an increase in the number of extreme wet days from 1946–99, although the spatial coherence of the trends was low for precipitation (Klein Tank and Können, 2003). Similar findings were observed in Africa but over a shorter period 1961–90 (Easterling et al., 2003). In China, the number of cold nights significantly decreased from 1961–2000 leading to a significant decrease in the diurnal temperature range (Qian and Lin, 2004); for precipitation, the number of rain days has decreased throughout most parts of China while the intensity has increased (Zhai et al., 2005). In Australia, the number of warm days and nights has increased while the number of cold days and nights has decreased since 1961; heavy rainfall has also increased in some areas although these trends were not significant (Plummer et al., 1999). The length of the frost-free season has substantially increased in Russia over the past five decades while a slight increase in heavy precipitation events was also observed (Groisman et al., 2003).

Similar patterns have been found in the Americas. In South America, no consistent changes were found in the indices based on daily maximum temperature while significant increasing trends were identified in the number of warm nights (Vincent et al., 2005); the rainfall indices indicate a change to wetter conditions during the last four decades (Haylock et al., 2005). In the Caribbean region, the percentage of days with very warm daytime and night-time temperatures has increased since the late 1950s, while the increase in heavy precipitation events has not been significant (Peterson et al., 2002). For the United States, the frequency of frost days decreased slightly from 1910–98 although a small downward trend was also observed in the number of warm days (Easterling et al., 2000). The same study indicated increasing trends in the number of heavy precipitation events.

Specifically in Canada, Bonsal et al. (2001) analysed annual and seasonal characteristics of temperature extremes from 1900–98. The results exhibited significant increasing trends in both the lower and higher percentiles of the daily minimum and maximum temperature distributions in southern Canada. There are fewer days with extreme low temperatures during winter, spring and summer and more days with extreme high temperatures during the winter and spring; however, no consistent trends were found in the number of extreme hot days during the summer. Spatial and temporal characteristics of heavy precipitation events were also examined in southern Canada for the same period of time (Zhang et al., 2001). The findings revealed no consistent trends in either the number or intensity of precipitation extremes during the last century. In a different study, using different datasets, an increase in heavy and very heavy precipitation events was found for British Columbia south of the 55°N from 1910–2001 (Groisman et al., 2005).

The lack of consistency between index definitions, adjustment methodologies, periods and procedures for computing the trends have made it difficult to compare results across regions and countries. In addition, different station density and area averaging methods can produce different results for larger areas (e.g., countries). Indicators of climatic extremes have been developed, through a number of international workshops, to facilitate consistency across borders and to promote international collaboration (Peterson et al., 2001). Since the year 2000, great efforts have been undertaken in Europe to generalize the indicator definitions and formulae, collect daily climatic data for the continent and analyse the indices using a consistent approach (online access of the European Climate Assessment & Dataset (ECA&D) is available at http://eca.knmi.nl/). A near-global analysis was also carried out by Frich et al. (2002) in order to provide a coherent analysis of the changes in extremes over the globe; however, it was found that large regions around the world had no digital daily data available for analyses. The World Meteorological Organization (WMO) CCL/CLIVAR Expert Team on Climate Change Detection, Monitoring and Indices (ETCCDMI) has recommended a comprehensive list of indices, accompanied by definitions, procedures and guidance, for large-scale regional studies (http://cccma.seos. uvic.ca/ETCCDMI/). They have also coordinated a series of international workshops for the preparation of climate change indices for those regions lacking data that are available internationally: these indices are used for a global analysis being prepared for the next Intergovernemental Panel on Climate Change (IPCC) assessment report.

The objective of this work is to present the trends in various indices of daily and extreme temperatures and precipitation in Canada for the periods 1950–2003 and 1900–2003, respectively. This study updates and extends the analysis of Bonsal et al. (2001) and Zhang et al. (2001) by using the latest version of the homogenized daily temperatures (Vincent et al., 2002) and the adjusted daily precipitation (Mekis and Hogg, 1999) for which some adjustment procedures have been recently refined and the datasets extended to 2003 (http://www.cccma.bc.ec.gc.ca/hccd/). Many of the indices used in this study are based on the definitions and procedures recommended by ECA&D and ETCCDMI to make Canadian results comparable with analysis conducted elsewhere in the world. Additional indices are also analysed to represent the climate characteristics of Canada better (for example, indices based on rain and snow separately). In this paper, the links between the temperature and precipitation indices are highlighted. Not only are the computed trends summarized for each station but national time series are also presented. Probability density functions are calculated and compared for the first half and second half of the century. The data and methodologies are described in Section 2. They are followed by the analyses in Section 3. Sections 4 and 5 contain the discussion and conclusion.

2 Data and methodologies

a Temperature Indices

The temperature indices are computed from homogenized daily maximum, minimum and mean temperatures for 210 stations across the country (Vincent et al., 2002). Homogeneity problems due to station relocation and changes in observing procedures were addressed using a technique based on regression models and surrounding stations (Vincent, 1998). The main causes of the identified inhomogeneities include changes in instrument exposure, changes in observing time, and co-located station observations that were joined to produce a longer time series. Several temperature indices were computed and analysed for the entire twentieth century as well as the past five decades (Mekis and Vincent, 2005; Vincent and Mekis, 2004); however, a set of only eight temperature indices was selected for this study. The definitions are presented in Table 1.

The indices describe cold events (frost days, cold days, cold nights), warm events (summer days, warm days, warm nights), and provide a measure of the temperature variability (diurnal temperature range, standard deviation of the daily mean temperature). They are calculated on an annual basis. Some are based on a fixed threshold (e.g., frost days, summer days), and their impact is easy to understand, however, they are not applicable to every region of the country (e.g., summer days are rare in the Arctic). Others are based on thresholds defined as percentiles calculated at each individual station (e.g., cold nights, warm days), which can be used to facilitate comparison between different climatic regions and stations. The diurnal temperature range is the annual average difference between daily maximum and minimum temperatures. The standard deviation is calculated using the daily mean temperature departures from the daily average for the 1961–90 period (Moberg et al., 2000). For the percentile-based indices, the annual value is missing if more than 5% of the daily data are missing in the given year. For other indices, the value for the month (and consequently the year) is considered to be missing if data for more than three consecutive days or more than five random days are missing in the month.

The indices based on percentiles represent the annual number of days above the 90th or below the 10th percentile level. The percentile values are calculated from the 1961–90 reference period. They are obtained for every day of the calendar year and do not necessarily represent extreme hot days in the summer or extreme cold days in the winter (Jones et al., 1999). The percentiles are computed using 150 values: a five-day window centred on each calendar day for 1961–90. Recently, it was determined that the procedure used to obtain the percentiles during the reference period could create artificial steps at the beginning and the end of this period, therefore a bootstrap resampling procedure was applied to estimate the percentiles better during the reference period (Zhang et al., 2005).

Table 2 shows the 1961–90 averages for six locations representing different climate regions in Canada (Fig. 1). There are more frost days in the Arctic (Eureka), in the high mountains (Barkerville), and in the interior part of the country (Regina), while summer days occur more often in the interior (Regina) and in southern Canada (Harrow). For the four indices based on the percentiles, an almost consistent value of 35 or 36 days is found throughout the table as a direct consequence of the definitions of these indices. The diurnal temperature range and standard deviation of the daily mean temperature (measures of temperature variability) are higher in the interior region (Regina) while they are lower on the west coast (Estevan Point).

b Precipitation Indices

The precipitation indices are computed from adjusted daily rain, snow and total precipitation amounts for 495 stations across Canada (Mekis and Hogg, 1999). Due to much higher spatial variability of precipitation compared to temperature, it was necessary to include more precipitation stations in order to include mesoscale precipitation events. Separate adjustments were applied to daily rain and snow. For each rain-gauge type, corrections to account for wind undercatch, evaporation and gauge specific wetting loss were implemented (for further details see Metcalfe et al. (1997)). For snowfall, ruler measurements were used during the entire period in order to include more stations and to minimize potential discontinuities

TABLE 1. Definitions of the eight temperature and ten precipitation indices used in the study. Tmax, Tmin and Tmean are daily maximum, minimum and mean temperatures respectively. P and R denote total precipitation and rain, respectively.

	Definitions	Units
Temperature Indices		
Frost days	Number of days with Tmin < 0°C	days
Cold days	Number of days with Tmax < 10th percentile	days
Cold nights	Number of days with Tmin < 10th percentile	days
Summer days	Number of days with Tmax > 25°C	days
Warm days	Number of days with Tmax > 90th percentile	days
Warm nights	Number of days with Tmin > 90th percentile	days
Diurnal temperature range	Mean of the difference between Tmax and Tmin	°C
Standard deviation of Tmean	Standard deviation of daily mean temperature from Tmean normal	°C
Precipitation Indices		
Annual snowfall precipitation	Annual accumulated liquid equivalent of snowfall amount	mm
Snow to total precipitation ratio	Annual accumulated snow to total precipitation ratio	%
Days with precipitation	Number of days with precipitation > trace	days
Days with rain	Number of days with rain > trace	days
Simple day intensity index of P	Annual total precipitation/number of days with P > trace	mm d^{-1}
Simple day intensity index of R	Annual total rain/number of days with rain > trace	mm d^{-1}
Max no of consecutive dry days of P	Max. number of consecutive dry days (trace days are excluded)	days
Highest 5-day precipitation amount	Maximum precipitation sum for 5-day interval	mm
Very wet days (≥ 95th percentile)	Number of days with precipitation ≥ 95th percentile	days
Heavy P days (≥ 10 mm)	Number of days with precipitation ≥ 10 mm	days

TABLE 2. Annual averages over 1961–90 for six stations in Canada. P and R denote total precipitation and rain, respectively.

	Estevan Point	Barkerville	Regina	Harrow	Sydney	Eureka	Units
Temperature Indices							
Frost days	33.1	239.9	202.9	128.4	163.6	302.3	days
Cold days	36.6	34.4	35.2	35.6	36.3	34.9	days
Cold nights	35.8	35.6	35.5	35.8	36.3	33.3	days
Summer days	0.1	8.6	58.0	68.5	25.0	0.0	days
Warm days	34.7	34.8	35.8	35.6	35.4	34.6	days
Warm nights	35.1	34.9	34.9	35.9	34.5	35.2	days
Diurnal temperature range	5.5	11.3	12.7	8.5	8.6	6.7	°C
Standard deviation of Tmean	2.0	4.7	6.0	4.4	3.9	5.3	°C
Precipitation Indices							
Annual snowfall precipitation	40.7	506.4	143.8	130.2	524.2	80.4	mm
Snow to total precipitation ratio	1.2	47.0	31.8	13.8	30.2	73.1	%
Days with precipitation	211.4	188.0	113.5	132.6	188.9	63.6	days
Days with rain	208.6	99.9	61.8	105.9	138.7	13.6	days
Simple day intensity index of P	15.6	5.7	3.8	7.3	9.1	1.5	mm d^{-1}
Simple day intensity index of R	15.7	5.8	4.9	7.9	8.7	1.9	mm d^{-1}
Max no of consecutive dry days of P	17.4	13.5	21.3	14.3	9.7	37.2	days
Highest 5-day precipitation amount	251.0	68.6	62.1	94.2	119.7	15.5	mm
Very wet days (≥ 95th percentile)	8.1	1.8	2.4	2.6	3.6	2.3	days
Heavy P days (≥ 10 mm)	101.2	30.1	10.0	30.5	56.3	0.5	days

introduced with the adoption of the Nipher shielded gauge in the mid-1960s. Snow density adjustments, based upon coincident ruler and Nipher observations, were applied to all ruler measurements to provide snow water equivalent and to convert snowfall amounts to snowfall precipitation amounts. The snow water equivalent adjustment factor map was recently updated (Mekis and Hopkinson, 2004). Trace adjustments were also added to the original observations depending on the station location and the measurement program (trace is defined as smaller than the minimum measurable amount). The problems related to trace measurements in Canada are summarized in Mekis (2005) and further research will follow to address these issues properly. Similar to temperature, the data for several stations were merged due to station closure and relocation.

As mentioned in Section 2a, a variety of precipitation indices were computed and examined for 1900–2003 and 1950–2003 (Mekis and Vincent, 2005; Vincent and Mekis, 2004). A set of ten precipitation indices was chosen for this study and the definitions are presented in Table 1. They illustrate the precipitation type, frequency, intensity and extremes. Some of the indices exclude the trace measurements as indicated in Table 1: for example, trace precipitation is excluded from the computation of the maximum number of consecutive dry days since it does not generally alleviate dry conditions. The annual snowfall precipitation and the snow to total precipitation ratio provide information on changes in solid precipitation, which is a very important climate characteristic for this country. The days with precipitation (rain and snow)

Fig. 1   Six stations representing different climatic regimes in Canada.

and the days with rain provide an indication of the number of precipitation and rain events while the simple day intensity index of precipitation and of rain describes the mean daily amount. The maximum number of consecutive dry days is used to characterize the length of dry spells. The highest five-day precipitation amount, the very wet days and the heavy precipitation days describe some extreme features of precipitation. For very wet days, the 95th percentile reference value was obtained from all non-zero total precipitation events for 1961–90. It is preferable to use indices based on percentile threshold rather than fixed threshold in Canada due to the huge scale differences in total precipitation between costal and Arctic regions. The monthly, and consequently the annual, values are considered to be missing when data for more than three consecutive days or more than five random days are missing in a given month.

The 1961–90 annual average values of the precipitation indices are given in Table 2 for the same six locations. The annual accumulated snowfall precipitation varies from 524.2 mm at Sydney on the east coast to 40.7 mm at Estevan Point on the west coast while the snow to total precipitation ratio varies from 73.1% to 1.2% at Eureka and Estevan Point respectively. The annual total precipitation and the amount of precipitation that fell as rain can also be computed from these values using simple multiplication and difference. The North is the driest area (Eureka) followed by the Prairies (Regina). There is a strong contrast in the difference in the precipitation frequency and intensity in the coastal areas compared to the rest of the country, especially in the Arctic. On average, there are 211.4 days with precipitation at Estevan Point with an average precipitation intensity of 15.6 mm d[−1] whereas there are only 63.6 days with precipitation (trace days are excluded) at Eureka with an average precipitation intensity of 1.5 mm d[−1]. The length of dry spells also characterizes the climate of the country. Annual average maximum length of dry spells of 21.3 and 37.2 days are observed in Regina and Eureka respectively. For the precipitation extreme indices, the highest five-day precipitation is observed at Estevan Point on the west coast with an annual average of 251.0 mm, along with an annual average of 8.1 very wet days and 101.2 heavy precipitation days. The second location with the highest averages of precipitation extremes is Sydney on the east coast although the magnitude is much lower.

c Trend Estimation
Frequently the best-fit linear trend is used to describe the linear change of a climatological time series. However, the estimated trend is often affected by outlier values or extremes such as those observed during La Niña and El Niño years. In addition, the statistical significance of the trend can be influenced by the non-symmetrical distribution of the data which often occurs in the indices of extreme events. In this work, an estimator of the slope proposed by Sen (1968) and based on Kendall's rank correlation was applied instead. The slope estimate is the median of the slopes calculated from all joining pairs of points in the series and the confidence interval is obtained from tabulated values in Kendall (1955). Since serial correlation is often present in many climatological time series, a procedure was also applied to take into account the first lag autocorrelation when present. A detailed description of the trend computation can be found in Zhang et al. (2000) and in Wang and Swail (2001). In the present study, the trends were computed for the periods 1950–2003 and 1900–2003 only if more than 80% of the values were present. The statistical significance of the trends was assessed at the 5% confidence level.

d National Time Series
Sometimes it is of interest to produce one single time series to represent a region or country in order to summarize the changes observed over the whole area. However, it is difficult to create such a series for Canada since it is a vast country with different climatic regimes, the station distribution is sparse in the north and the data do not necessarily cover the entire 1900–2003 period. Regional time series based on temperature and precipitation indices were obtained for the globe (Frich et al., 2002) and for Europe (Klein Tank and Können, 2003). In this work, a simple approach was used to facilitate the interpretation of the trends. The country was divided into 5° × 5° grid boxes. The arithmetic average was obtained from all stations available within a box and the national mean was computed from all box values. For the majority of the indices, the national anomaly series for 1950–2003 was calculated from the 1961–90 departures for the stations with more than 80% of the years in the 1950–2003 period; similarly, the national series for 1900–2003 was computed using the 1961–90 departures for the stations with more than 80% of the years in the 1900–2003 period. It is important to mention that the "national" series for 1900–2003 mainly represents the land mass south of 50°N in the Maritimes, Québec and Ontario, and south of 60°N in Manitoba, Saskatchewan, Alberta and British Columbia; only a few stations from the Yukon and Northwest Territories contributed to the national

series. This procedure is reasonable for different temperature and precipitation indices such as the frequency of cold days or the number of days with precipitation. However, since precipitation amounts can be very large in some areas compared to others, the percentage anomalies (departures from the 1961–90 average divided by the 1961–90 average and multiplied by 100) were used for the annual snowfall precipitation, simple day intensity indices and highest five-day precipitation.

e *Probability Density Function*
For each index, the probability density functions were calculated over two independent intervals, 1900–51 and 1952–2003, in order to determine if there was a change in the distribution of the given index over the past century. The stations involved are only those that have data available for more than 80% of the years for the entire 1900–2003 period. The probability density functions are obtained by simply computing the frequency of events across its range to produce a histogram. The results are then normalised to sum to one, providing the estimated probabilities rather than the frequency distribution. The probability density function plots are closely examined to determine if the distribution for the first half of the century differs from the distribution for the second half of the century. The non-parametric Wilcoxon Rank Sum test (Bethea et al., 1995) was used for testing the equality of the 1900–51 and 1952–2003 intervals' mean values. This test is less sensitive to the difference in variance and skewness of the distribution compared to the t-test. It was performed at the 5% confidence level.

3 Results
a *Trends in Temperature Indices, 1950–2003*
The number of stations with significant decreasing, non-significant, and significant increasing trends is given in Table 3 for each index. The results show that the number of cold events has significantly decreased while the number of warm events has significantly increased at many locations. It seems that changes are more evident in the night-time temperature than the daytime temperature since there are more stations with significant trends in both cold nights and warm nights. Figures 2 and 3 present the spatial distribution and magnitude of the trends. Many stations show a decrease of 10 to 20 days in the number of cold days and cold nights and an increase of 10 to 20 days in the number of warm days and warm nights over the last 54 years. The significant cold day and warm day trends occur mostly in western and central Canada while cold night and warm night trends are observed across the entire country. The national series (Table 4) indicates that there have been 11.9 fewer cold days and 8.0 more warm days since the beginning of the 1950s; the change is more pronounced in the night-time temperature with 15.1 fewer cold nights and 11.3 more warm nights occurring over the last 54 years. The number of frost days has decreased by 8.0 days and significant trends are evident at many locations in the western and eastern regions of the country (Fig. 3). The number of summer days has increased across the country with an average increase of 6.0 summer days during 1950–2003.

The diurnal temperature range and the standard deviation of the daily mean temperature do not indicate any strong or consistent changes over the second half of the century. Table 3 shows less than 20% of the stations with either significant positive or negative trends. Some stations show significant decreasing trends of 0.5° to 1.0°C in the diurnal temperature range over southern British Columbia, southern Manitoba, southern Ontario and Québec, and at a few locations in the Arctic (Fig. 3). The standard deviation of the daily mean temperature shows some increasing variability in the central and eastern regions while a decrease is also observed in western Canada although most of the trends are not significantly different from zero. The national series indicate no significant changes in either index during 1950–2003 (Table 4).

b *Trends in Precipitation Indices, 1950–2003*
Table 3 also includes the number of stations with significant trends for the selected ten precipitation indices. Almost 20% of the stations show significant trends in the annual total snowfall. Figure 4 indicates that the annual snowfall precipitation has significantly decreased, by at least 50 mm, during the last 54 years at some stations in southern Canada. In contrast, significant increases of at least 50 mm are also observed in the Arctic and in the eastern region of the country. The snow to total precipitation ratio confirms this result by showing decreasing trends of 10 to 20% in the south while a few significant increasing trends are found in the Arctic and on the east coast. The national trends do not show significant change in the annual snowfall precipitation and in the snow to total precipitation ratio (Table 4).

Significant changes are observed in the number of precipitation events where the number of days with total precipitation (rain and snow) and the number of days with rain have significantly increased by about 20 to 30 days at the majority of locations across the country (Fig. 4). These results include only days with measurable precipitation (greater than trace). It is important to mention that since the precision of minimum measurable precipitation has improved over time, the trends found in the number of precipitation events could be slightly overestimated. A few stations also show some decreasing trends in the days with precipitation, mainly in the central part of the country. The national series indicate that, on average, there are 21.4 more days with precipitation and 20.3 more days with rain in Canada today compared to the 1950s. Conversely, it seems that the mean daily amount of the precipitation has significantly diminished. The simple day intensity index of the annual total precipitation and of the annual total rainfall show significant decreasing trends of 1 to 3 mm d^{-1} at many locations across the country (Fig. 4). The national series indicate an overall significant decrease in the intensity of precipitation of 5.3% and an even larger decrease of 17.5% in the intensity of rain during the past 54 years.

The results for the extreme precipitation events are not as conclusive. Table 3 shows that there are fewer stations with significant trends in the indices characterizing extreme dry and wet events. The maximum number of consecutive dry

TABLE 3. Number of stations with significant negative, not significant, and significant positive trends for temperature and precipitation indices for 1950–2003 and 1900–2003 respectively (significant at the 5% level). The numbers in bold indicate that more than 20% of the stations have significant trends. P and R denote total precipitation and rain, respectively.

		1950–2003			1900–2003	
	negative	not significant	positive	negative	not significant	positive
Temperature Indices						
Frost days	**49**	110	0	**40**	25	0
Cold days	**39**	126	2	**32**	43	0
Cold nights	**71**	96	0	**62**	12	0
Summer days	0	118	**30**	4	49	9
Warm days	0	129	**38**	1	57	**17**
Warm nights	0	109	**58**	0	34	**40**
Diurnal temperature range	23	119	13	**34**	25	0
Standard deviation of Tmean	6	143	13	**15**	57	1
Precipitation Indices						
Annual snowfall precipitation	34	177	10	8	54	14
Snow to total precipitation ratio	**44**	161	4	**17**	48	7
Days with precipitation	8	101	**99**	1	20	**51**
Days with rain	3	68	**143**	1	19	**53**
Simple day intensity index of P	**58**	142	8	**40**	30	2
Simple day intensity index of R	**70**	143	1	**38**	32	3
Max no of consecutive dry days of P	34	172	2	**47**	25	0
Highest 5-day precipitation amount	5	223	11	1	65	9
Very wet days ($\geq 95^{th}$ percentile)	3	198	13	3	61	9
Heavy P days (≥ 10 mm)	8	187	19	4	52	**17**

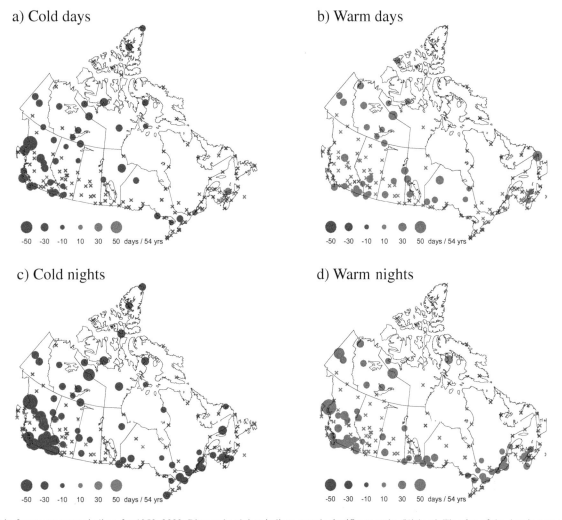

a) Cold days   b) Warm days

c) Cold nights   d) Warm nights

Fig. 2 Trends in four temperature indices for 1950–2003. Blue and red dots indicate trends significant at the 5% level. The size of the dots is proportional to the magnitude of the trend. Crosses denote non-significant trends.

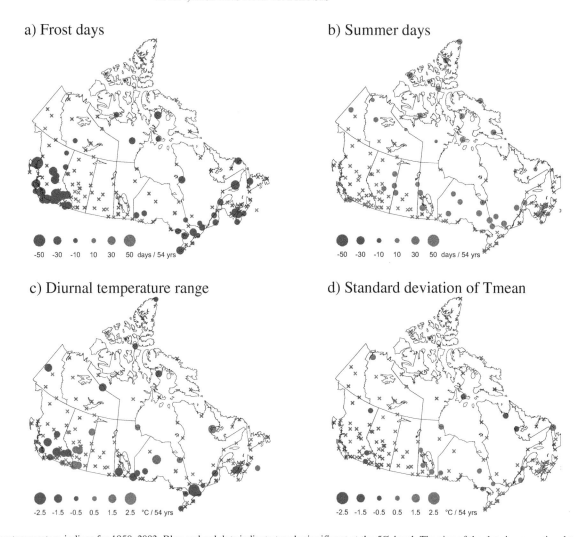

a) Frost days

b) Summer days

c) Diurnal temperature range

d) Standard deviation of Tmean

Fig. 3   Trends in four temperature indices for 1950–2003. Blue and red dots indicate trends significant at the 5% level. The size of the dots is proportional to the magnitude of the trend. Crosses denote non-significant trends. The grey dots indicate that the index is not characteristic for the location.

TABLE 4.   National trends for 1950–2003 and 1900–2003. The numbers in bold indicate the trends significant at the 5% level. The * indicates that the station percentage anomalies were used in the computation of the national series. P and R denote total precipitation and rain, respectively.

	1950–2003	1900–2003	Units
Temperature Indices			
Frost days	**−8.0**	**−15.6**	days
Cold days	**−11.9**	**−10.3**	days
Cold nights	**−15.1**	**−28.2**	days
Summer days	**6.0**	4.7	days
Warm days	**8.0**	**8.5**	days
Warm nights	**11.3**	**21.1**	days
Diurnal temperature range	−0.1	**−0.9**	°C
Standard deviation of Tmean	−0.1	−0.1	°C
Precipitation Indices			
Annual snowfall precipitation*	1.0	9.1	%
Snow to total precipitation ratio	−1.5	0.2	%
Days with precipitation	**21.4**	**42.9**	days
Days with rain	**20.3**	**28.8**	days
Simple day intensity index of P*	**−5.3**	**−31.9**	%
Simple day intensity index of R*	**−17.5**	**−30.0**	%
Max no of consecutive dry days of P	**−3.5**	**−6.8**	days
Highest 5-day precipitation amount*	4.7	1.9	%
Very wet days (≥ 95th percentile)	0.4	0.1	days
Heavy P days (≥ 10 mm)	**1.8**	**1.6**	days

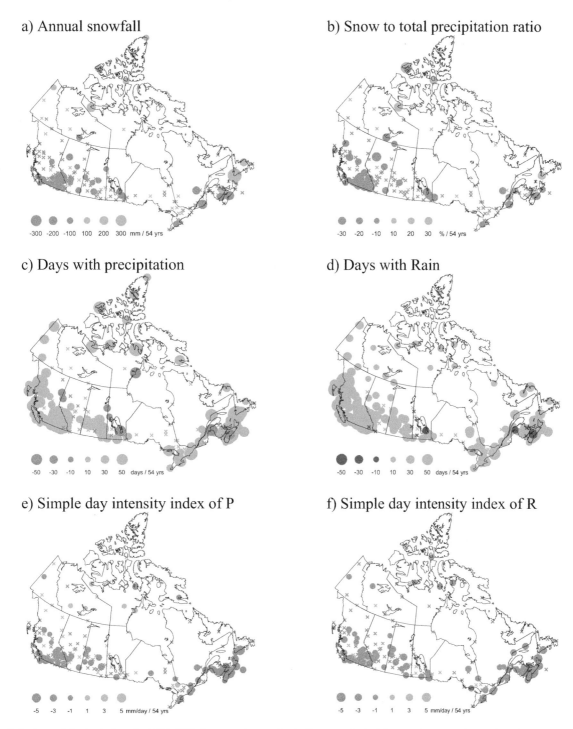

a) Annual snowfall

b) Snow to total precipitation ratio

c) Days with precipitation

d) Days with Rain

e) Simple day intensity index of P

f) Simple day intensity index of R

Fig. 4   Trends in six precipitation indices for 1950–2003. Brown and green dots indicate trends significant at the 5% level. The size of the dots is proportional to the magnitude of the trend. Crosses denote non-significant trends.

days exhibits a mixture of increasing and decreasing trends across the country (Fig. 5); although several stations exhibit a significant decrease in the number of consecutive dry days, which is in agreement with the observed increase in the frequency of days with precipitation. The national series show a significant decrease of 3.5 days in the annual maximum length of dry spells over the past 54 years (Table 4). There is also a mixture of increasing and decreasing trends in the high-

est five-day precipitation, very wet days and heavy precipitation days (Fig. 5). The national series show a significant increase of 1.8 days in heavy precipitation days (Table 4). Overall, these results suggest that the increase in the annual total precipitation observed during the second half of the century is due mostly to more days with precipitation with no consistent changes being found in most extreme precipitation indices.

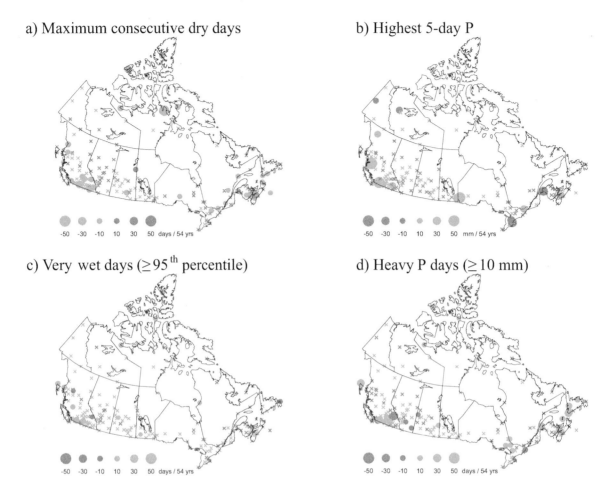

a) Maximum consecutive dry days

b) Highest 5-day P

c) Very wet days ($\geq 95^{th}$ percentile)

d) Heavy P days (≥ 10 mm)

Fig. 5   Trends in four precipitation indices for 1950–2003. Brown and green dots indicate trends significant at the 5% level. The size of the dots is proportional to the magnitude of the trend. Crosses denote non-significant trends.

c *Trends in Temperature Indices, 1900–2003*
In general, the changes observed during the 1900–2003 period are similar to those observed during 1950–2003. The occurrence of extreme cold events significantly decreased at most locations while the occurrence of extreme warm events significantly increased (Table 3). The changes are definitely more pronounced for night-time temperatures than for daytime temperatures: overall, there are 10.3 fewer cold days and 8.5 more warm days whereas there are 28.2 fewer cold nights and 21.1 more warm nights today in Canada compared to the beginning of the 1900s (Table 4). As mentioned in Section 2d, station data were used only if data were available for 80% of the years during 1900–2003. This condition was met by about 40 to 45% of the locations used in the 1950–2003 analysis and most of these stations are located in southern Canada.

The trends in the frequency of cold days and warm days are significant at many locations mainly in the western and eastern regions (not presented here) while the trends in the number of cold nights and warm nights are stronger and observed across the entire country. The probability density functions (PDFs) for the cold and warm events are presented in Fig. 6.

The figure shows a shift toward the occurrence of more warm days and more warm nights in the second half of the century associated with a shift toward the occurrence of fewer cold days and cold nights. The averages are significantly different for these four temperature indices. It is evident that the changes are pronounced in night-time temperature, particularly for cold nights where there is a visible change in the shape of the distribution.

The frequency of frost days significantly decreased during the twentieth century at the majority of stations in southern Canada (Fig. 7). The national trend also shows an average of 15.6 fewer frost days since the beginning of the century (Table 4). However, the decreasing trend is not strictly monotonic (Fig. 8): it shows little change between decreasing periods from the 1920s to the 1930s and from the 1970s to the 1980s. The results also indicate that the number of summer days has not changed consistently during the same period. The national trend shows an overall increase in the number of summer days which is interrupted by a short decrease from the 1940s to the 1960s.

For the diurnal temperature range and the standard deviation of the daily mean temperature, the findings are somewhat

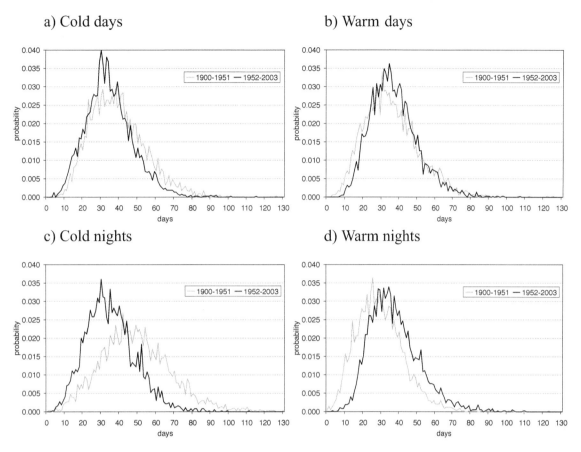

Fig. 6   Probability density functions for the periods 1900–51 and 1952–2003.

different from those observed during 1950–2003. Most stations across southern Canada display significant decreasing trends in the diurnal temperature range (Fig. 7) and the national series indicates a significant decrease of 0.9°C over the century. Furthermore, the time series shows an abrupt decrease in the 1950s (Fig. 8). It is well known that both the annual maximum and minimum temperatures have increased during the past century (Zhang et al., 2000), however, the annual minimum temperature was much cooler than the annual maximum temperature during the period prior to the 1940s (Vincent et al., 1999): this led to a considerable decrease in the diurnal temperature range at the beginning of the 1950s. The causes of the important increase in the night-time temperatures during the 1940s and 1950s are uncertain but they may be related to changes in cloudiness (Henderson-Sellers, 1989). Further investigation into the relationship between temperature and cloud cover is needed. The standard deviation of the daily mean temperature indicates a significant decreasing trend in the south-west part of the country (Fig. 7), however, the national series indicates no significant change during the past century.

d *Trends in Precipitation Indices, 1900–2003*
Overall, the changes seen during the entire twentieth century are similar to those observed in the last 54 years (Table 3).

Approximately 35% of stations included in the shorter interval qualified to be included in the century-long analysis. The annual snowfall precipitation exhibits a mixture of increasing and decreasing trends in the south (Fig. 9) with a non-significant increase for the entire country (Table 4). Many of the most significant and largest increases in annual snowfall are located in the Maritimes which suggests an increase in the number of east coast winter storms. The national series shows an almost consistent increase in annual snowfall precipitation from the beginning of the century up to the 1970s, which is followed by a substantial decrease until the 1980s and no change up to 2003 (Fig. 10). The increase in the national snow to total precipitation ratio is smaller than the increase in annual snowfall (Table 4) which means that the annual rainfall increased more than the annual snowfall. It is important to remember that these national series represent stations located mostly in the south.

The most notable changes are found in the number and intensity indices of the daily precipitation events. The number of days with precipitation (Fig. 9) and the number of days with rain have increased substantially everywhere in southern Canada. On average, today there are 42.9 more days with precipitation (snow and rain) and 28.8 more days with rain since the beginning of the century (days with trace precipitation are

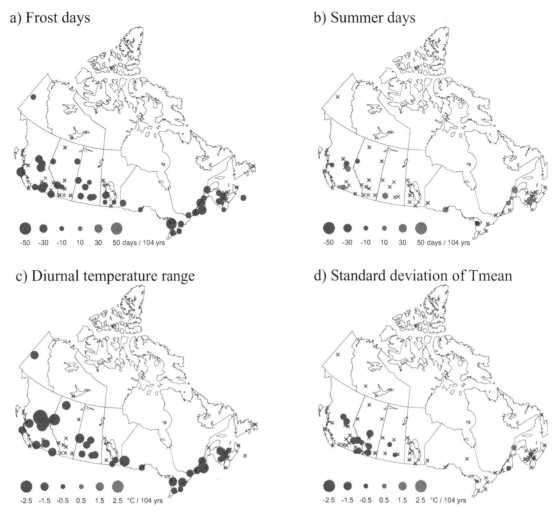

a) Frost days

b) Summer days

c) Diurnal temperature range

d) Standard deviation of Tmean

Fig. 7   Trends in four temperature indices for 1900–2003. Blue and red dots indicate trends significant at the 5% level. The size of the dots is proportional to the magnitude of the trend. Crosses denote non-significant trends.

excluded). As mentioned in Section 3b, the magnitude of the trends could be slightly overestimated since the precision of minimum measurable precipitation has increased over time. The number of days with snow has also increased significantly in the south during the 1900–2003 period (Mekis and Vincent, 2005). The national series shows an almost steady increase in days with precipitation (Fig. 10) and spatially, the trends are strong and significant at the majority of stations across Canada. The PDFs exhibit a statistically significant shift towards more days with precipitation and it also seems that the shape of the distribution has changed (Fig. 11). On the other hand, the mean daily amount of precipitation has decreased over the past 104 years; more than half of the stations show significant decreasing trends (Table 3). The national trend of the simple day intensity index for total precipitation and simple day intensity index for rain indicate significant changes of –31.9% and –30.0% respectively (Table 4). The station trends are significant almost everywhere in the south (not presented here). The PDFs indicate a small shift towards lower daily amounts during the second half of the century (Fig. 11).

Changes are observed in the index characterizing the occurrence of extreme dry events whereas no strong patterns are found in the number of extreme wet events. The maximum length of consecutive dry days has decreased by about 10 days at many southern stations since the beginning of the century (Fig. 9) which is in agreement with the increase in the number of days with precipitation. The national series shows a steady significant decrease of 6.8 days during the 104 years (Table 4 and Fig. 10), and the PDFs show a shift towards shorter dry periods in the second half of the century (Fig. 11). Compared to extreme dry events, fewer changes are observed in the frequency of extreme wet events. The national series show no significant increase in the highest five-day precipitation and in the very wet days (Table 4, Fig. 9 and Fig. 10). A significant increase of 1.6 days is found in the number of heavy precipitation days (≥10 mm), however, the PDFs show little change in the distribution (Fig. 11). Similar to 1950–2003, no consistent patterns are found in the number of extreme wet events and it seems that the increase in precipitation observed over the twentieth century is mainly due to more days with precipitation.

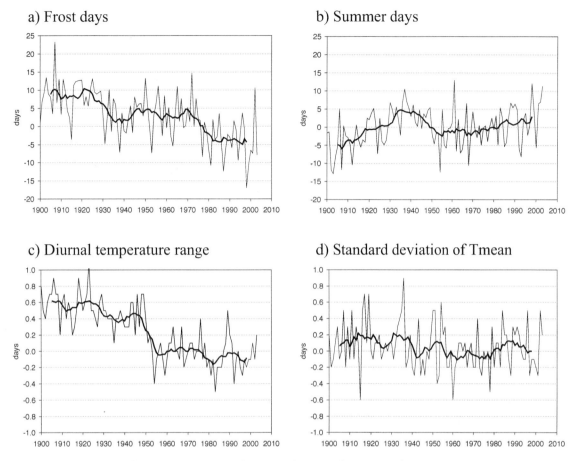

Fig. 8   National time series for 1900–2003 computed from the station anomalies. The bold line represents the 11-year running mean.

4 Discussion

Results from the analysis of extreme temperatures support the findings of Bonsal et al. (2001) which indicate that, from 1900 to 1998, most of southern Canada shows significant increasing trends in the lower and higher percentiles of the daily minimum and maximum temperature distributions. The warming observed in Canada over the past century is associated with fewer cold temperature events and more warm events. As mentioned earlier, these indices do not represent extreme hot days in the summer or extreme cold days in the winter but describe cold or warm events at any time during the year. The changes in the daytime and night-time temperatures are not symmetrical since the warming is more pronounced in the night-time as opposed to the daytime temperatures. This leads to a significant decrease in the diurnal temperature range over the past 104 years. Larger warming trends in the night-time indices have been observed in other regions of the globe. In particular, for Europe, an increase of 11.3 warm days and 13.5 warm nights occurred from 1946–99 (Klein Tank and Können, 2003). Similarly, in this study, an increase of 8.0 warm days and 11.3 warm nights was detected in Canada for 1950–2003. In addition, other indices based on daily minimum temperature have shown a greater warming than those based on the daily maximum temperature. The trends in Europe indicated 9.2 fewer frost days

and 4.3 more summer days, whereas in Canada there have been 8.0 fewer frost days and 6.0 more summer days during the past 54 years. These findings are also in agreement with those of the near-global analysis which suggests an overall increase in warm nights and an overall decrease in the frequency of frost days from 1946 to 1999 (Frich et al., 2002).

Extreme precipitation results support the findings of Zhang et al. (2001) who suggest that the observed increase in precipitation totals in Canada is due mostly to an increase in the number of small to moderate precipitation events and no consistent patterns in the extremes were found during 1900–98; however, in the current study, a small significant increase was also found in the number of days with heavy precipitation (≥10 mm) during 1900–2003. For the extreme precipitation indices, the spatial coherence of the trends is much lower than those observed for the extreme temperature indices. Even over relatively short distances, both positive and negative trends can be found at stations across the country. Significant increasing trends in precipitation extremes have been detected in other regions of the globe. For example, significant increases of 1.1 days and 1.6 days were found in the frequency of very wet days (≥ 95th percentile) and heavy precipitation days (≥10 mm) in Europe for 1946–99 (Klein Tank and Können, 2003). In Canada, this study shows increases of 0.4 days and 1.8 days for the same two indices but for the slightly different

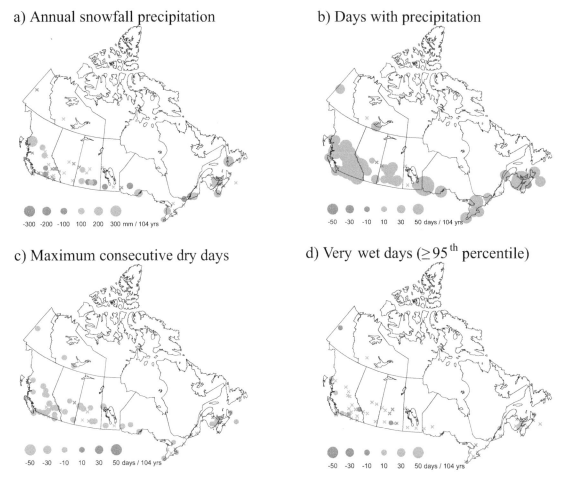

a) Annual snowfall precipitation

b) Days with precipitation

c) Maximum consecutive dry days

d) Very wet days ($\geq 95^{th}$ percentile)

Fig. 9   Trends in four precipitation indices for 1900–2003. Brown and green dots indicate trends significant at the 5% level. The size of the dots is proportional to the magnitude of the trend. Crosses denote non-significant trends.

period of 1950–2003. The near-global analysis (Frich et al., 2002) suggests an overall increase in the number of days with heavy precipitation, an overall increase in the highest five-day precipitation amount, and a general reduction in the maximum number of consecutive dry days (MCDD). In Canada, the reduction in MCDD and the increase in the days with heavy precipitation were confirmed from the above list. For the simple day intensity index, a mixed pattern of increasing and decreasing trends was found over the globe. The near-global trend indicates a slight increase in the intensity index while the trend for Canada shows a general significant decrease for the similar 1950–2003 period.

The observed temperature warming has changed some of the precipitation from a solid to liquid form which in turn can have a serious impact on some of the economic activities in Canada such as agriculture. A distinct pattern of warming in the western and southern regions of the country and cooling in the north-east was identified in the winter and spring during the second half of the twentieth century (Zhang et al., 2000). This led to a substantial change in the total snowfall precipitation which has significantly decreased, mainly in the west and in the Prairies since 1950. On the other hand, the annual snowfall precipitation has increased in the Arctic and

at a few stations in the east. The shift from solid to liquid precipitation is also confirmed by the decrease in snow to total precipitation ratio in the south. Over the century, the national annual snowfall trend shows an abrupt decrease from the 1970s to the 1980s which can be associated with the considerable decrease in the number of frost days over the same period of time. This interrelation provides further evidence of the warming influence on the precipitation type.

The changes observed in the daily and extreme temperature and precipitation indices can produce major impacts on our environment, society and economy. Fewer extreme cold days and frost days can influence the demand for heating buildings and for energy consumption; significant decreasing trends in the heating degree days were also observed at many stations in the southern and western regions of the country (Bonsal et al., 2001). With more summer days, the summer outdoor recreational season can be extended, but with less snowfall, the winter outdoor activities may be reduced especially in the south where ski resorts already have to rely on artificial snow-making. The warming in Canada is accompanied by a longer frost-free season and an augmentation in the growing degree days (Bonsal et al., 2001) which can be beneficial for agriculture; still the substantial decrease in the annual snowfall,

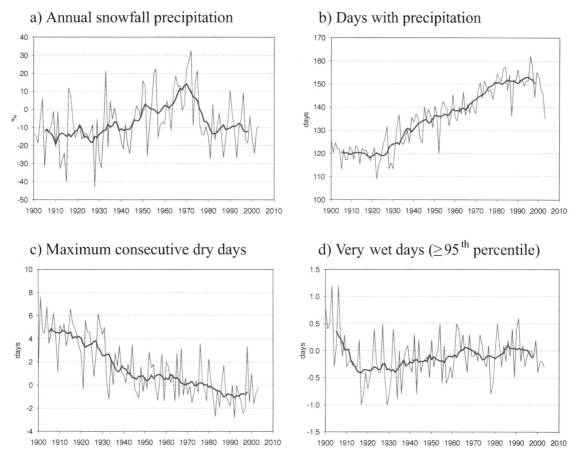

Fig. 10   National time series for 1900–2003 computed from a) the percentage station anomalies and b), c) and d) station anomalies. The bold line represents the 11-year running mean.

leading to less snow accumulation, can be crucial for the beginning of the growing season. Even if the frequency of days with precipitation has consistently increased in Canada, severe droughts still occurred during many summers in the 1930s and 1980s (Nkemdirim and Weber, 1999) and in the 2000s (Shabbar and Skinner, 2004). The highest five-day precipitation amount is a potential indicator of flood producing events and shows no consistent trends across the country. Since these indices demonstrate a high degree of spatial variability, more detailed research is needed on a regional basis.

5 Conclusion

This study examines the trends and variations in several daily and extreme temperature and precipitation indices over Canada during the twentieth century. It was found that the number of warm events increased significantly and the number of cold events decreased significantly during the two periods 1950–2003 and 1900–2003. A substantial decrease was also detected in the diurnal temperature range during the 1940s to 1960s due to much cooler night-time temperatures prior to the 1940s. The standard deviation of the daily mean temperature significantly decreased at several stations located in the west during the 104-year period. It is likely due to the greater change in the 'left side' of the distribution (fewer cold

daytime and night-time temperatures) than in the 'right side' of the distribution (more warm daytime and night-time temperatures) which is stronger in the west.

For precipitation indices, consistent increasing trends have been found for days with precipitation and days with rain, while significant decreases were detected in the simple day intensity index of precipitation and of rain for both periods, 1950–2003 and 1900–2003. Significant decreasing trends were also found in the MCDD across the country for both periods. The annual total snowfall increased from the beginning of the century until the 1970s and then a sharp decrease occurred from the mid-1970s to the 1980s which corresponds to a pronounced decreasing trend in the number of frost days. No consistent change has been found in the extreme precipitation indices with the exception of a significant increase in days with heavy precipitation for both periods. The inconsistency could be due to the high variability of the extreme precipitation events which increase the difficulty of detecting a significant trend or due to the sparse station network in the north. The lack of consistent patterns in extreme precipitation highlights the need for regional studies to explore local characteristics of precipitation extremes.

Examination of the century-long time series reveals considerable interdecadal variability. The causes of these variations

a) Simple day intensity index of P

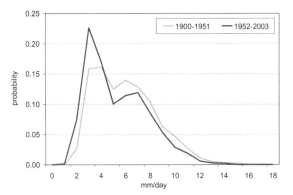

b) Days with precipitation

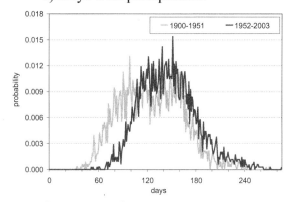

c) Maximum consecutive dry days

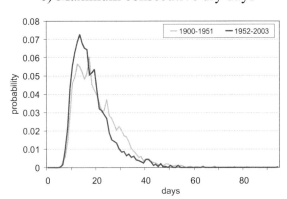

d) Heavy P days (≥ 10 mm)

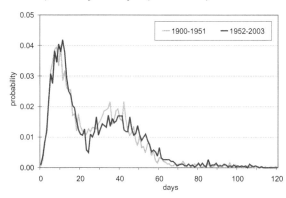

Fig. 11   Probability density functions for the periods 1900–51 and 1952–2003.

are beyond the scope of this study. The association between large-scale circulation anomalies and changes in extreme temperature and precipitation events requires further investigation. In conclusion, the results have improved our knowledge of daily and extreme temperature and precipitation trends and variability in Canada over the twentieth century. However, further research is needed to obtain a better understanding of the relationship between temperature and precipitation indices and to explain how the observed changes could affect our economy, ecological systems and our everyday lives.

Acknowledgements

The authors are grateful to Joan Klaassen and Xuebin Zhang of the Meteorological Service of Canada, and to three anonymous reviewers for their constructive comments and suggestions which led to an improved version of the manuscript.

References

BETHEA, R.M.; B.S. DURAN and T.L. BOULLION. 1995. *Statistical methods for engineers and scientists.* Third edition, revised and expanded. Marcel Dekker Inc., 652 pp.

BONSAL, B.R.; X. ZHANG, L.A. VINCENT and W.D. HOGG. 2001. Characteristics of daily and extreme temperatures over Canada. *J. Clim.* **14**: 1959–1976.

CUBASCH, U., G.A. MEEHL, G.J. BOER, R.J. STOUFFER, M. DIX, A. NODA, C.A. SENIOR, S. RAPER and K.S. YAP. 2001. Projections of future climate change. In: *Climate Change 2001: The Scientific Basic. Contribution of Working Group I to the Third Assessment Report of the Intergovernmental Panel on Climate Change.* Cambridge University Press, Cambridge UK and New York USA, 881 pp.

EASTERLING, D.R.; J.L. EVANS, P.YA. GROISMAN, T.R. KARL, K.E. KUNKEL and P. AMBENJE. 2000. Observed variability and trends in extreme climate events: a brief review. *Bull. Am. Meteorol. Soc.* **81**: 417–425.

————; L.V. ALEXANDER, A. MOKSSIT and V. DEEMMERMAN. 2003. CCI/CLIVAR workshop to develop priority climate indices. *Bull. Am. Meteorol. Soc.* **84**: 1403–1407.

FRICH, P.; L.V. ALEXANDER, P. DELLA-MARTA, B. GLEASON, M. HAYLOCK, A.M.G. KLEIN TANK and T. PETERSON. 2002. Observed coherent change in climatic extremes during the second half of the twentieth century. *Clim. Res.* **19**: 193–212.

GACHON, P.; A. ST-HILAIRE, T. OUARDA, V.T.V. NGUYEN, C. LIN, J. MILTON, D. CHAUMONT, J. GOLDSTEIN, M. HESSAMI, T.D. NGUYEN, F. SELVA, M. NADEAU, P. ROY, D. PARISHKURA, N. MAJOR, M. CHOUX and A. BOURQUE. 2005. A first evaluation of the strength and weakness of statistical downscaling methods for simulating extremes over various regions of eastern Canada. Subcomponent, Climate Change Action Fund (CCAF), Environment Canada, Final report, Montréal, Québec, Canada, 209 pp.

GROISMAN, P.Y.; B. SUN, R.S. VOSE, J.H. LAWRIMORE, P.H. WHITFIELD, E. FORLAND, I. HANSSEN-BAUER, M.C. SERREZE, V.N. RAZUVAEV and G.V. ALEKSEEV. 2003. Contemporary climate changes in high latitudes of the Northern Hemisphere: daily resolution. In: Proc. of the 14th Symposium on Global Change and Climate Variations, 9–13 February 2003, Long Beach, California. Am. Meteorol. Soc. 10 pp.

————; R. W. KNIGHT, D. R. EASTERLING, T. R. KARL, G. C. HEGERL, V. N. RAZU-VAEV and N. VYACHESLAV. 2005. Trends in intense precipitation in the climate record, *J. Clim.* **18**: 1326–1350.

HAYLOCK, M.R.; T. PETERSON, J.R. ABREU DE SOUSA, L.M. ALVES, T. AMBRIZZI, J. BAEZ, J.I. BARBOSA, V.R. BARROS, M.A. BERLATO, M. BIDEGAIN, G. CORONEL, V. CORRADI, A.M. GRIMM, R. JAILDO DOS ANJOS, D. KAROLY, J.A. MARENGO, M.B. MARINO, P.R. MEIRA, G.C. MIRANDA, L. MOLION, D.F. MONCUNILL, D. NECHET, G. ONTANEDA, J. QUINTANA, E. RAMIREZ, E. REBELLO, M. RUSTICUC-CI, J.L. SANTOS, I.T. VARILLAS, J.G. VILLANUEVA, L. VINCENT and M. YUMIKO. 2005. Trends in total and extreme South America rainfall 1960–2000. *J. Clim.* In press.

HENSERSON-SELLERS, A. 1989. North American total cloud amount variations this century. *Global Planet. Change*, **75**: 175–194.

JONES, P.D.; E.B. HORTON, C.K. FOLLAND, M. HULME, D.E. PARKER and T.A. BAS-NETT. 1999. The use of indices to identify changes in climatic extremes. *Clim. Change*, **42**: 131–149.

KENDALL, M.G. 1955. *Rank correlation methods*. Charles Griffin and Company, Second Edition, London. 196 pp.

KLEIN TANK, A.M.G. and G.P. KÖNNEN. 2003. Trends indices of daily temperature and precipitation extremes in Europe, 1946–99. *J. Clim.* **16**: 3665–3680.

MEKIS, É. 2005. Adjustments for trace measurements in Canada. *In*: Proc. 15th Conference on Applied Climatology, June 2005, Savannah, Georgia, USA, 6 pp.

———— and W.D. HOGG. 1999. Rehabilitation and analysis of Canadian daily precipitation time series. ATMOSPHERE-OCEAN, **37**: 53–85.

———— and R. HOPKINSON. 2004. Derivation of an improved snow water equivalent adjustment factor map for application on snowfall ruler measurements in Canada. *In*: Proc. 14th Conference on Applied Climatology, 11–15 January 2004, Seattle, Washington, Am. Meteorol. Soc. 5 pp.

———— and L.A. VINCENT. 2005. Precipitation and temperature related climate indices for Canada. *In*: Proc. 16th Symposium on Global Change and Climate Variations, 9–13 January 2005, San Diego, California. Am. Meteorol. Soc. 5 pp.

METCALFE, J.R.; B. ROUTLEDGE and K. DEVINE. 1997. Rainfall measurement in Canada: changing observational methods and archive adjustment procedures. *J. Clim.* **10**: 92–101.

MOBERG, A.; P.D. JONES, M. BARRIENDOS, H. BERGSTRÖM, D. CAMUFFO, C. COCHEO, T.D. DAVIES, G. DÉMARÉE, J. MARTIN-VIDE, M. MAUGERI, R. RODRIGUEZ and T. VERHOEVE. 2000. Day-to-day temperature variability trends in 160- to 275-year-long European instrumental records. *J. Geophys. Res.* **105**: 22849–22868.

NKEMDIRIM, L. and L. WEBER. 1999. Comparison between the droughts of the 1930s and the 1980s in the southern prairies in Canada. *J. Clim.* **12**: 2434–2450.

PETERSON, T.C.; C. FOLLAND, G. GRUZA, W. HOGG, A. MOKSSIT and N. PLUMMER. 2001. Report on the activities of the working group on climate change detection and related rapporteurs 1998–2001. Report WCDMP-47, WMO-TD 1071, Geneva, Switzerland, 143 pp.

————; M.A. TAYLOR, R. DEMERITTE, D.L. DUNCOMBE, S. BURTON, F. THOMPSON, A. PORTER, M. MERCEDES, E. VILLEGAS, R.S. FILS, A. KLEIN TANK, A. MARTIS, R. WARNER, A. JOYETTE, W. MILLS, L. ALEXANDER and B. GLEASON. 2002. Recent changes in climate extremes in the Caribbean region. *J. Geophys. Res.* **107**(D21): 4601, doi: 10.1029/2002JD002251.

PLUMMER, N.; M.J. SALINGER, N. NICHOLLS, R. SUPPIAH, K. HENNESSY, R. M. LEIGHTON, B. TREWIN, C.M. PAGE and J.M. LOUGH. 1999. Changes in climate extremes over the Australian region and New Zealand during the twentieth century. *Clim. Change*, **42**: 183–202.

QIAN, W. and X. LIN. 2004. Regional trends in recent temperature indices in China. *Clim. Res.* **27**: 119–134.

SEN, P.K. 1968. Estimates of the regression coefficient based on Kendall's tau. *J. Am. Stat. Assoc.* **63**: 1379–1389.

SHABBAR, A. and W. SKINNER. 2004. Summer drought patterns in Canada and the relationship to global sea surface temperatures. *J. Clim.* **17**: 2866–2880.

VINCENT, L.A. 1998. A technique for the identification of inhomogeneities in Canadian temperature series. *J. Clim.* **11**: 1094–1104.

————; X. ZHANG and W.D. HOGG. 1999. Maximum and minimum temperature trends in Canada for 1895–1995 and 1946–1995. *In*: Proc. 10th Symposium on Global Change Studies, 10–15 January 1999. Dallas, Texas. Am. Meteorol. Soc. pp. 95–98.

————; ————, B.R. BONSAL and W.D. HOGG. 2002. Homogenization of daily temperatures over Canada. *J. Clim.* **15**: 1322–1334.

———— and É. MEKIS. 2004. Variations and trends in climate indices for Canada. *In*: Proc. 14th Conference on Applied Climatology, 11–15 January 2004, Seattle, Washington. Am. Meteorol. Soc. 5 pp.

————; T.C. PETERSON, V.R. BARROS, M.B. MARINO, M. RUSTICUCCI, G. CARRAS-CO, E. RAMIREZ, L.M. ALVES, T. AMBRIZZI, M.A. BERLATO, A.M. GRIMM, J.A. MARENGO, L. MOLION, D.F. MONCUNILL, E. REBELLO, Y.M.T. ANUNCIAÇÃO, J. QUINTANA, J.L. SANTOS, J. BAEZ, G. CORONEL, J. GARCIA, I. TREBEJO, M. BIDE-GAIN, M.R. HAYLOCK and D. KAROLY. 2005. Observed trends in indices of daily temperature extremes in South America 1960–2000. *J. Clim.* **18**: 5011–5023.

WANG, X.L. and V.R. SWAIL. 2001. Changes of extreme wave heights in northern hemisphere oceans and related atmospheric circulation regimes. *J. Clim.* **14**: 2204–2221.

ZHAI, P.; X. ZHANG, H. WAN and X. PAN. 2005. Trends in total precipitation and frequency of daily precipitation extremes over China. *J. Clim.* **18**: 1096–1108.

ZHANG, X.; L.A. VINCENT, W.D. HOGG and A. NIITSOO. 2000. Temperature and precipitation trends in Canada during the 20th Century. ATMOSPHERE-OCEAN, **38**: 395–429.

————; W.D. HOGG and É. MEKIS. 2001. Spatial and temporal characteristics of heavy precipitation events over Canada. *J. Clim.* **14**: 1923–1936.

————; G. HEGERL, F.W. ZWIERS and J. KENYON. 2005. Avoiding inhomogeneities in percentile-based indices of temperature extremes. *J. Clim.* **18**: 1852–1860.

An Overview of the Second Generation Adjusted Daily Precipitation Dataset for Trend Analysis in Canada

Éva Mekis and Lucie A. Vincent

*Climate Research Division, Science and Technology Branch, Environment Canada
Toronto, Ontario*

ABSTRACT   *A second generation adjusted precipitation daily dataset has been prepared for trend analysis in Canada. Daily rainfall and snowfall amounts have been adjusted for 464 stations for known measurement issues such as wind undercatch, evaporation and wetting losses for each type of rain-gauge, snow water equivalent from ruler measurements, trace observations and accumulated amounts from several days. Observations from nearby stations were sometimes combined to create time series that are longer; hence, making them more useful for trend studies. In this new version, daily adjustments are an improvement over the previous version because they are derived from an extended dataset and enhanced metadata knowledge. Datasets were updated to cover recent years, including 2009. The impact of the adjustments on rainfall and snowfall total amounts and trends was examined in detail. As a result of adjustments, total rainfall amounts have increased by 5 to 10% in southern Canada and by more than 20% in the Canadian Arctic, compared to the original observations, while the effect of the adjustments on snowfall were larger and more variable throughout the country. The slope of the rain trend lines decreased as a result of the larger correction applied to the older rain-gauges while the slope of the snow trend lines increased, mainly along the west coast and in the Arctic. Finally, annual and seasonal rainfall and snowfall trends based on the adjusted series were computed for 1950–2009 and 1900–2009. Overall, rainfall has increased across the country while a mix of non-significant increasing and decreasing trends was found during the summer in the Canadian Prairies. Snowfall has increased mainly in the north while a significant decrease was observed in the southwestern part of the country for 1950–2009.*

RÉSUMÉ *[Traduit par la rédaction] Un ensemble de données quotidiennes de précipitations ajustées de deuxième génération a été préparé pour l'analyse des tendances au Canada. Les hauteurs quotidiennes des chutes de pluie et des chutes de neige ont été ajustées pour 464 stations en fonction de problèmes connus comme la sous-capture due au vent, l'évaporation et les pertes par mouillage pour chaque type de pluviomètre, l'équivalent en eau de la neige selon des mesures avec une règle, les observations de traces et les hauteurs accumulées de plusieurs jours. Les observations de stations situées à proximité ont parfois été combinées pour créer des séries chronologiques plus longues et donc plus utiles pour les études de tendance. Dans cette nouvelle version, les ajustements quotidiens constituent une amélioration par rapport à la version précédente parce qu'ils sont dérivés d'un ensemble de données étendu et d'une meilleure connaissance des métadonnées. Les ensembles de données ont été mis à jour pour inclure les années récentes, y compris 2009. L'effet des ajustements sur les hauteurs totales et les tendances des chutes de pluie et des chutes de neige a été examiné en détail. En raison des ajustements, les hauteurs totales de pluie ont augmenté de 5 à 10% dans le sud du Canada et de plus de 20% dans l'Arctique canadien, comparativement aux observations originales, alors que l'effet des ajustements sur les chutes de neige était plus important et plus variable à travers le pays. La pente des lignes de tendance de la pluie a diminué par suite de la plus forte correction appliquée aux pluviomètres plus anciens alors que la pente des lignes de tendance de la neige a augmenté, surtout le long de la côte ouest et dans l'Arctique. Finalement, les tendances annuelles et saisonnières des chutes de pluie et des chutes de neige basées sur les séries ajustées ont été calculées pour 1950–2009 et 1900–2009. Dans l'ensemble, les chutes de pluie ont augmenté dans le pays alors qu'on a trouvé un mélange de tendances non significatives à la hausse et à la baisse durant l'été dans les prairies canadiennes. Les chutes de neige ont augmenté principalement dans le nord alors qu'une diminution marquée a été observée dans la partie sud-ouest du pays pour 1950–2009.*

1 Introduction

Reliable climate datasets are crucial for climate monitoring and the detection of any climate change signal. However, climate observations require a great deal of processing before they are ready for analysis. Each observation needs to be recorded, transmitted, digitized, quality controlled and then examined by experts familiar with the instruments, observing practices and the climatology. These tasks are becoming even more complex because of the constantly changing observing network, which involves relocation and closure of sites and changes in instruments and practices. As a result, climate data have to be adjusted to address these issues and to ensure continuity of the records for climate monitoring and climate change studies. Methodologies required to adjust climate data have also been improving. When a new version of an adjusted dataset becomes available to the scientific community, it is vital to document the data properly along with the adjustments. Users can then understand the data better and determine if they are suitable for their own analyses.

It is widely recognized that gauge-measured precipitation has a systematic bias mainly caused by wind-induced undercatch, wetting losses (water adhering to the surface of the inner walls of the gauge that cannot be measured by the volumetric method) and evaporation losses (water lost by evaporation before the observation can be made). In 1985, the World Meteorological Organization (WMO) initiated the Solid Precipitation Measurements Intercomparison project to assess and document the national methods of measuring solid precipitation (Goodison et al., 1998). Corrections have been applied to a large number of stations located in the High Arctic: gauge-measured precipitation amounts for these stations have increased by about 10% for the summer and from 80 to 120% for the winter (Yang et al., 2005). In a study by Ding et al. (2007), it was noted that bias correction can change the magnitude and even the direction of a trend. Detailed documentation of the instruments, including their type, location and environment, is crucial since measurements can be affected by many changes during a station's history. The Meteorological Service of Canada (MSC) has used a number of different gauges for measuring rainfall over the past 150 years (Metcalfe et al., 1997). When the recently introduced rain-gauges were compared to the traditional gauges and to the WMO reference pit gauge (WMO, 1972), the results of the comparison indicated that the manual Type B gauge, in service since the 1970s, provided the most accurate measurements compared to the pit gauge data (Devine and Mekis, 2008).

In the mid-1990s, the first generation Adjusted Precipitation for Canada - Daily (APC1-Daily) dataset was prepared to provide a more accurate estimate of the precipitation amount and for the analysis of climate trends (Mekis and Hogg, 1999). Daily rainfall and snowfall were adjusted sezparately for 495 locations across the country. For each type of rain-gauge, corrections to account for wind undercatch as well as evaporation and wetting losses were implemented. For snowfall, ruler measurements were used throughout the entire period and a density correction based on a set of coincident ruler and Nipher gauge observations was applied to all ruler measurements. An adjustment was performed to account for trace observations of both rain and snow. Finally, neighbouring observations were sometimes joined and adjustments were applied based on a simple ratio computed using available periods of overlapping data. The APC1-Daily dataset was used in several studies, including temperature and precipitation trends in Canada (Zhang et al., 2000), changes in temperature and precipitation indices in Canada (Vincent and Mekis, 2006) and global changes in daily extreme temperature and precipitation (Alexander et al., 2006).

A number of improvements were introduced in the second generation datasets. The station list was revised to include stations with longer periods of observations. The rain-gauge adjustments were derived from more field experiments. The adjustment map for the snow water equivalent from ruler measurements was improved since it was based on 175 stations compared to the previous 63 stations. More background information was retrieved from the metadata for a better adjustment of trace observations. The accumulated amount flags, indicating that precipitation had fallen over a few days and was reported on the last day of the event, were taken into account. Joined stations were further tested using neighbouring stations to determine whether an adjustment was required.

The primary objective of this paper is to document all the procedures used to produce the second generation Adjusted Precipitation for Canada - Daily (APC2-Daily) dataset to ensure transparency to all users. To assist users in their analyses and the proper interpretation of results, it is important that all the information is merged into a single manuscript to provide an adequate understanding of the data. The second objective is to assess the impact of the adjustments on rainfall and snowfall amounts and the magnitude of the trends. The final objective is to present an updated analysis of annual and seasonal trends using this newest generation of datasets, which includes a longer period of time (to 2009). Section 2 presents the data and Section 3 describes the adjustments for known measurement issues and their impact on accumulated amounts and trends. Section 4 explains the adjustments for joined station observations. The trends in annual and seasonal rainfall and snowfall are presented in Section 5. Lastly, a summary is provided in Section 6.

2 Data

Daily rainfall gauge and snowfall ruler data were extracted directly from the National Climate Data Archive of Environment Canada. Adjusting rain and snow separately allows the correction of known problems such as instrument deficiencies and changes in observing procedures. All rainfall and snowfall measurements used in the APC2-Daily dataset were made by observers and no measurements were made by automatic systems. Station history files were searched for metadata information, such as installation dates of the rain-gauges, the introduction date of the 6-hourly and hourly measurement program and dates of station closure and/or relocation. Further metadata information was also provided by national experts in climate observing and climate practices.

Stations with long-term rainfall and snowfall measurements, covering as many of the most recent years as possible were chosen for inclusion in the second generation dataset. Since the 1990s, the climate observing network has changed considerably in Canada because of downsizing of the traditional network and increasing use of automated systems. Stations were often closed and/or relocated; hence, other stations were identified and added to the dataset to ensure the continuity of rainfall and snowfall observations with time. During station selection, consideration was given to Reference Climate Stations (RCS), which includes stations in the Global Climate Observing System (GCOS) Surface Network (GSN), because these stations are protected (i.e., will not close in the future). Station selection and the list of joined stations were finalized, using the expertise of regional climatologists to produce the best representation of precipitation for Canada.

The second generation dataset includes adjusted daily rainfall, snowfall and total precipitation for 464 locations (Fig. 1a). The adjusted daily total precipitation is the sum of the adjusted rainfall and adjusted snow water equivalent. Most of these stations can be found in the first generation dataset; however, some stations had to be removed because of large amounts of missing data at the end of the record or better quality data available for a neighbouring station. In addition, viable stations from the homogenized temperature dataset (Vincent et al., 2002) were added to provide the user community with both temperature and precipitation data at as many common locations as possible. In the second generation dataset, 53 stations were removed and 22 new locations were added. Although this newly developed database has 10% fewer stations than the first generation dataset, it has 23% more data in the last decade (Fig. 1b). The list of stations, along with their locations, can be obtained from the lead author.

3 Adjustments for known measurements issues

a Adjustments for Rainfall from Rain-Gauges
The MSC copper gauge, also called the Type A gauge, was originally used in Canada to measure daily rainfall. It was made entirely from copper, but the inside container was modified to soft plastic around 1965. In the 1970s, it was gradually replaced by the Type B gauge at all stations across the country. A complete description of the gauges can be found in Metcalfe et al. (1997) and Devine and Mekis (2008). Adjustments for wind undercatch, evaporation and retention (wetting at the receiver and measuring funnel area) for each type of gauge in the second generation dataset are similar to those applied to the first generation dataset (Mekis and Hogg, 1999); however, they are now based on more field experiments performed at various locations (Devine and Mekis, 2008). Table 1 provides

(a)

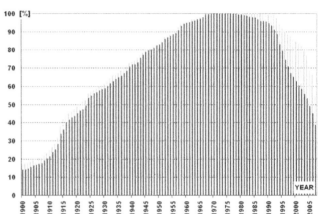

(b)

Fig. 1 a) Locations of the 464 stations used in the APC2-Daily dataset and b) percentage of stations with observations for every year: black bars indicate first generation stations (100% represents 495 stations) while grey bars indicate second generation stations (100% represents 464 stations).

a summary of required daily adjustments for the three major rain-gauges used in Canada during the last century.

Adjustments were applied using the following equation with the values given in Table 1:

$$R_a = (R_m + F_c + E_c + C_c) \times (1 + W_c),$$

where R_a is the adjusted daily rainfall (mm), R_m is the measured daily rainfall (mm), F_c is the funnel wetting correction (mm), E_c is the evaporation from container correction (mm), C_c is the container/receiver retention correction (mm), and W_c is the wind correction factor (%).

The sums of the adjustments in the last row of Table 1 show that the gauges have become more precise over time; therefore, less adjustment is needed for newer gauges. These adjustments were derived from a side-by-side gauge experiment by Routledge (1997) and further described in Devine and Mekis (2008). To avoid the loss of valuable information caused by small-scale corrections required for small rain events, precision of the adjusted daily precipitation dataset was increased from one to two digits after the decimal point.

TABLE 1.   Rain-gauge corrections

Type of correction	Unit	Notation	MSC, copper receiver	MSC, plastic receiver	Type B Gauge
1. Wind at Orifice level	%	W_c	0.04	0.04	0.02
2. Wetting at Funnel area	mm d^{-1}	F_c	0.13	0.13	0.08
3. Evaporation	mm d^{-1}	E_c	0.02	0.03	0.01
4. Wetting of Receiver or Container	mm d^{-1}	C_c	0.06	0.03	0.04
Sum of 2, 3 and 4	mm d^{-1}		0.21	0.19	0.13

b *Adjustments for Snowfall from Ruler Measurements*
Because of the availability of many sites with long-term ruler measurements, snow ruler data continue to be used for climate change studies. Historically, for all snow ruler measurements of freshly fallen snow, the liquid precipitation amount (water equivalent) was usually determined by assuming a fresh snow density of 100 kg m^{-3}. This estimate of fresh snow density can be improved because of the availability of coincident Nipher gauge and snow ruler measurements since the 1960s. The procedure was originally applied to the first generation dataset (Mekis and Hogg (1999) based on Metcalfe et al. (1994)) and involves the calculation of ratios of corrected solid Nipher gauge precipitation measurements to snowfall ruler depth measurements when both were operational. The snow water equivalent adjustment factor ρ_{swe} map (Fig. 2) has since been updated using 175 climatological stations with more than 20 years of concurrent observations (Mekis and Hopkinson, 2004; Mekis and Brown, 2010). For quality assurance, the snow water equivalent adjustment factor was verified using observations from independent stations with shorter records that were not originally included in the map production. The updated map allows estimates of ρ_{swe} for ruler-based snowfall observations to be obtained for all long-term climate stations in Canada. The spatial pattern is consistent with processes that influence the density of fresh snowfall and its initial settling, with values ranging from more than 1.5 in the Maritimes to less than 0.8 in south central British Columbia.

c *Adjustments for Flags*
Trace precipitation, flag "T", is less than the minimum measurable amount and is assigned a value of zero. Trace precipitation is important over vast parts of Canada, with the highest impact occurring in the Arctic, where precipitation amounts are very low and many trace events are recorded. Under these conditions, the sum of all trace amounts becomes a significant portion of the total precipitation.

Trace precipitation has not been observed consistently in time; both its definition and the minimum measurable amount have been modified (Mekis, 2005). During the conversion from the imperial to the metric system around 1977–78, all precipitation values were converted from inches to millimetres. Therefore, the minimum measurable amount was changed from 0.01 in (equivalent to 0.254 mm rounded to 0.3 mm) to 0.2 mm for rain measurements and 0.1 in (equivalent to 0.254 cm rounded to 0.3 cm) to 0.2 cm for snow ruler measurements.

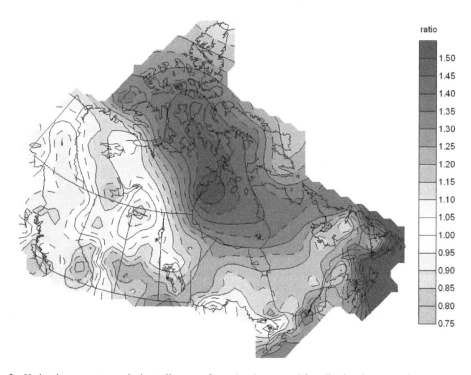

Fig. 2   Updated snow water equivalent adjustment factor (ρ_{swe}) map used for adjusting the snow ruler measurements.

The measurement of trace amounts has not always been practiced. Rainfall trace was mentioned as early as 1871 in an early observer's manual as well as in later manuals in 1878, 1893 and 1914 (Kingston, 1878). However, the next version of the *Instructions to Observers* manual does not contain any reference to the recording of traces of rain (Patterson, 1930). In 1947, trace precipitation was mentioned in the first edition of *Manual of Standard Procedures and Practices for Weather Observing and Recording MANOBS*, effective 1 January 1947 but only under Synoptic Reports (Meteorological Division, 1947). Finally, in Amendment No. 8 of the third edition of *MANOBS*, effective 9 June 1954, the first official mention of the trace flag appeared (Meteorological Division, 1951). Another definition appears in the booklet *How to Measure Rainfall* (Department of Transport, Meteorological Branch, 1955).

There are other reasons for trace observations not being consistent over time. New observers with different training or experience and the number of observations taken daily can affect the number of trace events reported annually. At synoptic stations (mostly airports), observations were taken twice daily prior to January 1941 and four times daily thereafter. Similarly, at climatological stations (volunteer stations), observations are taken twice daily but sometimes they are only taken once in the morning. For proper adjustment of the frequency of trace events, the number of daily and 6-hourly trace flag observations was compared with the Trace Occurrence Ratio calculated from overlapping synoptic and daily archive data (Mekis and Hogg, 1999).

Different adjustments were applied for rain and snow traces. For rainfall traces, amounts between 0.0 and 0.2 mm were considered equally probable and the average of 0.1 mm was applied. Adjustment for solid trace precipitation is more complex and is related to ice crystal events (Mekis and Hogg, 1999). Ratios of the number of ice crystal events to the number of solid precipitation trace events were computed using 6-hourly weather-type information (i.e., rain, ice crystal and snow traces) and mapped for many stations. This map was used to generate the trace adjustment factor for solid precipitation, with values ranging from 0.07 mm in the south to 0.03 mm in the High Arctic, which corresponds to amounts suggested by Metcalfe et al. (1994).

Even if trace amounts were carefully adjusted using all known metadata, it is possible that they could still cause artificial inhomogeneities in time series with low amount of precipitation. Inhomogeneities become less pronounced when a greater amount of precipitation is analyzed. To illustrate, precipitation series for the station at Campsie, Alberta, are presented in Fig. 3. The annual rain, snow and total precipitation calculated from events with less than 0.5, 5 and 10 mm are given in Figs 3a, 3b and 3c, respectively. There is a jump in the number of daily trace events in 1945 (Fig. 3d), which can be detected in the graph of total amounts calculated from events with less than or equal to 0.5 mm events of precipitation (Fig. 3a). However, this jump is not visible at higher thresholds.

It is important to note that although the value of trace precipitation is not zero after the adjustment is performed, it has to be excluded from the computations of some of the indices, such as the number of days with precipitation or the maximum consecutive dry days, because trace precipitation does not generally alleviate dry conditions.

Accumulated precipitation, flag "A", is recorded on the last day of a series of consecutive days with flag "C" (precipitation occurred) or flag "L" (precipitation may or may not have occurred). This situation can happen when an observer is absent for a few days (typically over a weekend) or for an extended period of time. However, the occurrence of an accumulated value does not happen very often and never occurs at some stations. Since daily observations can be used to prepare climate change indices or compute precipitation extremes, it is crucial that the accumulated value is taken into account to prevent further propagation of erroneous values. In APC2-Daily, flags "A", "C" and "L" are retained for further use and their corresponding value is replaced by the accumulated amount divided by the number of affected days; this is done to preserve the monthly total and minimize the impact on extreme values.

d *Impact of Adjustments for Known Measurement Issues*
To illustrate the impact of the adjustments, the relative change in annual total precipitation (%) and the trend for the entire period (mm decade^{-1}) are given after each adjustment for five stations representing different climate regions of Canada (Table 2 and Fig. 4). The two coastal stations, Port Hardy, British Columbia, and Gander, Newfoundland (Fig. 4a), have relatively high annual total precipitation (Table 2) whereas the Arctic station, Resolute, has low annual total precipitation with a high frequency of trace events (an average of 190 trace events a year). Island Falls is located in the relatively dry northwestern forest area while London is in a productive agricultural sector within the Great Lakes region.

In Figs 4b to 4f, four series are generated for each station. First, the annual total precipitation calculated from the original non-adjusted rainfall and snowfall (line 1), along with the best fit linear trend, and the annual total precipitation obtained after applying the rain-gauge correction (line 2) are plotted. Results indicate that the percentage change in the total amount of annual precipitation is similar for the five stations, varying from 3.5 to 4.9% (Table 2), and the trend slopes decreased because of the larger adjustments applied to the older rain-gauge for the periods of time indicated. When the snow water equivalent adjustment is performed (line 3), the total amount of annual precipitation increased at Resolute (30.4%) and Gander (20.8%), because of the larger snow water equivalent adjustment factor (ρ_{swe}) for these two stations (Fig. 2) and decreased slightly at Port Hardy, where the adjustment is less than one (Table 2). The slope of the trend depends on the actual snowfall amount and because ρ_{swe} is consistent throughout the entire period, this adjustment did not affect the trend much. The last adjustment is for the trace flags (line 4). The correction for trace values is always positive, an additional amount of precipitation is added to the observations whenever a trace flag occurs. The impact of trace corrections is dependent on the frequency of this observation at each station and its relative contribution to the annual total

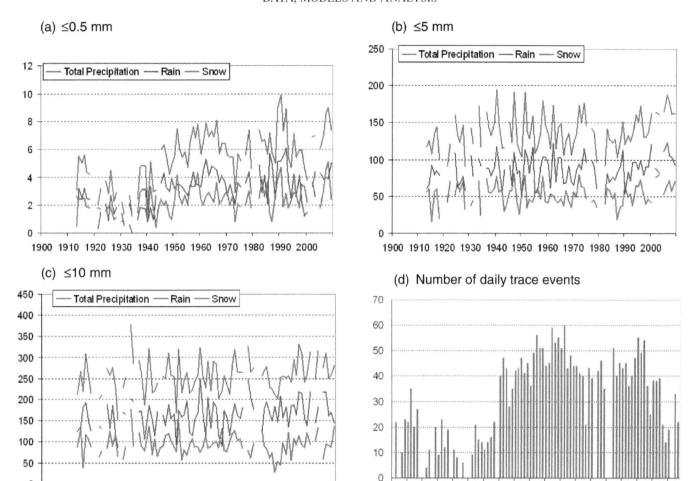

Fig. 3   Annual rain, snow and total precipitation (mm) calculated from events a) ≤0.5 mm; b) ≤5 mm and c) ≤10 mm at Campsie, Alberta. The number of trace events per year is given in d).

precipitation. In the Arctic, this correction increased the amount of precipitation by an additional 18.4% at Resolute (Table 2). The trend did not change considerably because trace adjustments depend on the frequency and distribution of trace observations with time.

A similar exercise was applied to all 464 stations. The magnitude of the adjustments for known rain and snow measurement issues, using the average ratio (for 1950–2009) of the adjusted annual rain (snow) value to the original rain (snow) value was mapped (Fig. 5). Results indicate that the adjustment for rain is larger in the Far North where the frequency

of trace measurements is considerable. Meanwhile, the adjustment for snow is more important in the Arctic and the East Coast since ρ_{swe} increases the water equivalent of snowfall considerably in these regions and decreases the water equivalent of snowfall in the western provinces. The adjustments increased solid precipitation by more than 50% in the North since the original amount was very small. Finally, when all the adjustments were applied, the total precipitation increased almost everywhere across the country except in the mountains in the west (Fig. 5c) mainly because of the lower adjustment required for snowfall.

TABLE 2.   Impact of adjustments for known measurement issues at five locations in Canada.

Station Name	Prov.	Period	Mean annual Prec. (mm)	(1) rain + snow measurements		(2) = (1) + rain-gauge corrections			(3) = (2) + snow density corrections		(4) = (3) + trace corrections	
				Amount change (%)	Trend (mm/10yr)	Amount change (%)	Trend (mm/10yr)	ρ_{swe}	Amount change (%)	Trend (mm/10yr)	Amount change (%)	Trend (mm/10yr)
Port Hardy	BC	1944–2008	1834	–	56.5	4.6	46.4	0.9	4.4	47.1	5.2	46.2
Island Falls	SASK	1931–2004	493	–	0.2	4.9	–1.7	1.1	6.8	–2.1	8.3	–2.7
London	ONT	1895–2001	974	–	7.1	4.9	5.5	1.2	9.4	4.8	11.0	5.6
Gander	NFLD	1937–2008	1159	–	50.4	4.3	46.2	1.5	20.8	58.4	23.3	57.3
Resolute	NU	1948–2007	159	–	8.3	3.5	7.7	1.4	30.4	11.4	48.8	13.6

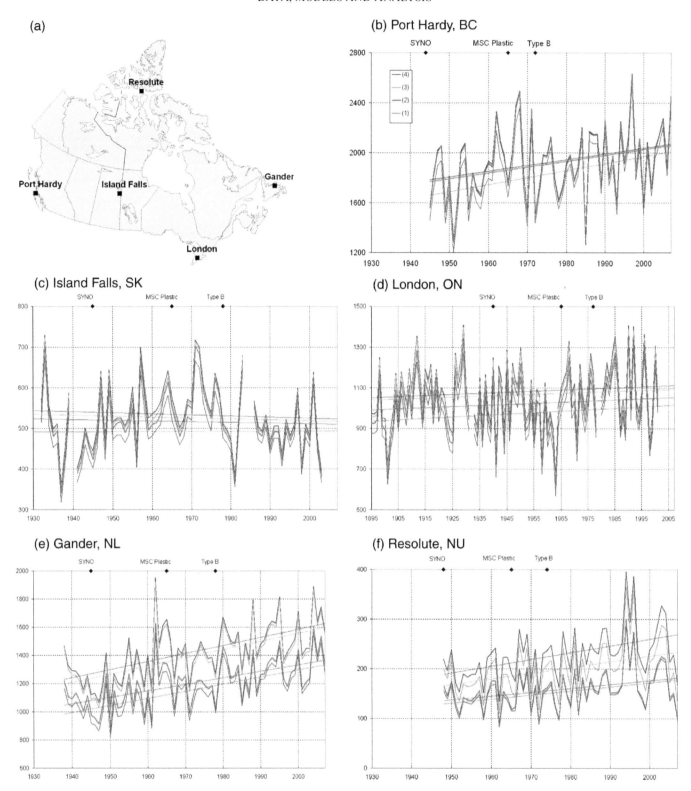

Fig. 4   Annual total precipitation (mm) for five stations: line (1) represents the precipitation total from the original data; line (2) is line (1) + rain-gauge corrections; line (3) is line (2) + snow water equivalent corrections; and line (4) is line (3) + trace corrections.

Rain and snow were adjusted not only to provide more accurate amounts but also to produce a better estimate of the trends. Figure 6 presents the difference in the annual rainfall and snowfall precipitation trends for 1950–2009 before and after the adjustments. This difference in trend was further divided by the mean for the entire period to take into account the climatology of the station (expressed as a percentage). Figure 6a shows that adjustments applied to the daily

(a) Rain

(b) Snow

(c) Total precipitation

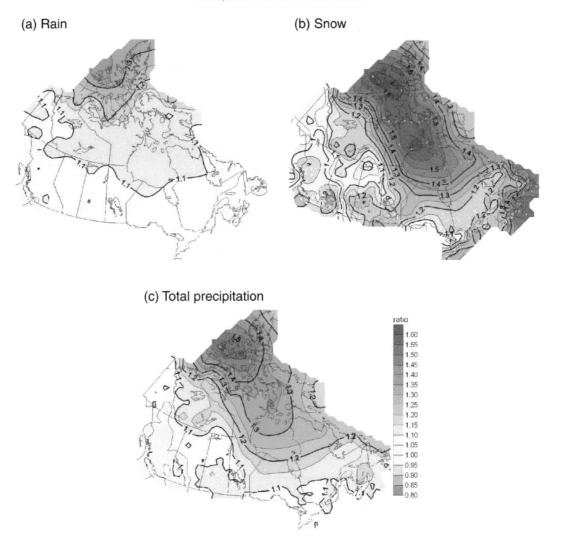

Fig. 5   Magnitude of adjustments for all known rain and snow measurement issues using the average ratio of the annual rain (snow, total) with all adjustments to the original rain (snow, total) measurements from 1950 to 2009.

rainfall decreased the trends by about 5% at many stations across the country. This is mainly because of the larger adjustment factor applied to the older rain-gauges. However, the adjustments increased the snow trends at many northern locations due to the larger adjustments required for trace observations (Fig. 6b).

4 Joining station observations

For many stations, it was necessary to merge observations from nearby stations to produce longer time series of rainfall and snowfall for trend analysis. A procedure to determine if joining precipitation observations created an artificial discontinuity at the joining dates was developed by Vincent and Mekis (2009) and was used to examine the 234 stations. Based on the results, about 35% of the stations were adjusted for rain and 58% of the stations were adjusted for snow (Fig. 7). The magnitude of the adjustment ratio varied from 0.78 to 1.29 for rainfall and 0.64 to 1.45 for snowfall. The adjustments

are based on neighbouring stations, which is a major improvement over the first generation dataset where adjustments were applied only if overlapping data were available.

Figure 8 presents the difference in annual rainfall and snowfall precipitation trends before and after all the adjustments were applied, including those for joining stations. The results show that the adjustment for joining does not have a uniform effect on the trends. For example, at Geraldton, Ontario, the slight positive difference in trend in snowfall (Fig. 6b) became negative after adjustments (Fig. 8b). Meanwhile, the slight positive difference in snowfall trend (Fig. 6b) at Beaverdell, British Columbia, became more positive after the adjustments (Fig. 8b).

As previously mentioned, the climate observing network in Canada has changed considerably since the 1990s and will continue to change. Station closures and relocation are ongoing issues. The list of joined stations along with the year of joining can be obtained from the lead author. Due to recent changes, observations were joined at 12.3% of all stations after 1990 in the second generation datasets.

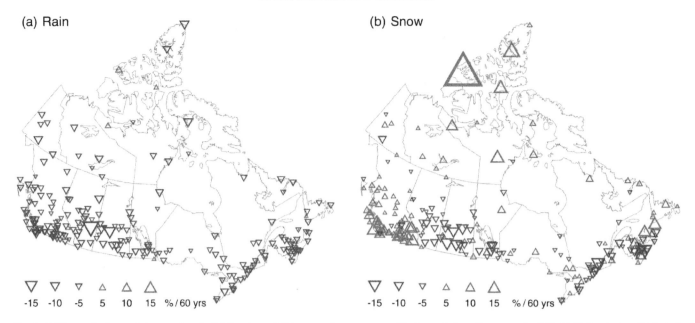

Fig. 6   Difference in the trends in annual total rainfall and snowfall (expressed as a percentage) before and after adjustments for 1950–2009. Panels a) and b) include rain-gauge, snow and trace adjustments. The size of the triangle is proportional to the magnitude of the change in trends.

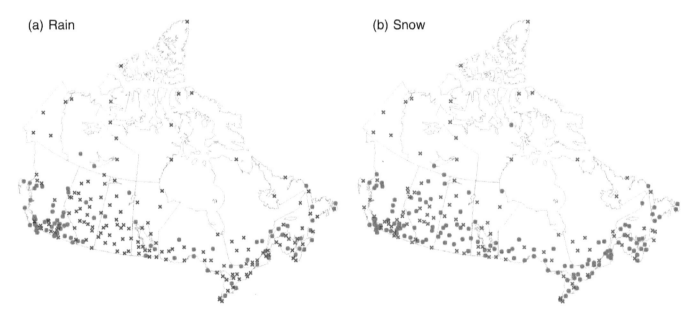

Fig. 7   Location of joined stations for a) rain and b) snow. Green crosses indicate joined stations without adjustments and red dots indicate joined stations with adjustments.

5 Trends in annual and seasonal rainfall and snowfall

Since the climate observing network was not established in northern Canada until the late 1940s, there are many regions in the North with very little data prior to 1945. For this reason, the trends were analyzed for two periods: 1950–2009 for the entire country and 1900–2009 for southern Canada (south of 60°N). Anomalies (expressed as a percentage) were computed at individual stations; these are departures from the reference period 1961–1990, which are further divided by the 1961–1990 mean. Trends were obtained for each station and for the national series representing all of Canada and southern Canada. To generate these last two series, the country was divided into 5°x5° grid boxes, and the national and southern Canada mean series were computed from the average of the stations within the individual boxes. The linear trend was estimated using the approach taken by Sen (1968) and the significance of the trend was determined using Kendall's test (Kendall, 1955). The trend was computed only if more than 80% of the values were present and the statistical significance was assessed at the 5% confidence level.

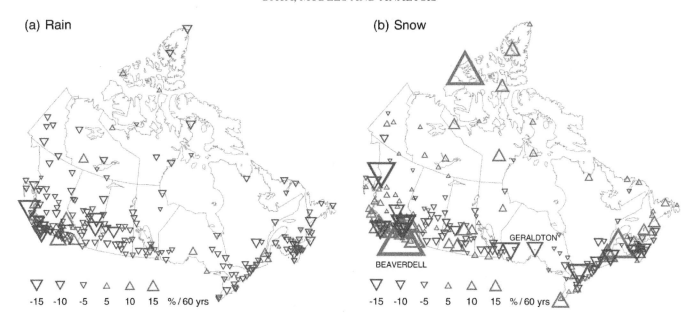

Fig. 8   Difference in trends of annual total a) rainfall and b) snowfall (expressed as a percentage) before and after all adjustments for 1950–2009 including adjustments for joining. The size of the triangle is proportional to the magnitude of the change in trends.

a *Trends for Canada, 1950–2009*

Annual rainfall in Canada increased by about 12.5% from 1950–2009 (Fig. 9a). Positive trends are observed from coast to coast and in the North (Fig. 10a). Even if most of the stations show an increase in annual rainfall during the past 60 years, significant increasing trends of 10 to 30% (depending on location) were found at only 26% of the stations. Overall, rainfall totals have increased in all seasons. The most pronounced increase is observed during the spring when 28% of the stations show significant increasing trends (Fig. 10b). During the summer (Fig. 10c), the pattern of change is less consistent, with many stations in the Canadian Prairies showing non-significant decreasing trends (southern Alberta, Saskatchewan and Manitoba). In the fall (Fig. 10d), significant increasing trends are found mainly in eastern regions including Ontario, Quebec and the Atlantic provinces (New Brunswick, Nova Scotia, Prince Edward Island and Newfoundland).

Annual snowfall in Canada increased slightly, by about 4%, from 1950 to 2009 (Fig. 9b). However, this increase in snowfall has not been consistent either temporally or spatially. The national time series for Canada shows a strong increase in annual snowfall from the 1950s to 1970, which is followed by a substantial decrease until the 1980s with no change up to 2009. Spatially, many stations located in the western provinces (British Columbia, Alberta and Saskatchewan) show significant decreasing trends whereas stations in the North show an increase in annual snowfall (Fig. 11a). With regard to seasonal trends, snowfall decreases in southwestern regions and increases in the North during winter (Fig. 11c) and, to a lesser extent, during spring (Fig. 11d). Furthermore, these changes are less pronounced in fall (Fig. 11b).

b *Trends for Southern Canada, 1900–2009*

Annual rainfall increased by 8.7% from 1900 to 2009 in southern Canada. The national time series for southern Canada indicate a strong decrease until 1920 followed by a steady increase from the 1920s to 2009 (Fig. 9c). This increase in annual rainfall is observed at most stations across the country (Fig. 12a). Although there are fewer stations in southern Canada with enough data to compute the trend from 1900 to 2009 than for the shorter 1950–2009 period, 54% of the long-term stations show a significant increase in annual rainfall with trends varying from 10 to 30% for the 110-year period. More stations were available on the seasonal time scale. Increasing rainfall totals are found across the country during all seasons: significant increasing trends were found in 42%, 31% and 35% of the stations during the spring (not presented), summer (Fig. 12b) and fall (not presented), respectively. A mix of non-significant increasing and decreasing trends is also observed in the Canadian Prairies (mainly southern Alberta, Saskatchewan and Manitoba); this pattern is more evident during summer and fall and less consistent during spring.

Annual snowfall increased slightly, by 6.8%, from 1900 to 2009 in southern Canada. The changes in snowfall are neither temporally nor spatially consistent. The annual snowfall time series for southern Canada shows a steady increase from the 1920s to 1970, followed by a considerable decrease to the 1980s with no major change up to 2009 (Fig. 9d). Spatially, a mix of increasing and decreasing trends is found annually (Fig. 12c). During winter (Fig. 12d), a mix of significant increasing and decreasing trends is observed in the southeastern provinces, with the significant decreasing trends observed at stations located along the more densely populated St. Lawrence River.

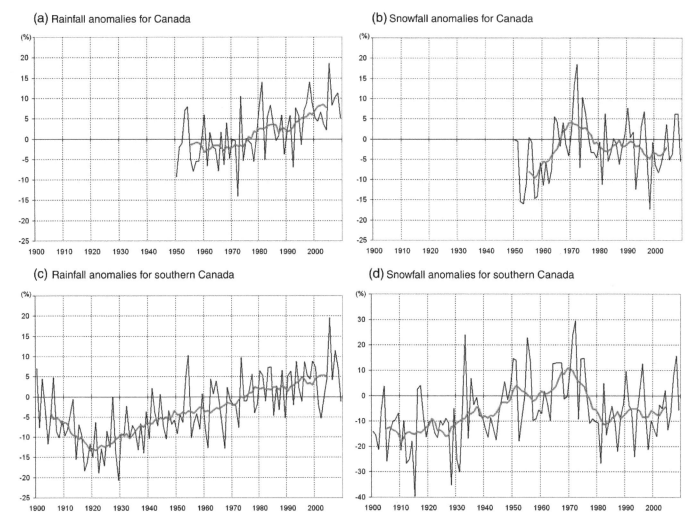

Fig. 9   Rainfall anomalies (expressed as a percentage) for a) Canada, 1950–2009 and c) southern Canada, 1900–2009; snowfall anomalies (expressed as a percentage) for b) Canada, 1950–2009 and d) southern Canada, 1900–2009. The red line represents the 11-year running mean.

6   Summary

Several improvements have been made to the APC2-Daily dataset:

- The station list was revised to include stations with longer periods of observations covering as much of the last 20 years (1990–2009) as possible with daily rain-gauge and snow ruler measurements. The continuity of snowfall observations had become more challenging as the number of stations with snow ruler measurements being taken by observers decreased because of automation.
- Rain-gauge adjustments were based on more field experiments and increased metadata.
- An updated snow water equivalent adjustment factor map, based on almost three times the number of stations as used previously with concurrent snow ruler and Nipher gauge observations, was used to obtain a better estimate of the water content of fresh snow at any location across Canada.
- More information was used to obtain a better adjustment of trace observations, including searches of historical files and

analysis of lower than minimum amount versus small amounts of measurable precipitation.

- The accumulated amount was taken into account by distributing it over the affected days. The distributed amount will be further refined using observations from surrounding stations.
- Since overlapping periods are not available at many merged stations, new adjustments were obtained for joined dates from standardized ratios between tested sites and their neighbours. Final adjustments were based on either the test results or overlapping observations.

The impact of the adjustments on precipitation amounts is summarized as follows. Rainfall amount has increased because of the adjustment for rain-gauges. Applying the snow water equivalent adjustment factor increased the snowfall precipitation amount on the East Coast and in northeastern regions and slightly decreased the amount on the West Coast and in western regions. Including trace event adjustments increased the total amount of precipitation; its overall effect depended on its frequency, which varies

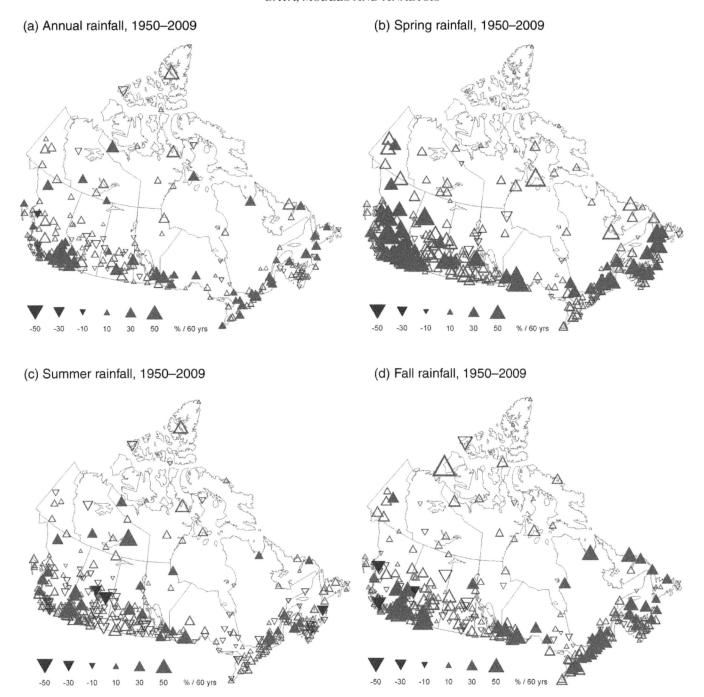

Fig. 10 Trends in annual and seasonal rainfall for 1950–2009. Upward and downward pointing triangles indicate positive and negative trends, respectively. Filled triangles correspond to trends significant at the 5% level. The size of the triangle is proportional to the magnitude of the trend.

throughout the country. Because of relatively low annual total precipitation and a high occurrence of trace events, rainfall and snowfall amounts increased the most in the North.

The impact of the adjustments on trends varies depending on the element and the climate characteristics of the region. Overall, since adjustments for older gauges are larger, rainfall trends decreased slightly after adjustment. Snow trends became more positive along the West Coast and in the Arctic because of snow water equivalent adjustments

applied to snow ruler measurements. Trace corrections did not significantly affect the trends.

Annual and seasonal rainfall and snowfall trends were examined for 1950–2009 and 1900–2009. Overall, rainfall increased across the country during all seasons for both periods; a mix of non-significant increasing and decreasing trends was also found in the summer for the Canadian Prairies. Snowfall increased mainly in the North while a significant decrease was found in the southwestern part of the country for 1950–2009.

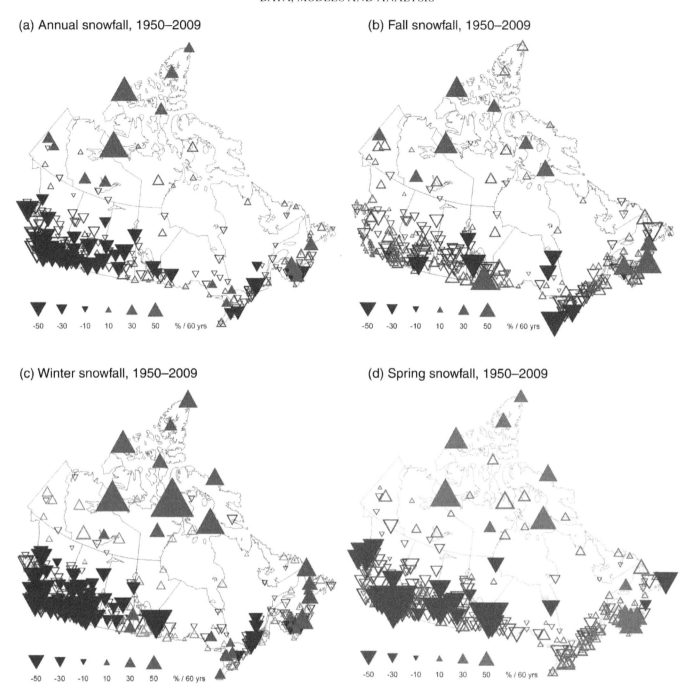

Fig. 11   Trends in annual and seasonal snowfall for 1950–2009. Upward and downward pointing triangles indicate positive and negative trends, respectively. Filled triangles correspond to trends significant at the 5% level. The size of the triangle is proportional to the magnitude of the trend.

Inhomogeneities due to "unknown" changes have not been resolved in this new version of adjusted precipitation. These include, for example, changes in observing practices, new observers, or any other undocumented change such as the relocation of the rain-gauge a short distance away. In several European studies, homogeneity testing and adjustments were performed on the total precipitation series and inhomogeneities were often caused by the instrument's relocation (Hanssen-Bauer and Førland, 1994; Tuomenvirta, 2001;

Wijngaard et al., 2003). In Canada, when a station is relocated, a new identification number is assigned to the new location; consequently, the change is known and data segments can be merged using the documented relocation date.

The main advantage of the procedures presented in this paper is the ability to directly adjust daily rain-gauge and snowfall ruler measurements when they are observed. The adjustments used for the second generation dataset seem to be reasonable when annual, seasonal and monthly totals are

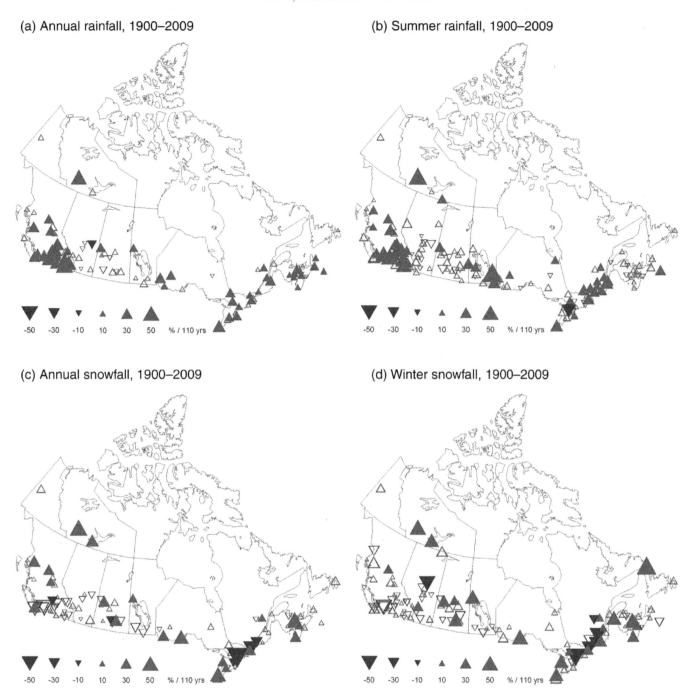

Fig. 12   Trends in annual and seasonal rainfall and snowfall for 1900–2009. Upward and downward pointing triangles indicate positive and negative trends, respectively. Filled triangles correspond to trends significant at the 5% level. The size of the triangle is proportional to the magnitude of the trend.

analyzed. However, because of the difficulties in the quality control of daily rainfall and snowfall for possible errors or outliers, caution should be used when daily precipitation is used in the analysis of climate indices and extremes. It is also important to note that the fresh snowfall water equivalent adjustment factor applied exhibits significant temporal and spatial variability; thus, this adjustment is not recommended for short time series or for events such as blizzards and blowing snow because of the local or short-term uncertainty involved in the computation of these events.

The ACP2-Daily dataset is now available to the scientific community as part of the Adjusted and Homogenized Canadian Climate Data (AHCCD) at http://www.ec.gc.ca/ dccha-ahccd/. The list of joined stations along with their exact location can be provided upon request.

Acknowledgements

The authors wish to acknowledge Ken Devine for his contribution to trace measurement history searches and his knowledge of measurement procedures. Regional climate experts Ron Hopkinson, Monique Lapalme, Bill Richards and Gary Myers provided invaluable help regarding station historical records and metadata information. Further, the authors are grateful to Ross Brown, who contributed to the improved snow ruler adjustment procedure and to Ewa Milewska and Xuebin Zhang of the Climate Research Division of Environment Canada for their constructive comments which led to an improved version of this manuscript.

References

ALEXANDER, L.V.; X. ZHANG, T.C. PETERSON, J. CAESAR, B. GLEASON, A.M. G. KLEIN TANK , M. HAYLOCK, D. COLLINS, B. TREWIN, F. RAHIMZADEH, A. TAGIPOUR, K. RUPA KUMAR, J. REVADEKAR, G. GRIFFITHS, L. VINCENT, D.B. STEPHENSON, J. BURN, E. AGUILAR, M. BRUNET, M. TAYLOR, M. NEW, P. ZHAI, M. RUSTICUCCI and J.L. VAZQUEZ-AGUIRRE. 2006. Global observed changes in daily climate extremes of temperature and precipitation. *J. Geophys. Res* **111**: D05109, doi:10.1029/2005JD006290

DEPARTMENT OF TRANSPORT, METEOROLOGICAL BRANCH. 1955. *How to measure rainfall* (Report No. 63-9056).

DEVINE, K.A. and É. MEKIS. 2008. Field accuracy of Canadian rain measurements. *Atmosphere-Ocean*, **46** (2): 213–227.

DING, Y.; D. YANG, B. YE and N. WANG. 2007. Effects of bias correction on precipitation trend over China. *J. Geophys. Res.* **112**, D13116, doi:10.1029/2006JD007938

GOODISON, B.E.; P.Y.T. LOUIE and D. YANG. 1998. *WMO solid precipitation measurement intercomparison* (Final report. Meteorological Organization, Instruments and Observing Methods Report No.67, WMO/TD-No.872). Geneva, Switzerland: WMO.

HANSSEN-BAUER, I. and E.J. FØRLAND. 1994. Homogenizing long Norwegian precipitation series. *J. Climate,* **7**: 1001–1013.

KENDALL, M.G. 1955. *Rank correlation methods* (2nd ed.). London, UK: Charles Griffin and Company.

KINGSTON, G.T. 1878. Instructions to observers connected with the Meteorological Service of the Dominion of Canada, Toronto, ON: Copp, Clark & Co.

MEKIS, É. 2005. Adjustments for trace measurements in Canada. In *Proc. 15th Conference on Applied Climatology*. Savannah, GA: Am. Meteorol. Soc.

MEKIS, É. and W.D. HOGG. 1999. Rehabilitation and analysis of Canadian daily precipitation time series. *Atmosphere-Ocean,* **37** (1): 53–85.

MEKIS, É. and R. HOPKINSON. 2004. Derivation of an improved snow water equivalent adjustment factor map for application on snowfall ruler measurements in Canada. In *Proc. 14th Conference on Applied Climatology*. Am. Meteorol. Soc, Seattle, WA.

MEKIS, É. and R. BROWN. 2010. Derivation of an adjustment factor map for the estimation of the water equivalent of snowfall from ruler measurements in Canada. *Atmosphere-Ocean,* **48** (4): 284–293.

METCALFE, J.R.; S. ISHIDA and B.E. GOODISON. 1994. *A corrected precipitation archive for the Northwest Territories* (Mackenzie Basin Study, Interim Report no. 2, pp. 110–117). Downsview, ON: Environment Canada.

METCALFE, J.R.; B. ROUTLEDGE and K. DEVINE. 1997. Rainfall measurement in Canada: Changing observational methods and archive adjustment procedures. *J. Climate,* **10**: 92–101.

METEOROLOGICAL DIVISION 1947. *Manual of standard procedures and practices for weather observing and reporting, MANOBS* (1st ed.). Toronto, ON: Department of Transport, Meteorological Division, Head Office.

METEOROLOGICAL DIVISION 1951. *Manual of standard procedures and practices for weather observing and reporting, MANOBS* (3rd ed.). Toronto, ON: Department of Transport, Meteorological Division, Head Office.

PATTERSON, J. 1930. *Instructions to Observers in the Meteorological Service of Canada*. Ottawa, ON: Department of Marine.

ROUTLEDGE, B. 1997. *Corrections for Canadian standard rain gauges* (Atmospheric Environment Service internal report, p. 8). (Available from Meteorological Service of Canada, 4905 Dufferin St., Downsview, ON, M3H 5T4, Canada.)

SEN, P.K. 1968. Estimates of the regression coefficient based on Kendall's tau. *J. Am. Stat. Assoc.* **63**: 1379–1389.

TUOMENVIRTA, H. 2001. Homogeneity adjustments of temperature and precipitation series – Finnish and Nordic data. *Int. J. Climatol.* **21**: 495–506.

VINCENT, L.A.; X. ZHANG, B.R. BONSAL and W.D. HOGG. 2002. Homogenization of daily temperatures over Canada. *J. Climate,* **15**: 1322–1334.

VINCENT, L.A. and É. MEKIS. 2006. Changes in daily and extreme temperature and precipitation indices for Canada over the twentieth century. *Atmosphere-Ocean,* **44** (2): 177–193.

VINCENT, L.A. and É. MEKIS. 2009. Discontinuities due to joining precipitation station observations in Canada. *J. Appl. Meteorol. Climatol.* **48** (1): 156–166.

WIJNGAARD, J.B.; A.M.G. KLEIN TANK and G.P. KÖNNEN. 2003. Homogeneity of the 20th century European daily temperature and precipitation series. *Int. J. Climatol.* **23**: 679–692.

WMO (WORLD METEOROLOGICAL ORGANIZATION). 1972. *Instructions for international comparisons of national precipitation gauges with a reference pit gauge* (revised, R/IOP Annex). Geneva, Switzerland: World Meteorological Organization.

YANG, D.; D. KANE, Z. ZHANG, D. LEGATES and B. GOODISON. 2005. Bias corrections of long-term (1973–2004) daily precipitation data over the northern regions. *Geophys. Res. Lett.* **32**: L19501, doi:10.1029/2005GL024057

ZHANG, X.; L.A. VINCENT, W.D. HOGG and A. NIITSOO. 2000. Temperature and precipitation trends in Canada during the 20th Century. *Atmosphere-Ocean,* **38**: 395–429.

Parametrization of Peatland Hydraulic Properties for the Canadian Land Surface Scheme

Matthew G. Letts, Nigel T. Roulet and Neil T. Comer
*Department of Geography, McGill University and Centre for Climate and
Global Change Research Burnside Hall, Montréal,
Québec*

Michael R. Skarupa
Department of Geography, Trent University, Peterborough, Ontario

and
Diana L. Verseghy
Atmospheric Environment Service, Downsview, Ontario

ABSTRACT   *A hydraulic parametrization is developed for peatland environments in the Canadian Land Surface Scheme (CLASS). Three wetland soil classes account for the typical variation in the hydraulic characteristics of the uppermost 0.5 m of organic soils. Review of the literature reveals that saturated hydraulic conductivity varies from a median of 1.0×10^{-7} m/s in deeply humified sapric peat to 2.8×10^{-4} m/s in relatively undecomposed fibric peat. Average pore volume fraction ranges from 0.83 to 0.93. Parameters have been designed for the soil moisture characteristic curves for fibric, hemic and sapric peat using the Campbell (1974) equation employed in CLASS, and the van Genuchten (1980) formulation. There is little difference in modelled soil moisture between the two formulations within the range of conditions normally found in peatlands. Validation of modelled water table depth and peat temperature is performed for a fen in northern Québec and a bog in north-central Minnesota. The new parametrization results in a more realistic simulation of these variables in peatlands than the previous version of CLASS, in which unrealistic mineral soil "equivalents" were used for wetland soil climate modelling.*

RÉSUMÉ   *Un paramétrage de nature hydraulique est développé pour des sols organiques du schéma CLASS («Canadian Land Surface Scheme»). Trois classes de nouveaux sols de tourbières rendent compte de la variation typique des caractéristiques hydrauliques pour des sols organiques du 0,5 premier mètre. Une revue de la littérature des paramètres hydrauliques révèle que la conductivité hydraulique saturée s'écarte de la médiane de $1,0 \times 10^{-7}$ m/s pour une tourbe humide saprique profonde jusqu'à $2,8 \times 10^{-4}$ m/s pour une tourbe fibrique relativement moins décomposée. La porosité moyenne varie de 0,83 à 0,93. Des paramètres ont été développés afin d'obtenir des courbes caractéristiques d'humidité du sol pour la tourbe fibrique, mésique et saprique, en utilisant le schéma de Campbell (1974) et celui de van Genuchten (1980). Aucune différence significative n'a été détectée dans les deux schémas lorsqu'on les utilise dans des conditions normales trouvées dans les tourbières. On a effectué la validation de la modélisation du niveau phréatique et de la température de la tourbe pour une tourbière carex (fen) dans le nord du Québec et une tourbe à sphaigne (bog) dans le centre nord du Minnesota. Le nouveau paramétrage produit une simulation plus réaliste que la version antérieure du schéma CLASS, qui utilisait des paramètres non réalistes du sol minéral pour modéliser le climat du sol humide.*

1 Introduction

a The Research Problem

The Canadian Land Surface Scheme (CLASS) is a soil-vegetation-atmosphere transfer scheme (SVAT) designed to simulate surface climate and hydrology (Verseghy, 1991). CLASS has proven to be effective at modelling the soil climate of the mineral soils for which it was originally designed. However, 12% of Canada's land surface, including much of the boreal region, is covered by peatlands (Tarnocai et al., 1995). The hydraulic and thermal properties of peat differ greatly from those of mineral soils. Previous versions of CLASS include soil moisture parametrizations for mineral soils, but not for organic soils (Verseghy, 1991). Accurate simulation of soil moisture and temperature is critically important for both micrometeorological and biogeochemical applications. Water availability and peat temperature near the surface affect the magnitude of the latent, sensible and ground heat fluxes (Liang et al., 1996). In this paper, a set of hydraulic parameters is developed and tested for peatland soils in CLASS.

There are several major differences between mineral and organic soils which suggest that mineral soil parametrizations would be deficient for soil climate modelling in wetland environments. Mineral soil porosity, θ_p, ranges from 0.4 to 0.6 (Dingman, 1996), while the porosity of peat is seldom less than 0.8 (Radforth et al., 1977). Near the surface of a peatland, hydraulic conductivity, k, can be very high, due to the large pore size within uncompressed, undecomposed plant material (Boelter, 1968). However, k decreases as much as five orders of magnitude by a depth of 0.4 to 0.8 m (Bradley, 1996), resulting in values similar to those of clays (Fig. 1). Soil-water characteristic curves differ from those of mineral soils. Much less suction, Ψ, is observed at a given volumetric water content, θ_l, than in mineral soils, except at saturation. Thermal properties, which are directly affected by hydrology, also differ substantially. Thermal conductivity, K, of organic soils is very low, ranging from 0.50 W/m K at saturation, to 0.06 W/m K under dry conditions (Farouki, 1981). Heat capacity is slightly higher in dry peat than in mineral soils, but

is much higher under saturated conditions as water occupies the majority of the porous soil volume.

To account for the variation of peat quality with depth, parametrizations are designed for three organic soil classes. Fibric peat is the highly permeable, undecomposed organic soil found near the surface of a peatland. Sapric peat, by contrast, is a deeply humified peat with very low hydraulic conductivity and higher suction. Sapric peat is found at the bottom of the peat profile. The use of these two soils, in conjunction with an intermediate, hemic peat, facilitates the maintenance of the high water table level characteristic of wetland environments.

b CLASS Hydrology

Verseghy (1991) outlines the soil climate algorithms in CLASS. Only the portions of the model relevant to this study are reviewed. CLASS hydraulic conductivity is given by:

$$k = k_s \cdot [\theta_l/\theta_p]^{2b+3} \tag{1}$$

where k (m/s) is the hydraulic conductivity within a given soil layer, θ_l (dimensionless) is the volumetric liquid water content, and b is the soil texture parameter derived from the soil-water characteristic curve (Campbell, 1974; Clapp and Hornberger, 1978). An alternative formulation of k is given by:

$$k = k_s \cdot \sqrt{\Theta}[1 - (1 - \Theta^{1/m})^m]^2, \tag{2}$$

where k_s is the saturated hydraulic conductivity and m is an empirical coefficient related to the pore size distribution parameter, n, by $m = 1 - 1/n$ (van Genuchten, 1980). Θ (dimensionless) is the normalized water content, expressed as:

$$Q = (\theta_l - \theta_{lim})/(\theta_p - \theta_{lim}), \tag{3}$$

where θ_{lim} is the residual soil-water content.

The Campbell (1974) soil characteristic curves, relating soil water suction, Ψ (cm) to volumetric water content, are defined by:

$$\Psi = \Psi_s[\theta_l/\theta_p]^{-b} \tag{4}$$

where Ψ_s is suction at saturation. The van Genuchten formulation of the soil water retention curve is:

$$\theta = \theta_{lim} + [(\theta_p - \theta_{lim})], \tag{5}$$

where α is a characteristic pore size parameter (cm^{-1}). For use in CLASS, this equation must be defined in terms of suction:

$$\Psi = (1/\alpha) \cdot [(1 - \theta_l^{1/m}) \theta_l^{1/m}]^{1/n} \tag{6}$$

We have added to CLASS a simple hydrological scheme that converts soil moisture to water table depth using the concepts of specific yield and specific retention. This also facilitated a comparison of modelled output with observed data since water table is commonly measured in peatlands. Specific yield, S_y (dimensionless), is the proportion of water yielded by gravitational drainage of a volume of saturated soil. S_y is much higher in fibric peat than in sapric peat or mineral soils, owing to the larger pore size. Subtracting S_y

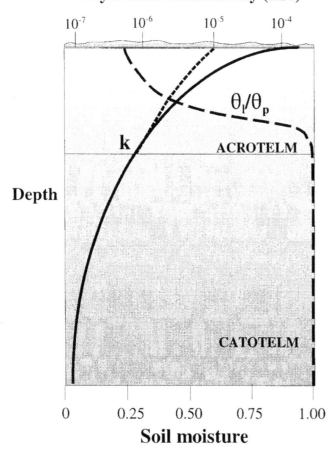

Hydraulic conductivity (m/s)

10^{-7} 10^{-6} 10^{-5} 10^{-4}

θ_l/θ_p

k ACROTELM

Depth

CATOTELM

0   0.25   0.50   0.75   1.00

Soil moisture

Fig. 1   Typical characteristics of two organic soil parameters. Saturated hydraulic conductivity, k_s, varies by over three orders of magnitude with depth in the peat profile. The stippled line to the left illustrates that k is reduced upon drying. In the summer months, soil wetness, θ/θ_p, falls in the acrotelm, but the deeper layers of the soil column remain saturated.

from θ_p yields specific retention, θ_r. Specific retention defines the point at which gravitational forces are balanced by suction. Using the new hydraulic parametrization in CLASS, soil moisture falls due to gravitational drainage until the specific retention value is reached. Beyond this point, additional reduction may occur as a result of evapotranspiration until soil moisture reaches θ_{lim}. To compute water table elevation the saturated soil volume, Θ_v, is calculated first:

$$\Theta_v = (\theta_l - \theta_r)/(\theta_p - \theta_r). \qquad (7)$$

If θ_l falls below θ_r, Θ_v is set to zero. The water table depth, h (m), is then calculated through the following equation:

$$h = \sum_{i=1}^{3} (z_i - z_i \bullet \Theta_{vi}), \qquad (8)$$

where z_i (m) is the depth of a given layer, i.

c Soil Thermal Regime in CLASS
Soil temperature is directly affected by the moisture content, and, therefore, by the hydraulic parametrization used. This is especially true in organic soils, where the high porosity of peat and the temporal variation in soil moisture leads to a great deal of variation in thermal conductivity and heat capacity with varying soil moisture. In CLASS, thermal conductivity, K (W/m K) is given as:

$$K = \prod_{i=1,j}^{3,n} \bullet K_{i,j}^{x_{i,j}}, \qquad (9)$$

where x is the volume fraction of a given soil component, j (Farouki, 1981). Specific heat capacity, C (W/m^{-3} K) is represented by:

$$C = \sum_{i=1,j}^{3,n} C_{i,j} \bullet x_{i,j}, \qquad (10)$$

where C_j is the specific heat capacity of soil component, j.

2 Method

There were two phases in this research. In the first phase, hydraulic and thermal parameters were determined for three classes of organic soils. Representative values of k_s, θ_p, S_y, C and K were determined from the literature on properties of fibric, hemic and sapric peat. In the second phase, CLASS was run using meteorological and hydrological input files from two North American peatlands (Comer et al., 2000). Predicted water table levels and peat temperatures were tested against measured values from these sites.

Statistical analysis of the modelled results follows the method of Willmott (1984). To obtain a comprehensive representation of the difference between observed and predicted values several statistics are reported. These include both systematic and unsystematic root mean square error (RMSE), observed and predicted average and standard deviations, as well as the index of agreement, d:

$$d = 1 - (N \bullet \mathrm{RMSE}^2) / \sum_{i=1}^{n} [(P_i - \bar{O}) - (O_i - \bar{O})]^2 \qquad (11)$$

where N is the number of observations, P is the predicted value and O is the observed value. The index of agreement, d, is a superior statistic to the coefficient of determination, r^2. The latter is insensitive to a variety of potential additive and proportional differences between observed and predicted values (Willmott, 1984).

3 Site descriptions

In this study, modelled water table depth is compared with observations from two distinct wetland sites. The first site, locally known as "Capricorn Fen," is a mineralpoor fen, located 5 km southeast of Schefferville, in northern Quebec (55°N, 67°W). This is an open fen (Moore et al., 1990), with hummock and hollow topography near the margin, and a flatter surface towards the centre. The peatland had some small shallow pools. The elevation of the water table in this fen was measured continuously (corrected for changes in surface elevation of the peat) at three sites with a potentiometric water level sensor (Roulet et al., 1991). The water table record for the central portion of the peatland was used as the test dataset. Moore et al. (1990) showed that water tables in this location are representative of a significant portion of the fen. Vegetation cover consists largely of *Sphagnum spp.* and *Carex spp.* This fen is located in the discontinuous permafrost zone, but is not underlain by permafrost.

The second site is Bog Lake Peatland, located in the Chippewa National Forest (47°32′N, 93°28′W) of north-central Minnesota (Shurpali et al., 1993). In this bog, vegetation is dominated by *Sphagnum papillosum*, though several emergent species were also noted, including some arrowgrass, sedge, leather-leaf, beak-rush, pitcher plant and tamarack species. The site consists of hummock and hollow topography. The water table in Bog Lake Peatland was measured using the same device as in the Capricorn fen, but the water table elevation was referenced to the 'average' hollow surface height (Shurpali et al.,1993).

4 Organic soil hydrology parametrization for CLASS

a Saturated Hydraulic Conductivity
The inclusion of varying hydraulic characteristics with depth is important in wetland hydrological modelling (Bradley, 1996). Hydraulic conductivity is the most variable of the peat parameters. Literature values of k_s were categorized into fibric, hemic or sapric peat classes based on descriptions of peat quality (Table 1). The hydraulic conductivities reported in Table 1 are obtained from both in situ pumping tests and laboratory permeability studies. Boelter (1965) found that laboratory-derived k_s were within the same order to one order of magnitude greater than field-derived k_s. However, the variance in k_s for similar classified peat soil shown in Table 1 is

TABLE 1.   Hydraulic conductivity observations for fibric, hemic and sapric peat.

k (m/s)	Peat quality observations	Source
Fibric peat		
1.28×10^{-4}	slightly decomposed herbaceous peat	Boelter (1968)
3.81×10^{-4}	undecomposed *Sphagnum* peat	Boelter (1968)
1×10^{-5}	undecomposed *Sphagnum* peat	Boelter (1968)
5×10^{-5}	slightly decomposed fen peat	Ivanov (1981)
1.5×10^{-4}	very slightly decomposed bog peat	Ivanov (1981)
4×10^{-5}	slightly decomposed bog peat	Ivanov (1981)
8.5×10^{-5}	von Post humification index = 2	Maelstrom, 1923 in Ingram et al. (1975)
1.7×10^{-4}	von Post humification index = 3	Maelstrom, 1923 in Ingram et al. (1975)
4.5×10^{-4}	*Sphapnum* peat, von Post humification index = 1	Sarasto, 1961 in Ingram et al. (1975)
2.8×10^{-4}	*Carex* peat, von Post humification index = 1.	Sarasto, 1961 in Ingram et al. (1975)
5×10^{-5}	slightly decomposed fen peat	Romanov (1968)
1×10^{-3}	von Post humification index = 1	Ryden (1990) in Magnussen (1994)
3×10^{-6}	von Post humification index = 1.5	Ryden (1990) in Magnussen (1994)
1×10^{-6}	von Post humification index = 2	Ryden (1990) in Magnussen (1994)
3×10^{-7}	von Post humification index = 3	Ryden (1990) in Magnussen (1994)
2.2×10^{-3}	von Post humification index = 1	Gafni (1986)
1.5×10^{-3}	von Post humification index = 2	Gafni (1986)
9×10^{-4}	von Post humification index = 3	Gafni (1986)
5×10^{-4}	surface *Sphagnum* peat	Boelter (1968)
1.6×10^{-3}	surface *Sphagnum* peat	Boelter (1968)
7.2×10^{-4}	surface *Sphagnum* peat	Boelter (1968)
1.4×10^{-3}	saturated fen peat, k averaged	Baird (1997)
Hemic peat		
1×10^{-6}	moderately-decomposed woody *Sphagnum* peat	Boelter (1968)
5×10^{-5}	moderately-decomposed woody peat	Boelter (1968)
6×10^{-6}	moderately-decomposed woody peat	Boelter (1968)
7×10^{-8}	moderately-decomposed woody peat	Boelter (1968)
8×10^{-6}	moderately-decomposed fen peat	Ivanov (1981)
5×10^{-7}	moderately-decomposed bog peat	Ivanov (1981)
2.2×10^{-4}	von Post humification index = 4.5	Maelstrom (1923) in Ingram et al. (1975)
2×10^{-6}	von Post humification index = 6	Maelstrom (1923) in Ingram et al. (1975)
2×10^{-6}	von Post humification index = 7	Maelstrom (1923) in Ingram et al. (1975)
8×10^{-6}	moderately decomposed fen peat	Romanov (1968)
6×10^{-7}	von Post humification index = 3.5	Ryden (1990) in Magnussen (1994)
1×10^{-6}	von Post humification index = 4	Ryden (1990) in Magnussen (1994)
1×10^{-7}	von Post humification index = 5	Ryden (1990) in Magnussen (1994)
6×10^{-8}	von Post humification index = 6	Ryden (1990) in Magnussen (1994)
1×10^{-9}	von Post humification index = 7	Ryden (1990) in Magnussen (1994)
5×10^{-4}	von Post humification index = 4	Gafni (1986)
2×10^{-4}	von Post humification index = 5	Gafni (1986)
3×10^{-5}	von Post humification index = 6	Gafni (1986)
6×10^{-6}	von Post humification index = 7	Gafni (1986)
4×10^{-7}	bulk density = 0.064–0.13 g/cc	Hanrahan (1954) in Radforth and Brawner (1977)
6×10^{-6}	bulk density = 0.172 g/cc	Irwin (1968)
5×10^{-6}	bulk density = 0.099 g/cc	Irwin (1968)
1×10^{-6}	bulk density = 0.096 g/cc	Irwin (1968)
4×10^{-6}	bulk density = 0.096 g/cc	Irwin (1968)
2×10^{-6}	bulk density = 0.096 g/cc	Irwin (1968)
2×10^{-6}	bulk density = 0.094 g/cc	Irwin (1968)
5×10^{-8}	bulk density = 0.11 g/cc, k averaged	Phalen (1961) in Radforth and Brawner (1977)
3×10^{-8}	mixed medium peat, $k = 1–5 \times 10^{-8}$ m/s	Tveiten (1956) in Radforth and Brawner (1977)
2×10^{-8}	grassy medium peat, $k = 1–3 \times 10^{-8}$ m/s	Tveiten (1956) in Radforth and Brawner (1977)
Sapric peat		
5×10^{-8}	decomposed	Boelter (1968)
5×10^{-7}	very decomposed	Ivanov (1981) in Ingram et al. (1975)
4×10^{-7}	von Post humification index = 8.5	Maelstrom (1923) in Ingram et al. (1975)
1×10^{-7}	von Post humification index = 9	Maelstrom (1923) in Ingram et al. (1975)
3×10^{-6}	*Sphagnum* peat, von Post humification index = 8	Sarasto (1961) in Ingram et al. (1975)
5×10^{-6}	*Carex* peat, von Post humification index = 8	Sarasto (1961) in Ingram et al. (1975)
5×10^{-7}	highly decomposed fen peat	Romanov (1968)
1×10^{-7}	von Post humification index = 7.5	Ryden (1990) in Magnussen (1994)
2×10^{-10}	von Post humification index = 8	Ryden (1990) in Magnussen (1994)
1×10^{-12}	von Post humification index = 9	Ryden (1990) in Magnussen (1994)
2×10^{-8}	dark peat, $k = 1–3 \times 10^{-8}$ m/s	Tveiten (1956) in Radforth and Brawner (1977)
2×10^{-7}	humic subsurface peat	Boelter (1968)
6.3×10^{-8}	clayey peat	Hemond and Chen (1990)

much larger than an order of magnitude, and the variance cannot be explained by the method of sampling.

Fibric peat is defined as having porosity, θ_p, greater than 0.90, bulk density, ρ_b, less than 75 kg m^{-3}, specific yield, S_y, in excess of 0.42 and volumetric water content, θ_l, less than 0.48 at 1 m suction (Boelter, 1968). Sapric peat, the most deeply humified organic soil, is characterized by $\theta_p > 0.85$, $\rho_b > 195$ kg m^{-3}, $S_y < 0.15$ and $\theta_l > 0.70$ at $\psi = 1$ m.

Measured k_s varies by several orders of magnitude within each class of peat. Since the k_s values are skewed within each class, the median was determined instead of the mean. The median values are 2.8×10^{-4} m s^{-1}, 2.0×10^{-6} m s^{-1} and 1.0×10^{-7} m s^{-1}, for fibric, hemic and sapric peat, respectively (Fig. 2).

b Suction versus Wetness Curves

Generalized mineral soil parameters are well established for both the Campbell soil-water characteristic curve (eg., Clapp and Hornberger, 1978; Cosby et al., 1984) and the van Genuchten formulation (van Genuchten, 1980). However, little attention has been paid to the parametrization of water retention curves for organic soils. da Silva et al. (1993) successfully used the van Genuchten equation in a laboratory study involving one peat type, but not all the fitted parameters were reported and θ_r was set to 0.0 rather than to $(\theta_p - S_y)$, making intercomparison with the results of this study difficult.

At a given suction, lower volumetric water content is generally observed in fibric peat than in sapric peat. The exception occurs at very low suction, when fibric peat holds slightly more water (Radforth et al., 1977). Such conditions are typical in wetland environments where the soil remains saturated at a shallow depth for much of the year. Soil-water versus suction values, derived for fibric, hemic and sapric peat based on data from Boelter and Blake (1964), Paquet et al. (1993), Berglund (1996) and Magnussen (1994), are plotted in Fig. 3. These values were obtained by using tension (low suctions) and pressure plates (higher suctions), standard techniques described in texts of soil physics (e.g., Hillel, 1998, pp. 166–168). Three parameters, θ_p, ψ_s and b relate suction to volumetric liquid water content, θ_l, in the Campbell scheme (see Eq. 4). The van Genuchten formulation requires α, θ_{lim}, θ_p and n (see Eq. 5). These parameters vary significantly with peat quality. Using the Campbell scheme, accurate modelling of θ_l at low suction is only possible at the expense of underestimation at high suction. This is a minor problem in wetland environments, where the deeper layers of the peat remain at or near saturation.

Porosity, which ranges from 0.81 to 0.95, is greater in fibric peat than in sapric peat (Radforth et al., 1977). Average values of 0.93, 0.88 and 0.83 were calculated for fibric, hemic and sapric peat, respectively (Fig. 2). Using these porosity parameters as constants, Campbell (1974) and van Genuchten (1980) water retention curves (Eqs 4 and 5) were fitted to the data displayed in Fig. 3. Saturated suction values are slightly higher for fibric peat than for sapric peat or mineral soils.

c Specific Yield

Specific yield is obtained by measuring the water released over time due to gravity drainage, but it is often measured as the difference in volumetric water content at 0 and 0.1 bars of tension (Boelter, 1968). Hillel (1998, pp. 481) points out that the assumption that there is a fixed value of drainable porosity and that soils drain instantly due to a change in water table is a gross approximation. Undecomposed moss peat, which contains approximately 93% water at saturation, releases up to 80% of this water to drainage (Radforth et al., 1977). Herbaceous peat contains less water at saturation, but contributes much less to drainage as a result of reduced pore size. In CLASS, specific yield is set at 0.66, 0.26 and 0.13, for fibric, hemic and sapric peat, respectively (Boelter, 1968). These values are subtracted from θ_p, to attain specific yield, θ_r.

d Thermal Conductivity and Heat Capacity

The thermal conductivity, K, of peat is lower than that of mineral soils, especially when drying occurs. Thermal conductivity is set to 0.25 W/m K for the organic matter fraction of Eq. 10 (Farouki, 1981). The thermal conductivities of water, air and ice are set to 0.57, 0.025 and 2.20 W/m K, respectively. Each of these components has a lower K than the average mineral soil value of 2.93 W/m K.

Heat capacity, C, is usually higher in organic soils than in mineral soils, though it may become lower under dry or frozen conditions. The heat capacity of organic matter is 2.51 W/m^3 K (van Wijk, 1963). The heat capacities of water, air and ice are set to 4.18, 0.00125 (Farouki, 1981) and 1.90 (van Wijk, 1963; Oke, 1987), respectively.

5 Impacts of the new parametrization

a Soil Moisture and Water Table Level

The previous version of CLASS (Version 2.6) did not include a complete organic soil parametrization. Instead, mineral soil parametrizations were used for peatland soil climate and energy balance simulations. In an effort to maintain high soil moisture levels, sand was used in the shallowest layer, with clay in the deepest layer. This approach led to underestimation of soil moisture and poor simulation of soil temperature. To alleviate this problem, drainage through the bottom layer was prohibited. This was unrealistic, because drainage occurs beneath wetlands, except when underlain by impermeable substrates or in receipt of groundwater input. The three varieties of peat included in the new parametrization (Version 2.6w) permit more realistic simulations to be performed without constraining drainage.

Figures 4(a) and (b) show the result of the water table simulation at Capricorn Fen, employing the parameter set of Fig. 2. The pattern of water table change is modelled effectively ($d = 0.92$, RMSE = 1.14 cm), though the water level is overestimated slightly in late summer, resulting in an average divergence between modelled and observed depths of

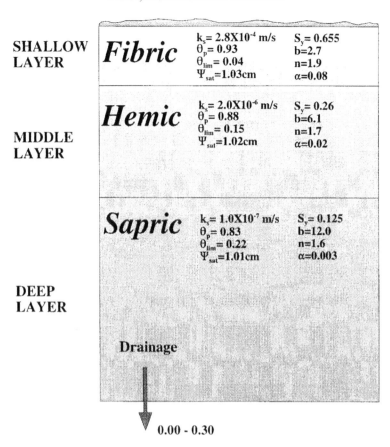

Fig. 2 The new CLASS hydraulic parametrization scheme. Included are characteristic values required for both the Campbell and van Genuchten water retention and hydraulic conductivity formulations. The drainage parameter is scaled between 0 and 1, where 1 represents a freely draining system and 0 is indicative of a fully impermeable underlying substrate.

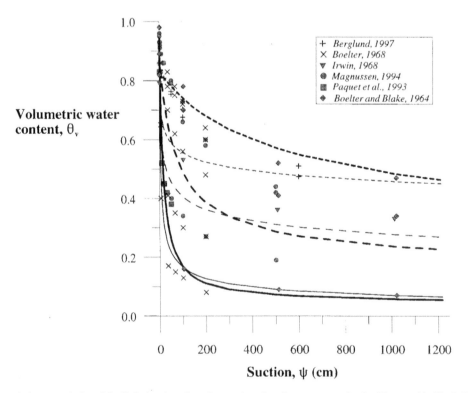

Fig. 3 Soil-water characteristic curves designed for fibric, hemic and sapric peat, based on the new parmetrization illustrated in Fig. 2. The bold curves are those of the van Genuchten formulation, while the lighter curves are those of the Campbell equations currently used in CLASS.

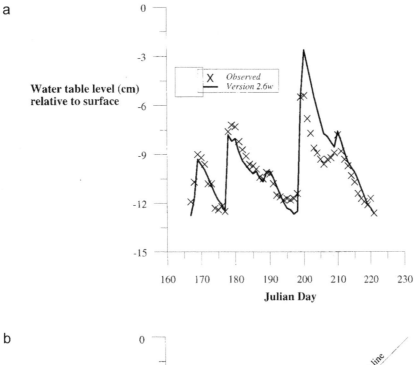

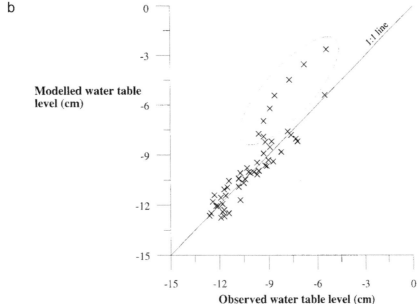

Fig. 4 (a) Observed and modelled water table depths at Capricorn Fen (1989). (b) A scatterplot of observed versus modelled water table levels at Capricorn Fen (1989). The stippled ellipse high-lights the overestimation of water table levels after the 60 mm rainfall event on Day 200.

0.38 cm (Table 2). This is largely due to the model's exaggerated response to the major rainfall event of Day 200. The small systematic root mean square error indicates that little further improvement is likely by altering the peat parameters.

The previous version of CLASS did not include a water table model. However, comparison of soil moisture plots demonstrates the improvement resulting from the new parametrization (Figs 5(a) and (b)). Using Version 2.6, drying occurs in layers 2 and 3, indicating that the water table would be located below 35 cm, much deeper than observed. In fact, excessive drying occurs in the two deepest layers with

drainage set to any value above zero. With drainage set to zero in Version 2.6, soil moisture is overestimated.

The result of the water table simulation at Bog Lake Peatland, using Version 2.6w, is shown in Figs 6(a) and (b). Analysis of soil moisture output reveals that Version 2.6 is far too dry when drainage is permitted, but is not capable of simulating the soil moisture reduction occurring in the dry period of late summer when drainage is set to zero. Using the new parametrization, the water table is modelled well ($d = 0.86$), though it is underestimated during a wet period in early summer (Table 2). There are two possible reasons

99

TABLE 2. Comparison of observed and modelled water table level and peat temperature.

	Average	Standard deviation	RMSE	RMSE$_u$	RMSE$_s$	d
			Water table level (cm)			
Capricorn Fen						
Observed	−9.95	1.73	–	–	–	–
Version 2.6w	−9.62	2.34	1.14	1.09	0.36	0.92
Bog Lake Peatland						
Observed	−8.38	6.24	–	–	–	–
Version 2.6w	−11.96	4.51	4.27	2.76	3.26	0.86
			Peat temperature (°C)			
Capricorn Fen						
Observed (10 cm)	14.17	4.41	–	–	–	–
Version 2.6 (layer 1)	12.63	4.51	3.04	2.60	1.59	0.88
Version 2.6w (layer 1)	11.83	3.79	3.23	2.26	2.31	0.86
Observed (20 cm)	9.29	2.41	–	–	–	–
Version 2.6 (layer 2)	11.80	4.00	3.86	2.78	2.68	0.69
Version 2.6w (layer 2)	9.93	2.21	1.48	1.34	0.61	0.90
Observed (2.0 m)	1.29	0.26	–	–	–	–
Version 2.6 (layer 3)	3.56	0.99	2.44	0.33	2.42	0.39
Version 2.6w (layer 3)	2.08	0.33	0.84	0.35	0.76	0.39

for this result. Firstly, the site receives water from the adjoining highlands and surrounding mineral soils (Shurpali et al., 1993). Secondly, CLASS is a one-dimensional model, and is not capable of receiving or releasing water through lateral flow. The one-dimensional form of CLASS only poses a problem when the water tables are high, as is the case immediately after a large rain storm or early in the spring during snowmelt. In these two situations, a significant proportion of the loss of water is through lateral runoff. However, if the water table is 10 cm or more below the surface of the peatland, the exchange of water becomes dominated by evapotranspiration and precipitation. The lack of lateral flow in this situation is due to the reduced hydraulic conductivity. As a result, modelled levels differ from observed values by an average of −3.58 cm, with a root mean square error of 4.27 cm.

The accuracy of this water table simulation is encouraging, especially given that identical physical parameters were used for both Bog Lake Peatland and Capricorn Fen. While the model performs well in these environments, equally good results would not be expected in rich fen environments or in highly ombrotrophic bogs. Soil climate simulation at rich fen sites is limited because CLASS is not presently capable of receiving groundwater input. Furthermore, there is a lack of understanding of the effect of non-vascular vegetation on the evaporation process. This could be problematic in particularly ombrotrophic bogs that may consist almost entirely of a *Sphagnum* sp. mat.

b Peat Temperature

Soil moisture and temperature are intimately linked, since wetness alters both thermal conductivity and heat capacity. The replacement of mineral soils with organic classes permits soil temperature simulation with more realistic values for these two parameters.

Figures 7(a) and (b) display observed and modelled peat temperature at Capricorn Fen. In the first layer, both the new and old parametrizations simulate the general trend of soil moisture very well until approximately Day 200 (Fig. 7(a); Table 2). Beyond this date, soil temperatures are underestimated. This coincides with the overestimation of soil moisture following a 60 mm precipitation event. Note that in hummock and hollow topography, it is more difficult to locate a representative surface elevation than in mineral soil environments. The measured values in Fig. 7(a) were taken at a depth of 10 cm within a hummock. In addition, soil temperature can vary as much as 5°C at similar depths between hummocks and hollows.

Version 2.6w provides far more accurate soil temperature output for layers 2 and 3. In Version 2.6, mineral soils are used resulting in overestimated thermal conductivity and underestimated heat capacity. As a result, soil temperatures rise far too quickly and are overly variable in the intermediate layer (Fig. 7(b); Table 2). The solution to this problem has proven beneficial for ongoing biogeochemical modelling efforts using the CLASS model.

c Sensitivity

The sensitivity of the model output to changes in porosity, saturation suction, hydraulic conductivity and fitted parameters for the soil-water characteristic curves was examined. For each parameter, with the exception of k_s, the parameter value was changed in steps of ±5% to a maximum 20% change from the base run. k_s was changed over a range of eight times (4 times below to 4 times above the standard run k_s). Only one parameter was changed at a time, therefore the compound effect of changes in multiple parameters was not assessed.

In the shallowest two layers, soil moisture and temperature outputs are insensitive to moderate changes in porosity,

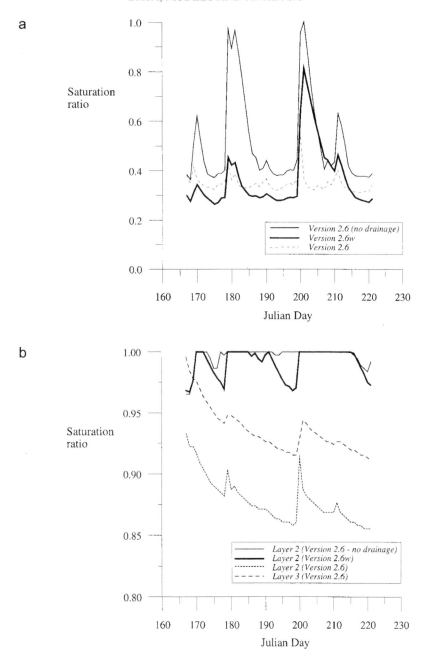

Fig. 5 (a) Modelled layer 1 soil moisture (Schefferville, Quebec, 1989). (b) Modelled daily average soil moisture in layers 2 and 3 (Schefferville, Quebec, 1989). Layer 3 is permanently saturated using Version 2.6 with zero drainage or Version 2.6w.

saturated suction, hydraulic conductivity and the fitting parameters of the soil-water characteristic curves. The deepest layer, which controls vertical drainage from the system, is highly sensitive to changes in hydraulic conductivity (Table 3), though the van Genuchten formulation is somewhat less sensitive to increasing k than the Campbell scheme. This is the result of the higher suction predicted at a given θ_l for sapric peat. Soil moisture levels also affect K and C, thus changing the peat temperature profile. The model is not sensitive to changes in other layer 3 parameters until soil moisture begins to fall in this layer. In wetland environments, this occurs only under conditions of extreme drought.

Using the parameters listed in Fig. 2, the Campbell and van Genuchten formulations produce very similar results. Hydraulic conductivity and suction values derived by the two formulations are identical at saturation (Figs 3 and 8). These two variables, along with the drainage parameter, determine the amount of seepage through the bottom of the profile. The third soil layer remains saturated throughout the simulation. It can, therefore, be concluded that while the implementation of the van Genuchten equation did not significantly alter results, differences between the two formulations would be greater in soils with lower hydraulic conductivity or in unusually dry conditions.

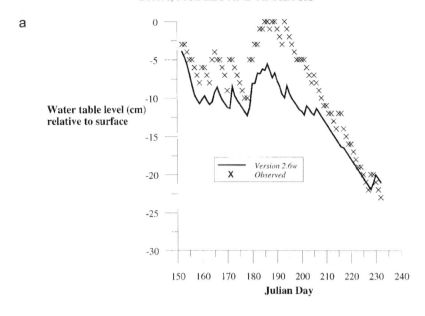

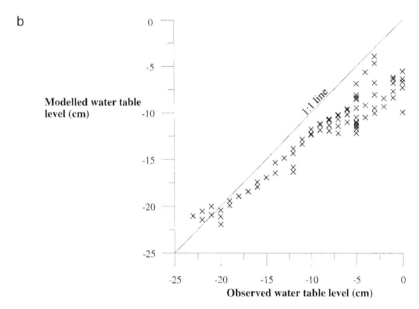

Fig. 6   (a) Observed and modelled water table levels at Bog Lake Peatland (1991). (b) A scatterplot of modelled water table levels at Bog Lake Peatland (1991).

Table 3.   Sensitivity of saturation ratio, water table level and peat temperature to varying hydraulic conductivity in the deepest layer. Outputs given are for Capricorn Fen on Julian Day 220, using Version 2.6w. The bolded values represent those of the standard runs.

k(m/s)	Saturation ratio			Peat temperature ($^{\circ}$C)			Water table (cm)
	L1	L2	L3	L1	L2	L3	
Campbell							
5×10^{-8}	0.33	1.00	1.00	11.60	8.29	2.54	−9.76
1×10^{-7}	**0.27**	**0.98**	**1.00**	**11.87**	**8.94**	**2.55**	**−12.07**
2×10^{-7}	0.23	0.89	1.00	12.33	9.92	2.79	−19.62
4×10^{-7}	0.24	0.70	0.98	9.98	5.43	2.23	−88.88
van Genuchten							
5×10^{-8}	0.31	1.00	1.00	11.59	8.27	2.54	−9.60
1×10^{-7}	**0.28**	**0.98**	**1.00**	**11.88**	**8.95**	**2.54**	**−11.97**
2×10^{-7}	0.22	0.90	1.00	12.24	9.83	2.81	−17.46
4×10^{-7}	0.08	0.70	0.99	9.08	6.36	2.65	−53.84

a

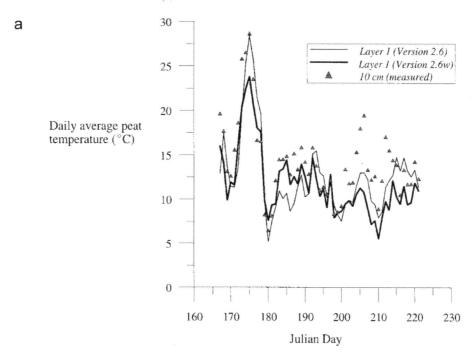

b

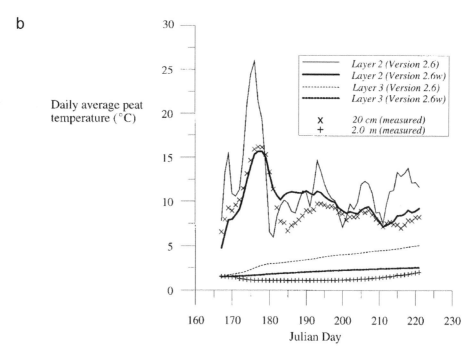

Fig. 7   (a) Modelled layer 1 peat temperature and 10 cm measured values (Schefferville, Quebec, 1989). (b) Modelled and measured peat temperatures in layers 2 and 3 (Schefferville, Quebec, 1989).

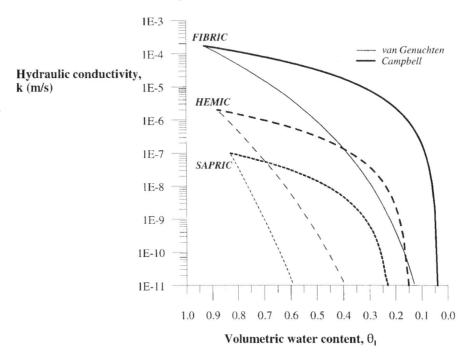

Fig. 8 Hydraulic conductivity versus volumetric water content for fibric, hemic and sapric peat. The bold curves are those of the van Genuchten formulation (Eq. 2), while the lighter curves are those of the Campbell scheme (Eq. 1).

6 Conclusions

Organic soil environments are under-represented in the SVAT modelling literature. This study represents a first effort to include an organic soil parametrization within such a model. The use of mineral soil "equivalents" to model the hydraulic and thermal regime of peatlands is unrealistic and should be discontinued. In CLASS Version 2.6, soil moisture is heavily underestimated unless drainage is set to zero. In Version 2.6w, the introduction of fibric, hemic and sapric peat classes allows the water table to be maintained without restricting drainage to unrealistically low levels at the bottom of the soil column.

The incorporation of organic soils into the CLASS model has improved the simulation of soil moisture and temperature in a northern Québec fen and in a north-central Minnesota bog. The results obtained from these peatlands are encouraging for both micrometeorological applications and biogeochemical simulations, which are heavily reliant upon accurate

soil climate modelling. There is, however, much room for improvement. CLASS eventually needs to be modified to be capable of receiving groundwater inputs and shedding water laterally. For peatlands, particularly the ombrotrophic cases examined in the present study, a simple exponential relationship between water table elevation and runoff (e.g., Verry et al., 1988) might be adequate to estimate the lateral loss of water. However, the modelling of groundwater input that occurs in rich fens will be problematic if detailed topography and geology wetlands are required.

Acknowledgements

We would like to thank Tim Moore for his comments on earlier drafts of this paper and the reviewers for their thoughtful suggestions. This study was funded by the Canadian Climate Research Network (Land Surface Processes node) and an NSERC Research grant to NTR.

References

BAIRD, A.J. 1997. Field estimation of macropore functioning and surface hydraulic conductivity in a fen peat. *Hydrol. Proc.* Vol. **11**: 287–295.

BERGLUND, K. 1997. Water retention data from Karungi, Sweden (1988) and Majnegarden, Sweden (1976). Swedish University of Agricultural Sciences, Department of Soil Sciences, Division of Hydrotechtonics. Uppsala, Sweden.

BOELTER, D.H. 1965. Hydraulic conductivity of peats. *Soil Sci.* **100**: 227–231.

———. 1968. Important physical properties of peat materials. *In*: Proc. Third International Peat Congress. Quebec. 18–23 August. Department of Energy, Mines and Resources Canada. National Research Council of Canada, pp. 150–154.

——— and G.R. BLAKE. 1964. Importance of volumetric expression of water contents of organic soils. *Soil Science Society of America Proceedings*, **28(2)**: 176–178.

BRADLEY, C. 1996. Transient modelling of watertable variation in a floodplain wetland, Narborough Bog, Leicestershire. *J. Hydrol.* **185**: 87–114.

CAMPBELL, J.D. 1974. A simple method for determining unsaturated conductivity from moisture retention data. *Soil Sci.* **117**: 311–314.

CLAPP, R.B. and G.M. HORNBERGER. 1978. Empirical equations for some soil hydraulic properties. *Water Resour. Res.* **14(4)**: 601–604.

COMER, N.T.; P.M. LAFLEUR, N.T. ROULET, M.G. LETTS, M. SKARUPA and D. VERSEGHY. 2000. A test of the Canadian Land Surface Scheme (CLASS) for a variety of wetland types. *Atmosphere-Ocean*, **38(1)**.

COSBY, B.J.; G.M. HORNBERGER, R.B. CLAPP and T.R. GINN. 1984. A statistical exploration of the relationships of soil moisture characteristics to the physical properties of soils. *Water Resour. Res.* **20(6)**: 682–690.

DA SILVA, F.F.; R. WALLACH and Y. CHEN. 1993. Hydraulic properties of Sphagnum peat moss and tuff (scoria) and their potential effects on water availability. *Plant Soil.* **154**: 119–126.

DINGMAN, S.L. 1996. *Physical Hydrology.* Prentice-Hall, Englewood, New Jersey, 575 pp.

FAROUKI, O.T. 1981. Thermal properties of soils. *CRREL Monograph.* Vol. 81, no. 1, 134 pp.

GAFNI, A. 1986. Field tracing approach to determine flow velocity and hydraulic conductivity in saturated peat soils. Ph.D. thesis. University of Minnesota, Minneapolis, Minnesota.

HEMOND, H.F. and D.G. CHEN. 1990. Air entry in salt marsh sediments. *Soil Sci.* **150**: 459–468.

HILLEL, D. 1998. *Environmental Soil Physics.* Academic Press, New York, 771 pp.

INGRAM, H.A.P.; D.W. RYCROFT and D.J.A. WILLIAMS. 1975. Anomalous transmission of water through certain peats. *J. Hydrol.* **22**: 213–218.

IRWIN, R.W. 1968. Soil water characteristics of some (southern) Ontario peats. *In*: Proc. Third International Peat Congress. Quebec. 18–23 August. Department of Energy, Mines and Resources Canada. National Research Council of Canada, pp. 219–223.

IVANOV, K.E. 1981. *Water Movement in Mirelands.* Academic Press, London, 276 pp.

LIANG, X.; E.F. WOOD and D.P. LETTENMAIER. 1996. Surface soil parameterization of the VIC-2L model: Evaluation and modification. *Global Planet. Change,* **13**: 195–206.

MAGNUSSEN, T. 1994. Studies of the soil atmosphere and related physical characteristics in peat forest soils. *Forest Ecol. Mgmt.* **67**: 203–224.

MOORE, T.R.; N.T. ROULET and R. KNOWLES. 1990. Spatial and temporal variations of methane flux from subarctic/northern boreal fens. *Global Biogeochem. Cycles,* **4**: 29–46.

OKE, T.R. 1987. *Boundary Layer Climates – Second Edition.* Methuen, London and New York, 435 pp.

PAQUET, J.M.; J. CARON and O. BANTON. 1993. In situ determination of the water desorption characteristics of peat substrates. *Can. J. Soil Sci.* **73**: 329–339.

RADFORTH, N.W. and C.O. BRAWNER (Eds.). 1977. *Muskeg and the Northern Environment in Canada.* University of Toronto Press, Toronto, Canada, pp. 82–147.

ROMANOV, V.V. 1968. Hydrophysics of Bogs. Israel Program for Translations, Jerusalem, 299 pp.

ROULET, N.T.; S. HARDILL and N. COMER. 1991. Continuous measurements of depth of water table (inundation) in wetlands with fluctuating surfaces. *Hydrol. Proc.* **5**: 399–403.

SHURPALI, N.J.; S.B. VERMA, R.J. CLEMENT and D.P. BILLESBACH. 1993. Seasonal distribution of methane flux in a Minnesota peatland measured by eddy correlation. *J. Geophys. Res.* **98(D11)**: 20,649–20,655

TARNOCAI, C.; I.M. KETTLES and M. BALLARD. 1995. Peatlands of Canada. Geological Survey of Canada map. Open File 3152.

VAN GENUCHTEN, M.T. 1980. A closed-form equation for predicting the hydraulic conductivity of unsaturated soils. *Soil Sci. Soc. Am. J.* **44**: 892–898.

VAN WIJK, W.R. 1963. *Physics of Plant Environment.* North Holland Publishing Company, Amsterdam, 372 pp.

VERRY, E.S.; K.N. BROOKS and P.K. BARTEN. 1988. Streamflow response from an ombrotrophic mire. *In*: Int. Symp. Hydrology of Wetlands in Temperate and Cold Regions, Joensuu, Finland, 6–8 June. Academy of Finland, Helsinki, pp. 52–59.

VERSEGHY, D. 1991. CLASS – A Canadian Land Surface scheme for GCMs. I. soil model. *Int. J. Climatol.* **11**: 111–133.

WILLMOTT, C.J. 1984. On the evaluation of model performance in physical geography. In: *Spatial Statistics and Models,* GAILLE, G.L. and C.J. WILLMOTT (Eds). D. Reidel Publishing Co., pp. 443–460

* Corresponding author: routlet@felix.geog.mcgill.ca

The 15-km Version of the Canadian Regional Forecast System

Jocelyn Mailhot[1], Stéphane Bélair[1], Louis Lefaivre[2], Bernard Bilodeau[1], Michel Desgagné[1],
Claude Girard[1], Anna Glazer[1], Anne-Marie Leduc[2], André Méthot[2], Alain Patoine[2],
André Plante[2], Alan Rahill[2], Tom Robinson[2], Donald Talbot[2], André Tremblay[1],
Paul Vaillancourt[1], Ayrton Zadra[1] and Abdessamad Qaddouri[1]

[1]*Meteorological Research Branch, Meteorological Service of Canada*
Dorval, Québec
[2]*Canadian Meteorological Centre, Meteorological Service of Canada*
Dorval, Québec

ABSTRACT *A new mesoscale version of the regional forecast system became operational at the Canadian Meteorological Centre on 18 May 2004. The main changes to the regional modelling system include an increase in both the horizontal and vertical resolutions (15-km horizontal resolution and 58 vertical levels instead of 24-km resolution and 28 levels) as well as major upgrades to the physics package. The latter consist of a new condensation package, with an improved formulation of the cloudy boundary layer, a new shallow convection scheme based on a Kuo-type closure, and the Kain and Fritsch deep convection scheme, together with a subgrid-scale orography parametrization scheme to represent gravity wave drag and low-level blocking effects. The new forecast system also includes a few changes to the regional data assimilation such as additional radiance data from satellites.*

Objective verifications using a series of cases and parallel runs, along with subjective evaluations by CMC meteorologists, indicate significantly improved performance using the new 15-km resolution forecast system. We can conclude from these verifications that the model exhibits a marked reduction in errors, improved predictability by about 12 hours, better forecasts of precipitation, a significant reduction in the spin-up time, and a different implicit-explicit partitioning of precipitation. A number of other features include: sharper precipitation patterns, better representation of trace precipitation, and general improvements of deepening lows and hurricanes. In mountainous regions, several aspects are better represented due to combined higher-resolution orography and the low-level blocking term.

RÉSUMÉ *Une nouvelle version à méso-échelle du système régional de prévision est devenue opérationnelle au Centre Météorologique Canadien le 18 mai 2004. Les principaux changements au système régional de modélisation incluent une augmentation de la résolution à la fois dans l'horizontale et dans la verticale (15 km de résolution horizontale et 58 niveaux verticaux plutôt que 24 km et 28 niveaux) et des améliorations majeures à la physique du modèle. Celles-ci comprennent un nouvel ensemble pour décrire les processus de condensation, avec une formulation améliorée de la couche limite nuageuse, un nouveau schéma de convection restreinte basé sur une fermeture à la Kuo et le schéma de convection profonde de Kain et Fritsch, ainsi qu'un paramétrage unifié des effets orographiques sous-maille, représentant la résistance sur l'écoulement due aux ondes de gravité déferlantes et les effets du blocage sur les vents de surface. Le nouveau système de prévision inclut aussi quelques changements au système régional d'assimilation de données, comme l'ajout de données de radiance satellitaires.*

Des vérifications objectives basées sur plusieurs séries de cas et au cours de la passe parallèle, de même que des évaluations subjectives par les météorologistes d'opérations du CMC, ont démontré une nette amélioration de la performance avec le nouveau système de prévision à 15 km. On peut conclure de ces vérifications que le modèle présente des erreurs nettement réduites, une prévisibilité accrue d'environ 12 heures, de meilleures prévisions des précipitations, une réduction significative du temps d'ajustement, et une répartition différente entre les précipitations implicites et explicites. Quelques autres caractéristiques incluent des zones de précipitations mieux définies, notamment pour la représentation de la trace de précipitations, ainsi que des améliorations au développement des dépressions et des ouragans. Dans les régions montagneuses, plusieurs aspects sont mieux représentés grâce à la combinaison d'une topographie plus réaliste et du terme de blocage.

1 Introduction

Since February 1997, the Canadian regional forecast system used operationally at the Canadian Meteorological Centre (CMC) for data assimilation and short-range weather forecasts, has been based on the Global Environmental Multiscale (GEM) model (Côté et al., 1998a, 1998b). The implementation of a 24-km version of the forecast system took place on 15 September 1998 (Bélair et al., 2000). With the main objective of improving summertime quantitative precipitation forecasts (QPF), it combined an increase in horizontal resolution (from 35 to 24 km) with the introduction of the Fritsch and Chappell (1980) deep convective scheme. In September 2001 a new surface modelling system was implemented based on a mosaic approach with four types of surfaces: vegetated land with the Interactions between Soil-Biosphere-Atmosphere (ISBA) scheme (Bélair et al., 2003a, 2003b), open water, sea ice with a thermodynamic ice model (Mailhot et al., 2002), and glaciers and ice sheets. Along with a sequential assimilation strategy for soil temperature and soil moisture variables, this led to significant improvements in the summertime QPF and low-level objective scores over land, through a better surface diurnal cycle and a marked reduction in temperature bias errors in the boundary layer (see Bélair et al. (2003a) for a discussion).

The purpose of this paper is to describe the main features of the new 15-km regional forecast system and to highlight some of the improvements that led to its operational implementation on 18 May 2004. The main changes to the mesoscale forecast system include an increase in both the horizontal and vertical resolutions and major improvements to the physics package. These are comprised of a new condensation package, with an improved formulation of the cloudy boundary layer and changes to the shallow and deep convection schemes, along with a subgrid-scale orographic drag parametrization scheme. The performance of the new modelling system has been compared with the current operational model for two two-month cycles during the winter and summer of 2002. The new forecast system was then run in real-time parallel mode at CMC where it was thoroughly evaluated both objectively and subjectively, starting in February 2004 until its operational implementation in May 2004.

An overview of the main changes to the regional forecast model is given in Section 2. Section 3 provides a summary of the objective evaluation for the winter and summer cycles and for the period of the parallel run. Highlights of the overall subjective evaluation are presented in Section 4. Finally, a summary and conclusions are provided in the final section of the paper.

2 Overview of the 15-km regional forecast system

The main features of the new regional forecast system (also referred to as GEM15 as opposed to GEM24 which is the operational version) are summarized in Table 1. To help with comparisons, the most significant changes between the GEM24 and GEM15 versions are highlighted in Table 2. In the regional version of GEM, the primitive hydrostatic equations are integrated on a global variable-resolution grid using the semi-implicit and semi-Lagrangian numerical techniques. The model central domain has a uniform resolution that covers the entire North American continent and adjacent oceans. The regional forecast system is run at CMC twice a day for 48 hours. The initial conditions at 00:00 and 12:00 UTC are pro-

TABLE 1.   Summary of the 15-km forecast system.

Regional data assimilation system
. 12-h spin-up cycle with 6-h trial field from regional GEM model;
. variable resolution in the horizontal, with uniform 1/3° resolution over North America;
. 28 η levels;
. incremental 3-D variational data assimilation;
. analysis on model grid;
. initialization by diabatic digital filter with 3-h span.

Dynamics
. Hydrostatic primitive equations;
. semi-implicit semi-Lagrangian (3-D) time integration;
. 58 hybrid η levels with top at 10 hPa;
. 48-h forecast at 15-km resolution (575×641), with a time step of 450 s;
. ∇^6 horizontal diffusion of momentum variables; enhanced diffusion on four uppermost levels (sponge layer).

Physics
. Planetary boundary layer based on TKE with statistical subgrid-scale cloudiness (MoisTKE);
. fully-implicit vertical diffusion;
. stratified surface layer, distinct roughness lengths for momentum and heat/moisture;
. thermodynamic sea-ice model and ISBA land surface scheme with snow physics;
. solar/infra-red radiation schemes with cloud-radiation interactions based on predicted cloud radiative properties;
. Kuo transient scheme for shallow convection;
. Kain-Fritsch deep convection scheme;
. Sundqvist condensation and cloud scheme;
. gravity wave drag and low-level blocking by subgrid-scale orography.

vided by a regional data assimilation system (RDAS), with 12-h cycles using trial fields from 6-h regional integrations and a three-dimensional variational technique (Laroche et al., 1999). In recent years, several changes to the current RDAS took place as discussed in Chouinard et al. (2001). Additional satellite data are also incorporated into the new regional system, including radiance data from the Advanced Microwave Sounding Unit-B (AMSU-B) instrument and the Geostationary Operational Environment Satellite-West (GOES-W) satellite, along with a better background check for satellite data. These modifications to the RDAS are essentially the same as those successfully implemented in June 2003 in the Canadian global forecast system (Chouinard and Hallé, 2003; Wagneur and Garand, 2003).

Besides the schemes described below, a complete set of physical parametrizations is included in GEM (Mailhot et al., 1998), in particular: a) vertical diffusion with a prognostic turbulent kinetic energy (TKE) and a turbulent mixing length (Benoit et al., 1989; Bélair et al. 1999); b) a grid-scale condensation scheme based on Sundqvist (1978) (see also Pudykiewicz et al. (1992)); and c) solar and infra-red radiation (Fouquart and Bonnel, 1980; Garand, 1983) fully interactive with clouds (Yu et al., 1997).

a Changes to the Dynamical Configuration
The horizontal resolution of the latitude-longitude grid has been increased from 24 km (0.22°) to 15 km (0.14°) within the uniform part of the model grid (refer to Fig. 1 of Bélair et al. (2000) for the grid of GEM24; GEM15 has essentially the

TABLE 2.   Summary of the changes to the regional forecast system.

	GEM24	GEM15
Regional data assimilation		
• new satellite data	–	AMSU-B, GOES-W radiances
• satellite background check	yes	improved
Dynamics		
• resolution	24 km (28 levels)	15 km (58 levels)
• grid points	354 × 415	575 × 641
• time step	720 s	450 s
• digital filter (span/period)	6 h / 6 h	3h / 3h
• horizontal diffusion	∇^2 (all variables)	∇^6 (momentum variables)
• sponge layer	none	4 uppermost levels
Physics		
• vertical diffusion	Dry diffusion	Moist diffusion
• shallow convection	Conres	Kuo transient
• deep convection	Fritsch-Chappell	Kain-Fritsch
• gravity wave drag	none	McFarlane
• SGS blocking	none	Zadra

same central domain at 15-km resolution). Since the regional GEM model is actually a global model with a variable-resolution grid, changes have also been introduced in the variable part of the grid where the resolution decreases at a rate of 10% per grid point in each direction. In the operational setup, there is no limit to the resulting degradation in horizontal resolution, while in the new model this is limited to 300 km (~3°). This was found to have a small but beneficial effect, especially for weather systems approaching the west coast of North America by the end of the 48-h integration period.

The GEM model resolution is also variable in the vertical, with a total of 58 levels extending to 10 hPa. The number of vertical levels has more than doubled (28 levels in the current model) and the levels are mostly concentrated in the boundary layer, the upper-level jet and the stratosphere. For comparison, there are ten levels below 850 hPa in the new model (seven in the operational model) and five levels between 200 and 300 hPa in the new model (only two in the operational model). The number and positioning of the levels are quite similar to those in the next configuration of the global model currently under development (Bélair et al., 2001) and this is important since the regional and global configurations are closely linked in terms of data assimilation. The addition of levels near the top of the model allows for the introduction of a sponge layer (characterized by enhanced horizontal diffusion at the four uppermost levels) to prevent undesirable wave energy reflection from the rigid top of the model, and will also aid the assimilation of new types of satellite observations in the upper atmosphere in the future.

The time step of the new model has been reduced according to the increased horizontal resolution and set to 450 seconds (eight time steps per hour compared to five for the operational model). The larger number of time steps is also the rationale used to decrease the span and period of the digital filter from six to three hours, without loss of information. The digital filter (Fillion et al., 1995) is applied at the begin-ning of the model integration to filter out spurious gravity waves that could still be present in the analysed fields. The combined increases in the total number of grid points and time steps make the new model about eight times more expensive to run. This is compensated for by the increased power of the new IBM p690 (960 CPUs) supercomputer that allows the new model to run in about the same wall clock time as the operational one (typically 45–50 minutes).

A few changes were also made for the horizontal diffusion that damps the spurious numerical noise at the shortest scales. The operational model applies ∇^2 horizontal diffusion to all variables, while the new model applies a higher-order, more scale-selective, ∇^6 diffusion operator to the momentum variables only. In addition, since the semi-Lagrangian advection scheme already has some intrinsic numerical diffusion, the diffusion coefficient does not need to be too large and has been reduced somewhat.

The geophysical fields (orography, land-sea mask, soil and vegetation characteristics, etc.) also benefited from the increased horizontal resolution with a better definition of their main features. These fields are generated directly on the model grid based on global high-resolution databases, such as the 1-km resolution US Geological Survey vegetation data (see Bélair et al., 2003a).

b *Changes to the Physics Package*

An improved version of the physics package has been developed for the GEM15 forecasting system. Essentially the same modifications are also being tested for the new mesoscale version of the global model (see Bélair et al., 2001). The new condensation package consists of a combination of four different schemes to represent: 1) boundary-layer clouds (stratus, stratocumulus, and small cumulus clouds); 2) overshooting cumulus clouds; 3) deep convective clouds (Kain and Fritsch, 1990); and 4) non-convective clouds (Sundqvist scheme). As discussed by Bélair et al. (2005), this multi-scheme approach is able to represent realistically the wide variety of clouds observed in large-scale weather systems. The first three of these schemes are new features in the model and a short summary is given here (refer to Bélair et al. (2005) for more details). It is also worth noting that, despite some preliminary efforts to test a more detailed cloud microphysics scheme as a possible replacement for the Sundqvist scheme and to upgrade the Yu et al. (1997) cloud optical properties describing the cloud-radiation interactions, more development was deemed necessary on those aspects and it was decided to keep these schemes in the final configuration of the physics package.

1 THE CLOUDY BOUNDARY LAYER

An improved formulation of the cloudy boundary layer using a unified moist TKE approach following the strategy of Bechtold and Siebesma (1998) has been tested. In the new scheme (referred to as MoisTKE), the vertical diffusion is calcultated for the conservative thermodynamic variables (ice-liquid potential temperature and total water content). The MoisTKE scheme generates low-level fractional clouds, including mixed-phase clouds, that interact with the radiation scheme. This formulation is appropriate for a low-order turbulence model such as our current TKE scheme and allows a general description of stratiform clouds and shallow non-precipitating cumulus convection regimes using a single parameter Q_1, representing the normalized saturation deficit. Statistical relations appropriate to the various boundary-layer cloud regimes were obtained by Bechtold and Siebesma (1998) based on observations and large-eddy simulations, permitting the definition of the subgrid-scale cloud fraction and cloud water content in terms of Q_1 only. Preliminary tests of this formulation on a case of Arctic boundary-layer clouds over a polynya observed during the FIRE Arctic Cloud Experiment (FIRE.ACE) indicated much improved performance of the model (Mailhot and Bélair, 2002).

2 KUO-TRANSIENT SHALLOW CONVECTION SCHEME

A new convection scheme to represent overshooting cumulus cloud activity has been developed (see Mailhot et al., 1998; Bélair et al., 2005 for details) based on a Kuo-type closure (the so-called Kuo-Transient). Since boundary-layer moistening due to surface fluxes is in quasi-equilibrium with drying by condensation and vertical transport from shallow cumulus activity, the closure hypothesis assumes a direct relation with the vertical diffusion tendency of specific humidity. Therefore, the Kuo-Transient shallow convection scheme is intimately linked with the TKE vertical diffusion scheme. The scheme is also designed to represent overshooting cumulus clouds that can precipitate; this feature was found to be quite important in situations of streamers over open waters in winter, when the convective instability is not vigorous enough to trigger deep convection.

3 KAIN-FRITSCH DEEP CONVECTION SCHEME

An optimized version of the Kain-Fritsch (KF) deep convective scheme has been evaluated as a replacement for the Fritsch-Chappell (FC) scheme. The KF scheme is an extension of the FC scheme that also relates the intensity of deep convection to the convective available potential energy (CAPE) and triggers convective activity when low-level upward motion becomes sufficient to overcome the convective inhibition. It has an improved representation of the one-dimensional entraining/detraining plume model for the updrafts and moist downdrafts, and more detailed microphysics (including glaciation effects). An important advantage is that the KF scheme has more general closure assumptions that are less dependent on the model resolution and that appear to perform well over a wide range of horizontal resolutions. This is a crucial aspect on a variable-resolution grid and, indeed, the KF scheme allowed the removal of the double-scheme strategy for deep convection currently used in the operational model (FC scheme on the uniform grid and Kuo scheme on the variable grid) without noticeable detrimental effects.

One parameter of the KF scheme that needed some tuning is the trigger function parameter. A large value hinders the activation of the implicit convective scheme and, as a result, most of the precipitation is produced by the explicit scheme at the resolved scales. This often leads to significant overestimation of the precipitation amounts on a 15-km grid. Therefore, a fine balance has to be found between the implicit and explicit schemes in order to generate the correct precipitation amounts. An additional problem arises from the fact that the data assimilation does not necessarily result in moisture fields that are consistent with the physics parametrization. Based on experimentation with a limited number of test cases, a lower value of the trigger function parameter (=0.05) has been chosen for the first six hours of integration to avoid excessive, localized maxima in the precipitation patterns, and a more optimal value (=0.12) is used afterwards. As revealed in the QPF verifications, this generally results in better forecasts of precipitation amounts in the new system, but occasional large precipitation amounts may occur in the first six hours of the forecast. Other important parameters of the KF scheme are the convective timescale needed to release the CAPE (set to 45 min) and the average radius of the convective updrafts (set to 1500 m).

4 THE SUBGRID-SCALE OROGRAPHIC DRAG

The gravity-wave drag and blocking effects due to subgrid-scale orography are parametrized by the McFarlane (1987) and Zadra et al. (2003) schemes, respectively. They aim to reduce the winds when the flow encounters mountainous terrain. Even if the model resolution has increased, there is still a significant subgrid-scale part of the orography that has to be taken into account. The gravity-wave drag emulates the breaking of mountain waves in the troposphere and lower stratosphere. Generally, this is not considered to be too important in a regional model, since it mostly affects the winds at relatively high levels (~100 hPa) and this is why it has been neglected until now.

The blocking term is a parametrization based on the formulation of Lott and Miller (1997) that represents the drag on low-level winds blocked on the flanks of mountains. It was introduced in the global configuration of GEM in December 2001 by Zadra et al. (2003) and was found to improve the large-scale flow both for short- and medium-range forecasts, especially in winter when the surface winds are strong. The blocking term reduces the low-level winds in mountainous regions and also has a positive impact in the regional configuration. It was found that the modified wind fields and the related low-level convergence often result in a displacement and enhancement of the precipitation patterns upstream of the mountain ranges. The temperature bias in the lower troposphere is also reduced due to low-level warming indirectly caused by the blocking term, as discussed in Zadra et al. (2003).

3 Objective evaluation

a *Winter and Summer Cycles*

Two series of cases were used to develop and verify the GEM15 configuration: a winter cycle based on 15 cases (23 December 2001 to 27 January 2002) and a summer cycle based on 18 cases (13 August to 24 September 2002). Each case is integrated for 48 h and is separated from the next one by 60 h, to ensure that their results are independent. During the development work it was not always possible to assess the effects of each of the changes separately, but this will be indicated whenever possible. Otherwise, the evaluation that led to the final configuration of GEM15 was based on performance improvement resulting from a combination of the various components and their synergistic interactions.

Objective evaluation of the 0- to 48-h forecasts is done using the North American upper-air radiosonde network for the zonal and meridional wind components, temperature, geopotential height and dewpoint depression. Only the verifications for the 48-h forecasts are shown here, but they are typical of all forecast ranges. For wintertime, verification curves (Fig. 1) indicate significant improvements using the 15-km resolution model compared to the operational model. The large improvements in the root-mean-square (rms) errors for the wind, temperature, geopotential height, and dewpoint depression are impressive throughout the troposphere, and especially at the jet level where the vertical resolution has been increased. In fact, it is noteworthy that, overall, the rms errors of GEM15 at 48-h are almost as good as those of GEM24 at 36-h (not shown), indicating a significant gain in the model predictability. In the stratosphere (above 250 hPa), the improvements are even more significant for both rms and bias errors, but they do not have much impact in the context of a short-term regional model. One of the main objectives of the new physics package was to correct the known tropospheric biases of the operational model, the model being generally too warm in the mid-troposphere due to the incorrect release of latent heat by condensation and too cold and moist in the boundary layer. Clearly, the changes to the condensation package have almost completely eliminated the mid-level biases for temperature (with corresponding beneficial impacts for the bias of geopotential heights). The boundary-layer biases have also been significantly reduced as a result of the inclusion of the moist vertical diffusion (MoisTKE) scheme. The only deterioration in the scores is seen in the bias error of dewpoint depression above 500 hPa, where, in any case, there is not much moisture. The summertime verifications (Fig. 2) again favour GEM15 although the improvements are less spectacular than for winter. Still there is a noticeable positive impact in the rms errors of geopotential heights and temperatures. For dewpoint depression, the rms errors are reduced but again the bias errors worsen above 500 hPa, the new model being too dry. Note also the improvements in the stratosphere resulting mostly from the increased vertical resolution in the upper levels. Finally, the less striking scores obtained in summer may well illustrate several inherent difficulties with verifications: the synoptic observation network has a relatively low resolution, while summertime weather is characterized by large variability at small scales.

Surface verifications are made using data from North American synoptic stations for temperature and dewpoint depression. Both models in winter (Fig. 3) exhibit a cold bias of the order of ~1.5°C over North America, the bias being slightly larger for the 15-km model. Stratification by regions (not shown) indicates that the difference in bias comes mostly from the western part of the continent. The two models have a similar moist bias (less than 0.5°C in dewpoint depression). The summer verifications are nearly identical in the two models and are not shown; both models are too cold (though by less than 1°C) and slightly too dry. In summary, the surface scores are rather similar for GEM24 and GEM15 and this may not be too surprising considering that there is no major change related to surface processes in the new forecast system.

Precipitation is probably the most difficult field to verify and there are seldom enough cases in the verifications. In principle, a database of 15 independent cases in winter and 18 summer cases would seem to be sufficient for verification. In order to have access to a larger number of stations for verification, the US Standard Hydrometeorological Exchange Format (SHEF) surface observation network covering most of the continental US at much higher spatial density, is also used. The conventional North American synoptic network reports 12-h accumulations of precipitation at 00:00 UTC and 12:00 UTC. On the other hand, the SHEF observations are only available once a day at 12:00 UTC and report 24-h accumulations. Since half of the model runs are started at 12:00 UTC (the other half being started at 00:00 UTC), the verifications of QPF will be done on 24-h accumulations for the three periods of 0–24 h, 12–36 h and 24–48 h forecasts. In order to draw sound conclusions, the number of observations needs to be considered together with the number of weather systems (number of cases), the variety of atmospheric circulations in the sample, and so on. Verification indices based on less than 100 observations are probably inconclusive and will not be considered in the discussion.

Verifications are done for the widely accepted bias and threat scores (see the appendix in Bélair et al. (2000) for the

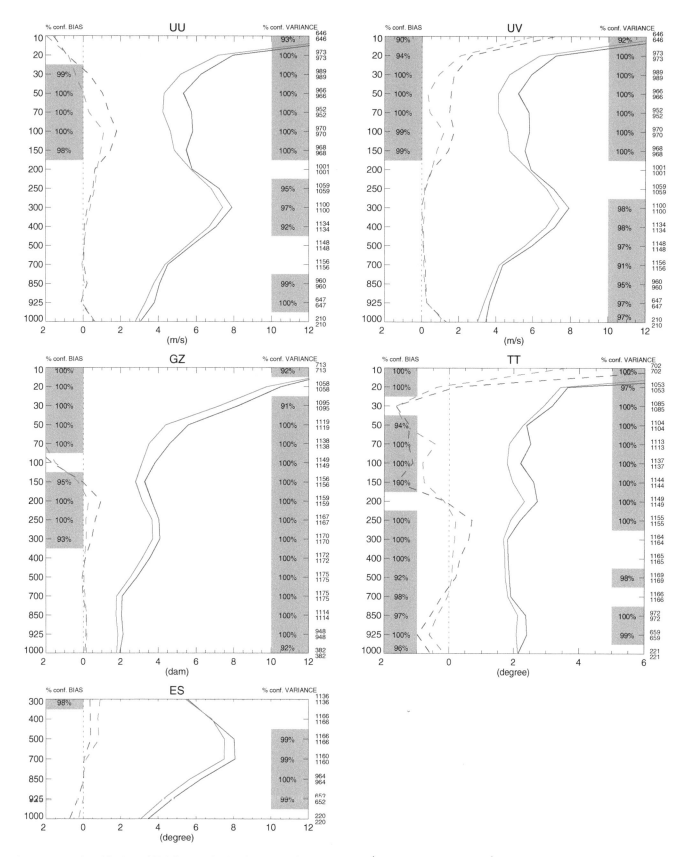

Fig. 1   Upper-air verifications of 48-h forecasts for the winter cycle of zonal (UU in m s⁻¹) and meridional (UV in m s⁻¹) winds, geopotential height (GZ in dam), temperature (TT in °C), and dewpoint depression (ES in °C). Bias (dashed) and rms (solid) errors (forecast minus observation) for operational (blue) and new (red) models. Green tags indicate the level of statistical significance separating the two curves. The left vertical axis is pressure in hPa (number of observations at each level on the right) and the units are indicated on the horizontal axis.

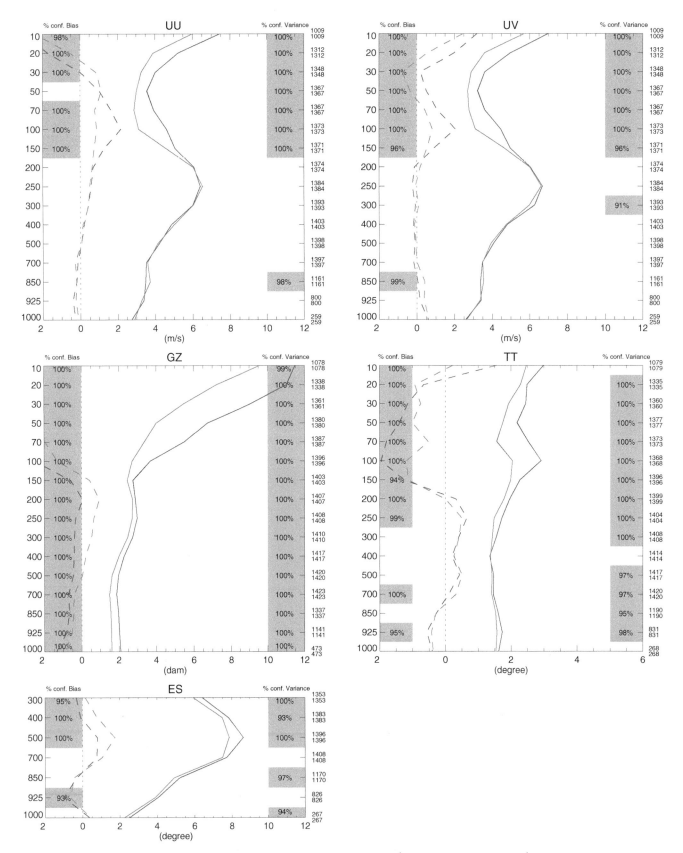

Fig. 2   Upper-air verifications of 48-h forecasts for the summer cycle of zonal (UU in m s⁻¹) and meridional (UV in m s⁻¹) winds, geopotential height (GZ in dam), temperature (TT in °C), and dewpoint depression (ES in °C). Bias (dashed) and rms (solid) errors (forecast minus observation) for operational (blue) and new (red) models. Green tags indicate the level of statistical significance separating the two curves. The left vertical axis is pressure in hPa (number of observations at each level on the right) and the units are indicated on the horizontal axis.

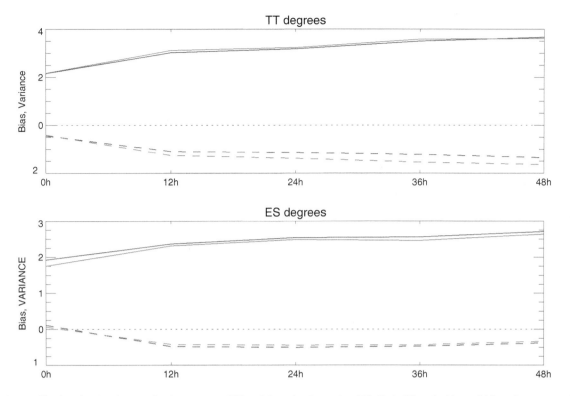

Fig. 3 Surface verifications for the winter cycle of temperature (TT) and dewpoint depression (ES) (dashed lines for bias; solid lines for rms errors). The operational model is in blue, the new model is in red.

definition of the scores). For the winter cycle based on the high-resolution SHEF network over the US (Fig. 4), both the biases (slightly reduced) and the threat scores (significantly larger) indicate improvements with GEM15. For the corresponding scores using the North American synoptic network (not shown), the biases are generally similar in the two models but the threat scores deteriorate for larger amounts (more than 5 mm of accumulation in 24 h) in the new model. Therefore, wintertime QPF verifications using the SHEF or the synoptic networks lead to apparently differing conclusions. The explaination for this is not clear but there are several differences between the two observing networks, such as a larger number of SHEF observations, different geographical coverage and possibly contrasting precipitation regimes over the two verification areas. For instance, there is likely more convective precipitation over the southern part of the US than over the rest of the North American continent, even in winter. One possible reason could also be that the verifications using data from the synoptic network are not statistically significant due to the relatively small number of observations (note that there are between three and five times more observations with the SHEF network in the precipitation classes of 5 10 mm and 10–25 mm per day). In summer, the SHEF verifications (Fig. 5) generally indicate that the bias is slightly increased with GEM15 but the threat scores are clearly improved for almost every precipitation class. In addition, the conclusions for summertime are more consistent over the two verification areas (results from the synoptic network are not shown). It is worth emphasizing that this improved QPF performance

results from the combined changes in the dynamics (resolution) and the condensation package. Indeed, early experimentation indicated that changes in the dynamics alone did not bring about improvements in the QPF (deterioration of the threat scores was noted for several precipitation classes). The improved performance also highlights that a satisfactory balance has been achieved between the KF and the explicit condensation schemes, since there is no signal from the bias scores of overestimation of QPF for the larger classes.

Time series of precipitation rates (not shown) indicate that the spin-up time for QPF (the time needed for the precipitation to reach quasi-equilibrium) is reduced to about 8 h in the new system (compared to ~12 h in the operational model) for both summer and winter. Moreover, the implicit-explicit partition of precipitation is somewhat modified with averaged precipitation rates in summer accounting for 28% and 72% of the total for the implicit and explicit schemes, respectively, in GEM24, while the corresponding values are 19% and 81% in the new model. Thus, there is an increased role in the 15-km resolution model for the explicit precipitation scheme arising partly from the replacement of the FC deep convection scheme by KF, and partly from the enhanced resolution. These results are in general agreement with previous studies (e.g., Molinari and Dudek, 1986; Bélair et al., 2000; Bélair and Mailhot, 2001) and emphasize the impact of higher horizontal resolution on the partitioning of condensation and the needs for realistic explicit cloud schemes in mesoscale models. As expected, the differences between the two models in winter (not shown) are much smaller, as most of the precipitation is then generated by the explicit scheme (90–95% of the total).

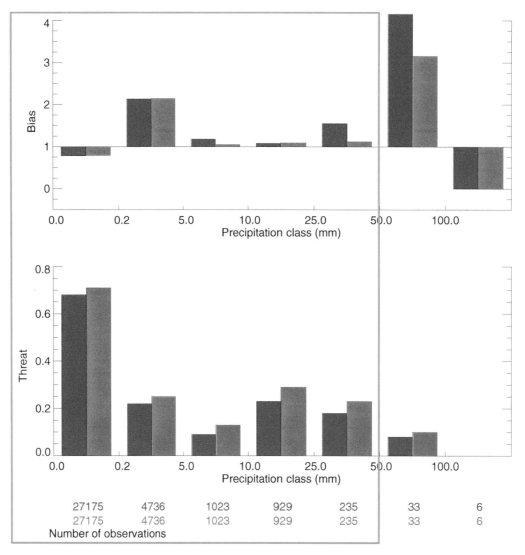

Fig. 4   Precipitation verification scores for the winter cycle, 24–48 h forecasts, versus the high density US SHEF network observations. The framed area high-lights the results based on more than 100 observations. The operational model is in blue, the new model is in red.

Some verification of radiation fluxes has been done for the two cycles based on observations from the Clouds and the Earth's Radiant Energy System (CERES) Atmospheric Radiation Measurement (ARM) Program Validation Experiment (CAVE) network. This network consists of about 30 North American surface stations equipped with radiometers measuring shortwave (SW) and longwave (LW) downward radiative fluxes and providing an estimate of the cloud fraction. As summarized in Table 3, both models are found to underestimate the LW surface fluxes in winter by some 20 W m^{-2} under clear or cloudy skies, while this radiation deficit is reduced to about 10 W m^{-2} under summer clear-sky conditions (almost no bias in cloudy skies). The surface SW radiative fluxes are overestimated both for summertime and wintertime by about 25 W m^{-2} and 35 W m^{-2} in GEM24 and GEM15, respectively, the errors being mostly present in cloudy situations. Therefore, the clouds of the new model appear to be even more transparent in the SW than the operational ones, suggesting that revisions to the current

cloud optical radiative properties will be needed in the future to handle cloud-radiation interactions better.

b *Parallel Run*
Objective verification scores using radiosonde observations for geopotential height, temperatures and winds at standard levels (850, 500 and 250 hPa) were calculated for the entire period of the official parallel run (24 February to 18 May 2004). Here again, the scores for the new forecast system showed a general improvement, with a remarkable decrease in both the rms and bias errors. Indeed, rms errors for the 48-h forecasts of heights and temperatures are found to be about 10% lower than the operational model at all levels during the entire parallel run. This is quite an achievement since, as noted by CMC meteorologists, it represents one of the largest rms error reductions in the regional forecast system during the last ten years. The reduction in bias errors is also significant, especially for temperatures. Examples are shown for the

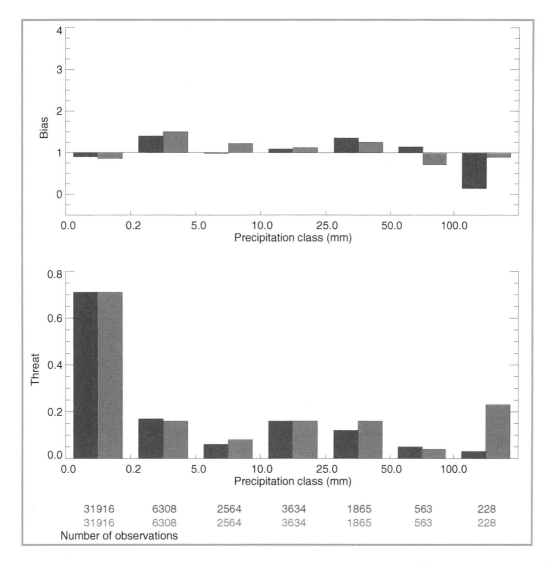

Fig. 5   Precipitation verification scores for the summer cycle, 24–48 h forecasts, versus the high density US SHEF network observations. The framed area highlights the results based on more than 100 observations. The operational model is in blue, the new model is in red.

geopotential heights at mid-levels (Fig. 6) and temperatures at lower levels (Fig. 7) and the situation is similar at all levels for both fields. The rms errors of the wind vectors also decreased, though not as much as for heights and temperatures. The largest gains are found at 250 hPa near the upper-level jet (not shown). For wind speed, the bias was also less, again especially at higher levels. Therefore, the verification scores for the parallel run are in general agreement with those of the winter cycle and indicate a marked improvement in the model performance during the cold seasons.

Objective verifications of QPF using data from both the North American synoptic network and the high density US SHEF network show a net improvement using GEM15 over the operational GEM24 (for example, see Fig. 8 based on SHEF observations). During the parallel run, the operational system had a positive bias (bias score larger than unity) for all precipitation thresholds, indicating a tendency to over-predict the number of events for all precipitation categories. Note that

TABLE 3.   Verification of downward surface radiative fluxes (LW and SW) based on CAVE network observations (model minus observed values in W m^{-2}) for the winter and summer cycles. Differences are also stratified into clear sky (both model and observed cloud fraction less than 10%) and cloudy sky (more than 10%).

	Winter (clear / cloudy)	Summer (clear / cloudy)
LW		
GEM24	−21 (−23/−19)	−11 (−11/−3)
GEM15	−22 (−21/−18)	−12 (−9/−3)
SW		
GEM24	26 (4/41)	25 (13/5)
GEM15	32 (9/58)	35 (15/20)

this was also the case with the winter cycle (cf. Fig. 4). As shown in Fig. 8, biases are generally less using the 15-km resolution model, though a slight tendency to over-forecast the higher amounts was noted in some areas. This was further

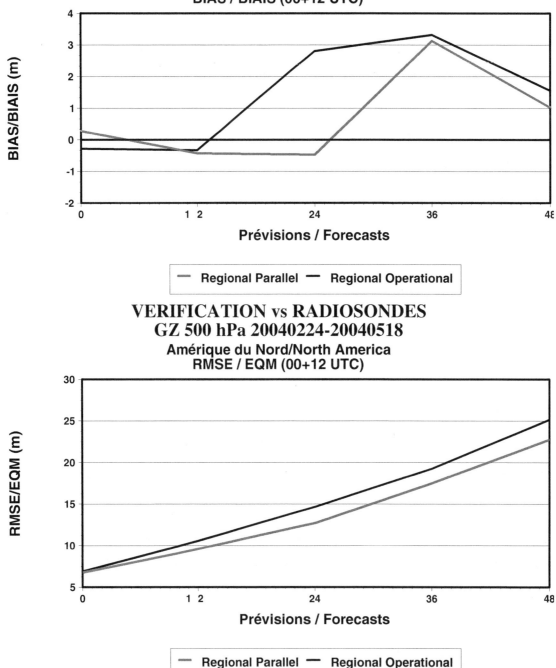

Fig. 6   The 500-hPa geopotential height bias and rms errors (in m) for the operational (black) and parallel (red) runs, versus the North American radiosonde observation network, for the period 24 February–18 May 2004.

corroborated, for instance, by the subjective verifications over areas with complex orography. Moreover, objective verifications against observations over a window, including the mountainous areas of the western United States, also indicat- ed a somewhat higher bias for precipitation classes exceeding 25 mm in 24 h (not shown). Equitable threat scores generally improved in those ranges up to ~15 mm of accumulated pre- cipitation in 24 h, especially in the 24–48 h forecast period.

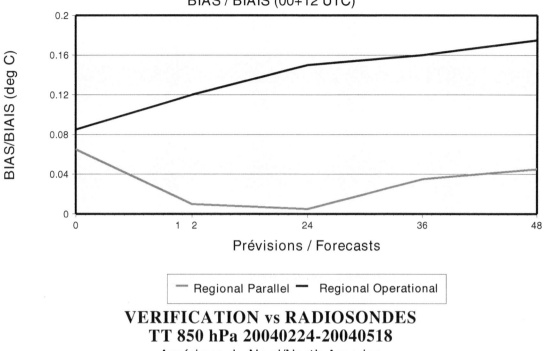

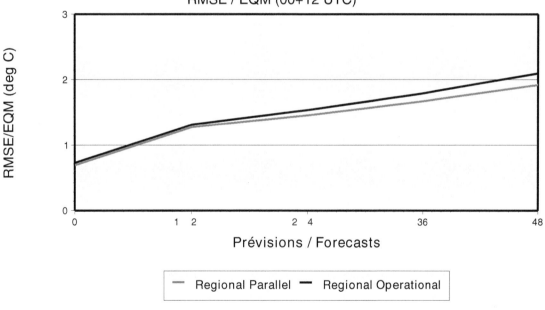

Fig. 7   The 850-hPa temperature bias and rms errors (in °C) for the operational (black) and parallel (red) runs, versus the North American radiosonde observation network, for the period 24 February–18 May 2004.

The only appreciable deterioration occurred for amounts larger than 25 mm at longer term (Fig. 8), but the threat score values are less than 0.2 and are probably not significant anyway. It is also worth emphasizing that, in contrast to the results using the winter cycle, the threat scores from the synoptic network (not shown) are in general agreement with those from

the SHEF network, showing significant improvements in almost all precipitation categories.

4 Highlights of the subjective evaluation

CMC operational meteorologists have carefully evaluated various versions of the new regional prediction system. An

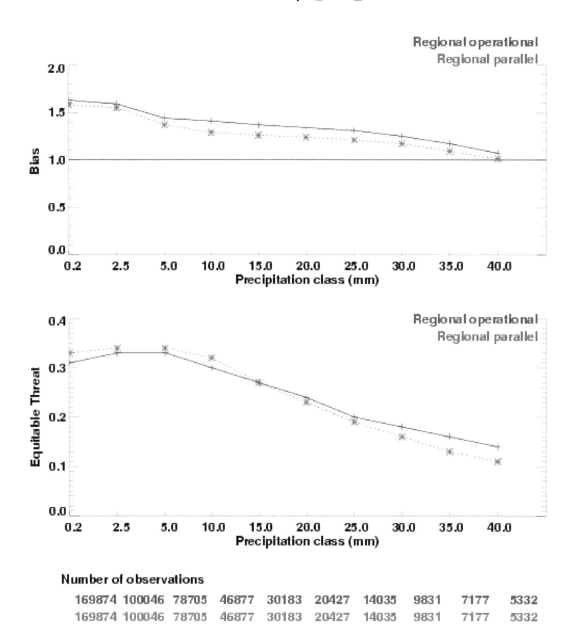

24-hour precipitation forecast verification against observations
SHEF network data observed at 12z
24-48 hour forecast All of USA
20040224-20040518 par_15km_2004

Fig. 8   Bias and equitable threat scores for the operational (blue) and parallel (red) runs, 24–48 h forecasts, versus the high density US SHEF network observations, for the period 24 February–18 May 2004.

overall summary of the subjective evaluation of the summer and winter cycles, as well as benchmark cases and the real-time parallel run, resulted in a consensus that the new 15-km regional forecast system is a clear improvement over the operational system. Here are some highlights of their comprehensive evaluation.

GEM15 is superior to GEM24 in forecasting the speed, intensity and phasing of 500-hPa short waves. When relatively large differences between the two models were present, the new model usually compared better with the verifying analysis. In most cases it was estimated that the predictability improved by about 12 h using GEM15. Note that this subjective estimate is in very good agreement with the conclusions obtained from the objective verification scores for the winter cycle and the parallel run. A negative height bias noted in GEM24 over the Rockies has been corrected and, in particular, upper ridges that are systematically too weak in GEM24 are better represented in GEM15. In situations of low pressure systems forming or reforming over Alberta in the lee of the mountains, GEM24 typically moves them too far north and forecasts too low a pressure. In such cases, GEM15 usually has a weaker low farther south and exhibits better agreement with the verifying analysis (Fig. 9). Chinooks in the lee of the Rockies are better represented in GEM15 with a good definition of the wind patterns descending the eastern mountain slopes.

Better QPF verifications are obtained, in general, with the new model. Due to higher horizontal resolution, the zones of maximum precipitation are often smaller than those of GEM24, both in winter and summer and for both stratiform and convective precipitation regimes. The precipitation bands are often sharper with generally larger maximum values. GEM15 clearly performed better in cases where precipitation moves from the west into the interior of British Columbia. In such cases, GEM24 has a strong tendency to push the precipitation too far east. A positive precipitation bias in upslope flow and a better definition of the rain shadow in mountainous areas were also noted with the new model. The forecasts of precipitation types for both models are rather similar and GEM15's colder surface temperature does not seem to have any detrimental impact. However, GEM15 tends to generate small disorganized freezing precipitation zones in the mountain valleys of British Columbia and over the western high plains more frequently.

Wider areas of trace amounts of precipitation (less than 0.2 mm) are produced by GEM15 in cases of air mass convection and light snow falling from stratocumulus, and larger precipitation amounts are produced in cases of deep convection in a warm air mass that generally exhibit better agreement with the available observations (Fig. 10). The same comparisons also indicate that GEM15 performs better with respect to the position of the axes of convective precipitation. Based on radar observations, the apparently noisier patterns of GEM15 are usually considered to be more realistic than those of GEM24. The new model was significantly better for several hurricanes over the Gulf of Mexico while the improvement was smaller over the Atlantic coast. Subjective evaluation also revealed a slight improvement in the forecast clouds in the new model, in particular for low-level clouds over open waters such as the convective streamers produced over the Great Lakes and the Gulf of St. Lawrence.

GEM24 tends to release too much latent heat at low levels, leading to excessive deepening of lows over the Atlantic and to QPF problems for the Maritimes and Newfoundland. GEM15 was found to be generally better in those situations. The new model handles convective instability in a different way that is generally judged to be much better. For instance, in a case of a deepening low over the eastern United States

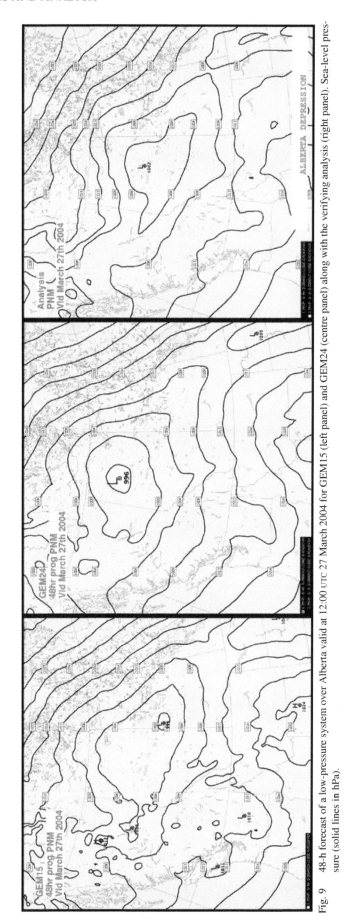

Fig. 9 48-h forecast of a low-pressure system over Alberta valid at 12:00 UTC 27 March 2004 for GEM15 (left panel) and GEM24 (centre panel) along with the verifying analysis (right panel). Sea-level pressure (solid lines in hPa).

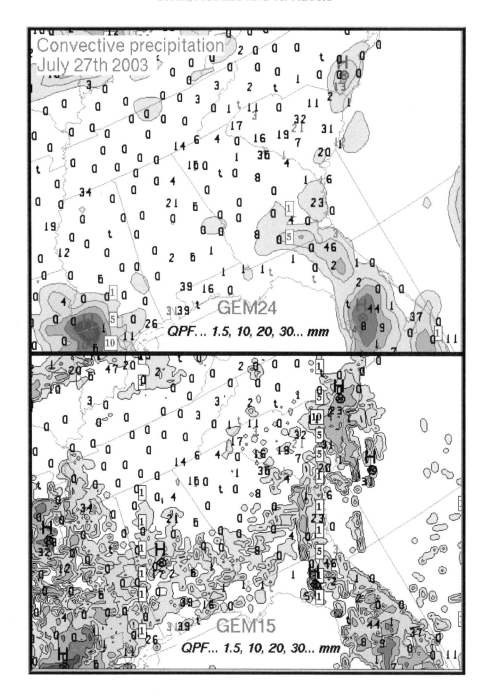

Fig. 10   24-h accumulation of precipitation (mm) over the south-eastern United States on 27 July 2003 for GEM24 (upper panel) and GEM15 (lower panel) with the surface raingauge observations.

characterized by strong, active thunderstorms, the position and central values of the low exhibited much better agreement using GEM15 (Fig. 11).

5 Summary and conclusions

An overview of the main features of the new mesoscale forecast system implemented at CMC in May 2004 has been presented. The changes include an increased horizontal resolution from 24 to 15 km, and improved resolution in the vertical with

58 levels. Major upgrades to the physics package consist of new condensation schemes that better describe the cloudy boundary layer (MoisTKE scheme), overshooting cumulus clouds (Kuo-Transient scheme) and deep convection (Kain and Fritsch scheme). A subgrid-scale orography parametrization scheme to represent gravity wave drag and low-level blocking effects has also been included. The regional data assimilation incorporates several changes such as additional radiances from AMSU-B and GOES-W. Increased horizontal resolution also

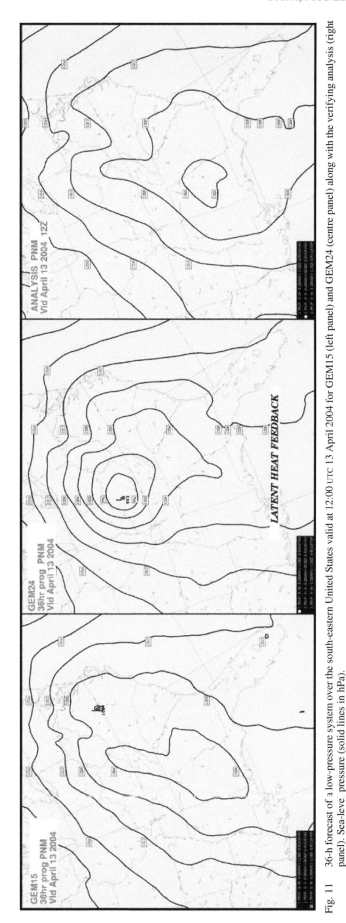

Fig. 11   36-h forecast of a low-pressure system over the south-eastern United States valid at 12:00 UTC 13 April 2004 for GEM15 (left panel) and GEM24 (centre panel) along with the verifying analysis (right panel). Sea-level pressure (solid lines in hPa).

results in a better definition of the geophysical fields in the new version of the mesoscale modelling system.

Objective verifications using a series of cases and the parallel run, along with subjective evaluations by CMC operational meteorologists, indicated quite significant improvements with the new regional forecast system. A remarkable reduction in model errors was obtained throughout the troposphere in winter and, to a lesser extent, in summer. Subjective estimates suggest that the predictability improved by nearly 12 hours. The winter-time and summertime QPF show significantly better scores in the 15-km model for most precipitation classes, in particular for day-2 forecasts. Moreover, the new system has a much reduced spin-up time for the precipitation and a modified partitioning between the implicit and explicit precipitation. Other features include precipitation patterns organized in sharper bands, more realistic trace amounts of precipitation, and better handling of convection for several cases of deepening lows and hurricanes. In mountainous regions, several phenomena are better described such as: Chinook winds, rain shadows, and re-formation of low pressure systems in the lee of mountains, which is a result of the combination of higher-resolution orography and the subgrid-scale blocking term.

The verifications also revealed a few weaknesses, some of which were already present in the operational model. Evaluation with the CAVE network indicated a deficit in downward LW radiation, particularly in winter, and too large a SW radiative flux under cloudy skies. There is a slightly larger cold bias present at the surface in winter due to noctur-nal cooling under calm winds and dry conditions, that appears to be related to the deficit in LW radiation and possibly to a lack of thin low-level cloudiness in those situations. There are also some indications for potential over-prediction of precip-itation amounts on the windward side of mountains and dur-ing the first six hours of the integration. Therefore, the study emphasizes the need for further experimentation and tuning with the blocking term and the balance of implicit-explicit precipitation, and for more detailed cloud microphysics schemes and improved cloud optical radiative properties.

6 Acknowledgments

The implementation of the new 15-km modelling system is the result of several years of continuing collaborative effort between staff from both the Meteorological Research Branch (MRB) and CMC. Besides the extensive list of authors, many other persons contributed to various aspects and at various stages of this project. In particular, thanks are due to Clément Chouinard, Jean Côté, Yves Delage, Luc Fillion, Louis Garand, Stéphane Laroche, Vivian Lee, Michel Roch, Lubos Spacek, and Michel Valin from MRB and to the following staff from the Development, Operations, and Informatics Branches of CMC: Maryse Beauchemin, Doug Bender, Henri-Paul Biron, Amin Erfani, Richard Hogue, Marc Klasa, Manon Lajoie, Robert Mailhot, Gérard Pellerin, Suzanne Roy, Jean-Pierre Toviessi, Gilles Verner, Richard Verret and Nicolas Wagneur. The support of MRB and CMC manage-ment, particularly Gilbert Brunet and Jean-Guy Desmarais, throughout the project is gratefully acknowledged.

References

BECHTOLD, P. and P. SIEBESMA. 1998. Organization and representation of boundary layer clouds. *J. Atmos. Sci.* **55**: 888–895.

BÉLAIR, S.; J. MAILHOT, J. W. STRAPP and J. I. MACPHERSON. 1999. An examination of local versus nonlocal aspects of a TKE-based boundary-layer scheme in clear convective conditions. *J. Appl. Meteorol.* **38**: 1499–1518.

——; A. MÉTHOT, J. MAILHOT, B. BILODEAU, A. PATOINE, G. PELLERIN and J. CÔTÉ. 2000. Operational implementation of the Fritsch-Chappell convective scheme in the 24-km Canadian regional model. *Weather Forecast.* **15**: 257–274.

—— and J. MAILHOT. 2001. Impact of horizontal resolution on the numerical simulation of a midlatitude squall line: Implicit versus explicit condensation. *Mon. Weather Rev.* **129**: 2362–2376.

——; ——, A. TREMBLAY, A.-M. LEDUC, A. MÉTHOT, M. ROCH and P. VAILLANCOURT. 2001. A mesoscale model for global medium-range weather forecasting in Canada. *In*: Preprints 9th Conference on Mesoscale Processes, Am. Meteorol. Soc. 30 July–2 August 2001, Fort Lauderdale FL, pp. 242–245.

——; L.-P. CREVIER, J. MAILHOT, B. BILODEAU and Y. DELAGE. 2003a. Operational implementation of the ISBA land surface scheme in the Canadian regional weather forecast model. Part I: Warm season results. *J. Hydrometeorol.* **4**: 352–370.

——; R. BROWN, J. MAILHOT, B. BILODEAU and L.-P. CREVIER. 2003b. Operational implementation of the ISBA land surface scheme in the Canadian regional weather forecast model. Part II: Cold season results. *J. Hydrometeorol.* **4**: 371–386.

——; J. MAILHOT, C. GIRARD and P. VAILLANCOURT. 2005. Boundary-layer and shallow cumulus clouds in a medium-range forecast of a large-scale weather system. *Mon. Weather Rev.* **133**: 1938–1960.

BENOIT, R.; J. CÔTÉ and J. MAILHOT. 1989. Inclusion of TKE boundary-layer parameterization in the Canadian regional finite-element model. *Mon. Weather Rev.* **117**: 1726–1750.

CHOUINARD, C.; C. CHARETTE, J. HALLÉ, P. GAUTHIER, J. MORNEAU and R. SARRAZIN. 2001. The Canadian 3D-VAR analysis scheme on model vertical coordinate. *In*: Preprints 14th Conference on Numerical Weather Prediction, Am. Meteorol. Soc. 30 July–2 August 2001, Fort Lauderdale FL, pp. 14–18.

—— and J. HALLÉ. 2003. The assimilation of AMSU-B radiance data in the Canadian Meteorological Centre global data assimilation system: their difficulties relative to the assimilation of AMSU-A radiances. *In*: Proceedings 13th International TOVS Study Conference (ITSC-13), 29 October–4 November 2003, Ste-Adèle QC, pp. 1–14. (Available online at http://cimss.ssec.wisc.edu/itwg/)

CÔTÉ, J.; S. GRAVEL, A. MÉTHOT, A. PATOINE, M. ROCH and A. STANIFORTH. 1998a. The operational CMC-MRB global environmental multiscale (GEM) model. Part I: Design considerations and formulation. *Mon. Weather Rev.* **126**, 1373–1395.

——; J.-G. DESMARAIS, S. GRAVEL, A. MÉTHOT, A. PATOINE, M. ROCH and A. STANIFORTH. 1998b. The operational CMC-MRB global environmental multiscale (GEM) model. Part II: Results. *Mon. Weather Rev.* **126**: 1397–1418.

FILLION, L.; H. L. MITCHELL, H. RITCHIE and A. STANIFORTH. 1995. The impact of a digital filter finalization technique in a global data assimilation system. *Tellus,* **47A**: 304–323.

FOUQUART, Y. and B. BONNEL. 1980. Computations of solar heating of the earth's atmosphere: A new parameterization. *Contrib. Atmos. Phys.* **53**: 35–62.

FRITSCH, J. M. and C. F. CHAPPELL. 1980. Numerical prediction of convectively driven mesoscale pressure systems. Part I: Convective parameterization. *J. Atmos. Sci.* **37**: 1722–1733.

GARAND, L. 1983. Some improvements and complements to the infrared emissivity algorithm including a parameterization of the absorption in the continuum region. *J. Atmos. Sci.* **40**, 230–244.

KAIN, J.S. and J.M. FRITSCH. 1990. A one-dimensional entraining/detraining plume model and its application in convective parameterization. *J. Atmos. Sci.* **47**: 2784–2802.

LAROCHE, S.; P. GAUTHIER, J. ST-JAMES and J. MORNEAU. 1999. Implementation of a 3D variational data assimilation system at the Canadian Meteorological Centre. Part II: The regional analysis. ATMOSPHERE-OCEAN, **37**: 281–307.

LOTT, F. and J. M. MILLER. 1997. A new subgrid-scale orographic drag parameterization: Its formulation and testing. *Q. J. R. Meteorol. Soc.* **123**: 101–127.

MAILHOT, J.; S. BÉLAIR, R. BENOIT, B. BILODEAU, Y. DELAGE, L. FILLION, L. GARAND, C. GIRARD and A. TREMBLAY. 1998. Scientific Description of RPN Physics Library - Version 3.6 - Recherche en prévision numérique, 188 pp. (Available from RPN, 2121 Trans-Canada Highway, Dorval, PQ H9P 1J3, Canada; also online at http://www.cmc.ec.gc.ca/rpn/physics/physic98.pdf.)

—— and ——. 2002. An examination of a unified cloudiness-turbulence scheme with various types of cloudy boundary layers. *In*: Preprints 15th Symposium on Boundary Layer and Turbulence, 15–19 July 2002, Wageningen, The Netherlands, Am. Meteorol. Soc. pp. 215–218.

——; A. TREMBLAY, S. BÉLAIR, I. GULTEPE and G. A. ISAAC. 2002. Mesoscale simulation of surface fluxes and boundary layer clouds associated with a Beaufort Sea polynya. *J. Geophys. Res.* **107**(C10): 8031, doi:10.1029/2000JC000429.

MCFARLANE, N. A. 1987. The effect of orographically excited gravity-wave drag on the circulation of the lower stratosphere and troposphere. *J. Atmos. Sci.* **44**: 1175–1800.

MOLINARI, J. and M. DUDEK. 1986. Implicit versus explicit convective heating in numerical weather prediction models. *Mon. Weather Rev.* **114**: 1822–1831.

PUDYKIEWICZ, J.; R. BENOIT and J. MAILHOT. 1992. Inclusion and verification of a predictive cloud water scheme in a regional weather prediction model. *Mon. Weather Rev.* **120**: 612–626.

SUNDQVIST, H. 1978. A parameterization scheme for non-convective condensation including prediction of cloud water content. *Q. J. R. Meteorol. Soc.* **104**: 677–690.

WAGNEUR, N. and L. GARAND. 2003. Operational assimilation of GOES water vapor imager channel at MSC. *In*: Proceedings 13th International TOVS Study Conference (ITSC-13), 29 October–4 November 2003, Ste-Adèle QC, pp. 654–659. (Available online at http://cimss.ssec.wisc.edu/itwg/)

YU, W.; L. GARAND and A. P. DASTOOR. 1997. Evaluation of model clouds and radiation at 100 km scale using GOES data. *Tellus,* **49A**, 246–262.

ZADRA, A.; M. ROCH, S. LAROCHE and M. CHARRON. 2003. The subgrid scale orographic blocking parametrization of the GEM model. ATMOSPHERE-OCEAN, **41**: 155–170.

The Canadian Fourth Generation Atmospheric Global Climate Model (CanAM4). Part I: Representation of Physical Processes

Knut von Salzen[1], John F. Scinocca[1], Norman A. McFarlane[1], Jiangnan Li[1], Jason N. S. Cole[1], David Plummer[1], Diana Verseghy[2], M. Cathy Reader[3], Xiaoyan Ma[†], Michael Lazare[1] and Larry Solheim[1]

[1]*Canadian Centre for Climate Modelling and Analysis, Environment Canada, Victoria, British Columbia*
[2]*Climate Processes Section, Environment Canada, Downsview, Ontario*
[3]*School of Earth and Ocean Sciences, University of Victoria, Victoria, British Columbia*

ABSTRACT   *The Canadian Centre for Climate Modelling and Analysis (CCCma) has developed the fourth generation of the Canadian Atmospheric Global Climate Model (CanAM4). The new model includes substantially modified physical parameterizations compared to its predecessor. In particular, the treatment of clouds, cloud radiative effects, and precipitation has been modified. Aerosol direct and indirect effects are calculated based on a bulk aerosol scheme. Simulation results for present-day global climate are analyzed, with a focus on cloud radiative effects and precipitation. Good overall agreement is found between climatological mean short- and long-wave cloud radiative effects and observations from the Clouds and Earth's Radiant Energy System (CERES) experiment. An analysis of the responses of cloud radiative effects to variations in climate will be presented in a companion paper.*

RÉSUMÉ   *[Traduit par la rédaction] Le Centre canadien de la modélisation et de l'analyse climatique (CCmaC) a mis au point la quatrième génération du modèle canadien de circulation générale de l'atmosphère (CanAM4). Le nouveau modèle comprend des paramétrisations physiques passablement modifiées comparativement à son prédécesseur. En particulier, le traitement des nuages, des effets radiatifs des nuages et des précipitations a été modifié. Les effets directs et indirects des aérosols sont calculés à l'aide d'un schéma d'aérosols en bloc. Nous analysons des résultats de simulation pour le climat général du jour présent en mettant l'accent sur les effets radiatifs des nuages et les précipitations. Nous trouvons un bon accord général entre la moyenne climatologique des effets radiatifs des nuages pour les courtes et les grandes longueurs d'onde et les observations de l'expérience CERES (Clouds and Earth's Radiant Energy System). Une analyse de la réponse des effets radiatifs des nuages aux variations du climat sera présentée dans un article connexe.*

1 Introduction

The fourth generation of the Canadian Atmospheric Global Climate Model (CanAM4), developed by the Canadian Centre for Climate Modelling and Analysis (CCCma), is based on the third generation Atmospheric General Circulation Model (AGCM3 McFarlane et al., 2006; Scinocca and McFarlane, 2004; Scinocca, McFarlane, Lazare, Li, and Plummer, 2008), which has been and continues to be widely used in climate research activities in Canada. However, CanAM4 differs substantially from AGCM3 in its treatment of a number of physical processes. In particular, CanAM4 includes prognostic representations of stratiform (layer) clouds and aerosols. In addition, treatments of radiative transfer, convection, and turbulent mixing

have been completely revised. These changes enable modelling experiments for established and newly emerging questions in climate research such as the role of aerosol/cloud interactions and feedbacks between atmospheric, biogeochemical, and oceanic processes in the climate system. CanAM4 is part of the Canadian Earth System Model (CanESM) and has recently been applied in numerous studies, including studies on carbon cycle feedbacks (Arora et al., 2011), seasonal forecasting (Merryfield et al., Unpublished manuscript; Fyfe et al., 2011) and attribution of climate change (Gillett, Arora, Flato, Scinocca, and von Salzen, 2012). Model results were provided to the Fifth Coupled Model Intercomparison Project (CMIP5; Taylor, Stouffer, and Meehl, 2011).

[†]Current affiliation: The Atmospheric Sciences Research Center, State University of New York at Albany, New York, USA.

Accurate simulations of global climate and climate change depend upon realistic model representations of clouds and their effects on radiative transfer and on amounts and distribution of heat, moisture, and chemical tracers in the atmosphere. For instance, all past climate assessment reports by the Intergovernmental Panel on Climate Change (IPCC) concluded that cloud feedbacks remain the largest source of uncertainty in climate sensitivity estimates (e.g., Forster et al., 2007). In CanAM4 and other models, simulated clouds are affected by radiation, transport, and a multitude of thermodynamic, microphysical, and chemical processes. Variations in simulated cloud processes and properties are substantial in space and time and are non-trivially related to radiative forcings from solar radiation, greenhouse gases, and short-lived climate forcers. In addition, interactions between atmospheric processes and land/ocean processes have considerable impacts on clouds. Effects of processes or model parameters on cloud properties are the focus of a considerable amount of scientific research.

The approach taken here is to analyze fundamental climatological model results, with an emphasis on clouds and precipitation. This provides a baseline for a subsequent analysis of responses of cloud radiative effects and precipitation to variations in climate in the second part of this paper.

This study is partly motivated by results from the Cloud Feedback Model Intercomparison Project (CFMIP). According to Williams and Webb (2009), a developmental version of this model (CCCma AGCM4) produced a particularly skillful representation of clouds for different cloud regimes, including those which are considered most important in a warming climate. Here, a more recent version of the model and improved satellite datasets are used.

The outline of the paper is as follows. A description of the modelling approach is given in Sections 2 and 3. Climatological mean results and variability for cloud radiative effects and precipitation are discussed in Section 5. Finally, Section 6 provides a brief summary.

2 Model description

a *Overview of model changes*

A summary of the main model features in CanAM4 and a comparison with the approach taken in the previous version of the model, AGCM3, are presented in Table 1. These features will be described in detail in the following sections.

b *Advection of thermodynamic quantities and chemical tracers*

As in AGCM3, horizontal advection is performed spectrally in CanAM4 while vertical advection employs rectangular finite elements defined for a hybrid vertical coordinate as described by Laprise and Girard (1990). Spectral transforms required for evaluation of quadratic products associated with advection are carried out on the usual quadratic Gaussian grid, which eliminates aliasing of unresolved spectral components arising from projecting quadratic terms into the spectral domain.

A second set of spectral transforms is used in order to facilitate evaluation of the physical processes on a reduced linear grid associated with the chosen spectral truncation. The linear grid has a resolution of 128×64 for T63. Performing the physics tendency calculations on the linear grid rather than the quadratic grid results in considerable computational savings because the linear grid contains less than half the number of grid points.

Table 1. Summary of main model features.

	AGCM3	CanAM4
Horizontal resolution	T47 or T63, double transform	As in AGCM3
Vertical grid	31 levels, top ≈1 hPa (≈50 km), $\Delta z \approx 100$ m near surface	As in AGCM3, 35 levels (enhanced near tropopause)
Topography	Optimal spectral (Holzer, 1996)	As in AGCM3
Advection	Hybridized moisture variable, spectral and semi-Lagrangian advection, global mass fixer	As in AGCM3, but hybridized tracer variables, local mass fixer, physics filtering (Lander and Hoskins, 1997)
Tracer horizontal diffusion	Leith	Laplace
Boundary layer turbulence	Based on bulk Richardson number (Abdella and McFarlane, 1996), dry non-local thermodynamic mixing	As in AGCM3, revised mixing lengths, moist non-local thermodynamic mixing
Land surface scheme	CLASS 2.7 (Verseghy 2000; Verseghy, McFarlane, and Lazare, 1993), predicted soil temperature and humidity	As in AGCM3
Gravity wave drag	Anisotropic orographic low-level drag (Scinocca and McFarlane, 2000)	As in AGCM3
Radiative transfer	4-band solar, 6-band terrestrial (Fouquart and Bonnel, 1980; Morcrette, 1984)	Correlated-k distribution and McICA
Clouds	Diagnostic cloud properties (McFarlane, Boer, Blanchet, and Lazare, 1992)	Prognostic cloud liquid water and ice, statistical cloud scheme, interactive with aerosols
Moist convection	Mass flux scheme for shallow and deep convection (Zhang and McFarlane, 1995)	As in AGCM3, mass flux scheme for deep convection, extended for aerosol chemistry. Separate shallow convection scheme (von Salzen, McFarlane, and Lazare, 2005)
Cloud base closure for deep convection	Diagnostic (Zhang and McFarlane, 1995)	Prognostic (Scinocca and McFarlane, 2004)
Atmospheric composition	Specified GHGs and aerosols	Specified GHGs, prognostic bulk aerosols

As in AGCM3, the vertical domain of CanAM4 extends from the surface to the stratopause region (1 hPa, approximately 50 km above the surface), but the vertical resolution is slightly higher in CanAM4 across the tropopause where more uniform resolution is employed. In the vertical, the domain is spanned by 35 layers. Layer depths increase monotonically with height from approximately 100 m at the surface to 3 km in the lower stratosphere.

CanAM4 employs two strategies that are used in combination to deal with the artifacts of overshoots and undershoots associated with spectral advection (Gibbs effect), which can induce negative concentrations of positive definite tracers. In the first, the hybrid variable approach adopted for the moisture variable in AGCM3 has been generalized and extended to tracers in CanAM4. The second strategy follows an approach first suggested by Lander and Hoskins (1997) in which both input fields and the output tendencies from the physics package are spatially filtered. Further details regarding the implementation of the second strategy may be found in Scinocca et al. (2008).

According to the hybrid variable approach, the quantity that is transported is the transformed variable

$$S = \begin{cases} \dfrac{q_0}{[1 + p \ln(q_0/q)]^{1/p}}, & \text{if } q < q_0, \\ q & \text{if } q \geq q_0, \end{cases} \quad (1)$$

with inverse transformation for $S < q_0$,

$$q = q_0 \exp\left[\frac{1 - (q_0/S)^p}{p}\right], \quad (2)$$

where q is the physical variable (i.e., specific humidity and chemical tracer mixing ratios), and q_0 and p are constants. This is a generalization of the hybrid variable transformation proposed by Boer (1995).

The use of this transformed variable alleviates, to a considerable extent, the undesirable overshoots and undershoots that can occur when spectrally transporting rapidly varying tracers (Merryfield, McFarlane, and Lazare, 2003). In particular, unphysical negative values of positive definite tracers are largely suppressed. One drawback of the hybrid transform approach involves mass conservation. Because the transformed variable S is now advected, it is spectrally conserved rather than the original variable q. However, by judiciously choosing q_0 and p, the degree of time- and globally averaged non-conservation can be effectively controlled. Therefore, q_0 and p are empirically assigned separately for each tracer according to a criterion that time-averaged global non-conservation errors do not exceed a certain threshold under present-day climate conditions. At each time step, global mass conservation of each hybrid tracer is also enforced in CanAM4 by applying local corrections for any residual mass changes which result from the transport of its hybrid form for static choices of q_0 and p in Eq. (1). As is evident from

theory and numerical experiment, the sign and magnitude of the correction necessarily depends on the given tracer distribution so that the correction will generally be different at each time step if the tracer concentrations are changing with time. The method is described in Appendix A.

c Aerosols

Different types of natural and anthropogenic aerosols are considered in CanAM4, including sulphate, black and organic carbon, sea salt, and mineral dust. Parameterizations for emissions, transport, gas-phase and aqueous-phase chemistry, and dry and wet deposition account for interactions with simulated meteorological variables in CanAM4 for each individual grid point and time step.

Similar to an earlier version of the model (Lohmann, von Salzen, McFarlane, Leighton, and Feichter, 1999), various processes are considered for sulphate. Monthly mean emissions of gas phase sulphur dioxide (SO_2) from anthropogenic and biomass burning sources are from Lamarque et al. (2010). In addition, climatologically representative emissions for non-explosive volcanoes (Dentener et al., 2006) are included. The model also accounts for production of SO_2 from oxidation of dimethylsulphide (DMS). For DMS, monthly mean emissions from the terrestrial biosphere (Spiro, Jacob, and Logan, 1992) and emissions based on climatological concentrations of DMS in sea water (Kettle et al., 1999) are used.

Under clear-sky conditions daytime production of sulphate (SO_4^{2-}) aerosol occurs via the oxidation of SO_2 by the hydroxyl radical (OH). At nighttime, SO_4^{2-} is produced by oxidation of SO_2 by the nitrate radical (NO_3). Chemical reactions in CanAM4 are summarized in Table 2.

The model accounts for in-cloud oxidation in deep and shallow convection, and stratiform clouds with hydrogen peroxide (H_2O_2) and ozone (O_3) as oxidants (von Salzen et al., 2000). The oxidation rates depend on the pH of the cloud water (Table 2), which is calculated from an ion balance for the dissolution of the chemical species SO_2, ammonia (NH_3), nitric acid (HNO_3), and carbon dioxide (CO_2) in could droplets (von Salzen et al., 2000). For in-cloud oxidation in deep convection, the cumulus cloud fraction is calculated from the precipitation flux (Slingo, 1987).

Three-dimensional monthly averaged concentrations of OH, O_3, H_2O_2 and NO_3 from the Model for Ozone and Related Chemical Tracers (MOZART; Brasseur et al., 1998) are used for aqueous and gas-phase chemical processes. Additionally, concentrations of NH_3 and ammonium (NH_4^+) from Dentener and Crutzen (1994) are used for the calculation of pH-dependent reaction rates in the clouds.

Primary particle emissions for black and organic carbon aerosol from anthropogenic and biomass burning sources are based on Lamarque et al. (2010). In addition, emissions of precursors for secondary organic aerosol are considered using monthly mean emissions (Dentener et al., 2006). Hydrophobic and hydrophilic black and organic carbon aerosol are considered, with a specified lifetime of $\tau = 24$ h for conversion

Table 2. Chemical reactions included in the model.

Reaction	Rate Coefficient or Equilibrium Constant[b]	Unit	Reference[a]
	Gas Phase Reactions		
$DMS + OH \rightarrow SO_2$	$9.6 \times 10^{-12} \exp(-234/T)$	$cm^3 \ molec^{-1} \ s^{-1}$	1
$DMS + NO_3 \rightarrow SO_2$	$1.9 \times 10^{-13} \exp(500/T)$	$cm^3 \ molec^{-1} \ s^{-1}$	2
$SO_2 + OH \rightarrow H_2SO_4$	$a_0 \times m/(1 + a_0 \times m/a_\infty) \times 0.6^{[1./(1.+\log(a_0 m)/a_\infty)^2]}$	$cm^3 \ molec^{-1} \ s^{-1}$	3
	with $a_0 = 3 \times 10^{-31} \times (300/T)^{3.3}$ and $a_\infty = 1.5 \times 10^{-12}$		
	Equilibrium Reactions		
$SO_2(g) + H_2O(aq) \leftrightarrow SO_2(aq)$	$1.23 \times 3120(1/T - 1/298)$	$M \ atm^{-1}$	4
$SO_2(aq) \leftrightarrow H^+ + HSO_3^-$	$35.53(1/T - 1/298)$	M	4
$HSO_3^- \leftrightarrow H^+ + SO_3^{2-}$	6.72×10^{-5}	$M \ atm^{-1}$	4
$O_3(g) + H_2O(aq) \leftrightarrow O_3(aq)$	$29.44(1/T - 1/298)$	$M \ atm^{-1}$	4
$H_2O_2(g) + H_2O(aq) \leftrightarrow H_2O_2(aq)$	6.4×10^8		4
	Aqueous Phase Reactions	$M \ atm^{-1}$	
$S(IV) + H_2O_2 \rightarrow S(VI) + H_2O$	$8 \times 10^4 \exp[-3650(1/T - 1/298)] \ (0.1 + [H^+])^{-1}$	$M \ s^{-1}$	5
$S(IV) + O_3 \rightarrow S(VI) + O_2$	$4.4 \times 10^1 \exp(-4131/T) + 2.6 \times 10^3 \exp(-966/T)[H^+]^{-1}$	$M \ s^{-1}$	6

[a]References: 1. Atkinson, Baulch, Cox, Hampson, and Troe (1989); 2. DeMoore et al. (1992); 3. Pham, Müller, Brasseur, Granier, and Mégie (1989); 4. Chameides (1984); 5. Martin (1984); 6. Maahs (1983)
[b]T: temperature, m: air density in molec cm^{-3}.

of hydrophobic to hydrophilic aerosol, representing aerosol aging processes (Croft, Lohmann, and von Salzen, 2005).

For dry sea salt aerosol, two separate log-normally distributed size modes with median radii $R_g = 0.209$ μm and $R_g = 1.75$ μm and geometric standard deviation $\sigma_g = 2.03$ are considered. Concentrations of hygroscopically grown sea salt particles in the first layer above the sea surface are calculated as a function of wind speed based on a relationship that was obtained from least-squares fitting of observations from the Joint Air-Sea Interaction (JASIN) experiment (Fairall, Davidson, and Schacher, 1983). Above this layer and over land, aerosol concentrations are determined by gravitational settling, dry and wet deposition, and transport.

The source flux for coarse ($R_g = 1.9$ μm, $\sigma = 2.15$) and accumulation mode ($R_g = 0.39$ μm, $\sigma = 2$) mineral dust is calculated based on the work of Marticorena and Bergametti (1995). It is proportional to the cube of the surface wind friction speed, including a gustiness adjustment, above a threshold value. It is additionally dependent on the fraction of exposed soil, the clay (accumulation mode) or silt (coarse mode) fraction in the originating soil and has a dependence on soil moisture similar to Fécan, Marticorena, and Bergametti (1999). A proportionality constant was determined by fitting the simulated deposition to the Dust Indicators and Records of Terrestrial and Marine Paleoenvironments (DIRTMAP) II database of climatological ice-core and ocean sediment dust deposition data (Kohfeld and Harrison, 1999). Similar to sea salt, once in the atmosphere, the dust is subject to transport, gravitational settling, and dry and wet deposition.

The parameterization of dry deposition in CanAM4 is based on specified deposition velocities for each type of aerosol and a further dependency on local conditions at the surface (Croft et al., 2005; Lohmann et al., 1999). Wet deposition fluxes from below- and in-cloud scavenging of aerosols depend on local rates of precipitation and conversion of cloud water to rainwater, respectively (Croft et al., 2005).

The approach of Dufresne, Quaas, Boucher, Denvil, and Fairhead (2005) is used in CanAM4 to calculate the cloud droplet number concentration as a function of sulphate aerosol concentration. Following their approach, parameters in the parameterization of cloud droplet number were tuned in order to match simulated and observed results for the cloud droplet effective radius, giving

$$N_c = 60(SO_4^{2-})^{0.2}, \qquad (3)$$

with N_c in units of droplets cm^{-3} and the concentration of sulphate, SO_4^{2-}, in units of μgm^{-3}. A lower bound of 1 droplet cm^{-3} is used for N_c. See Ma, von Salzen, and Cole (2010) for details.

The treatment of the radiative effects of aerosols in CanAM4 is described in Section 2f.

d *Layer clouds*
1 STATISTICAL CLOUD SCHEME
Similar to other models that employ statistical cloud schemes, statistical cloud properties in CanAM4 are diagnosed from grid-cell mean thermodynamic quantities. Following the basic framework that was first outlined by Mellor (1977), quasi-conserved thermodynamic variables are used, that is, liquid/ice water static energy ($h = c_p T_l + gz$) and total (non-precipitating) water specific humidity ($q_t = q_v + q_l + q_i$). Here, $T_l = T - (L_v q_l + L_s q_i)/c_p$ is the liquid water temperature, c_p the specific heat capacity at constant pressure, g the gravitational acceleration, z the height, T the temperature, L_v the latent heat of vapourization, and L_s the latent heat of sublimation. The specific humidities of liquid water, ice, and water vapour are q_l, q_i, and q_v, respectively. The variables h and q_t are directly derived from prognostic temperature, humidity, cloud condensate, and geopotential height in CanAM4.

Following Mellor (1977), the use of a truncated Taylor series for the saturation specific humidity leads to expressions for cloud parameters in terms of the variable

$$s = \frac{1}{2}(aq_t' - bT_l'),$$

with

$$a = \left(1 + \frac{(1 - f_i)L_v + f_i L_s}{c_p} \frac{\partial q_s}{\partial T}\bigg|_{T=\overline{T_l}}\right)^{-1},$$

$$b = a \frac{\partial q_s}{\partial T}\bigg|_{T=\overline{T_l}},$$

where the saturation specific humidity is q_s, and the ice fraction $f_i = q_i/(q_l + q_i)$. Bars indicate averaging over the grid cell volume and primed variables refer to deviations from the mean. In the model, f_i is calculated based on cloud liquid water and ice concentrations from the previous time step. If no cloud existed during the previous time step, an empirical, temperature-dependent fractional probability of liquid water (Rockel, Raschke, and Weyres, 1991) is used to initialize the cloud microphysical calculations.

Based on the assumption of a joint Gaussian probability distribution for h and q_t in each grid cell of the model, the mean αth moment of the cloud liquid/ice specific humidity and the variance of the cloud condensate distribution can be expressed in terms of s, that is,

$$\overline{q_c^\alpha} = \int_{-a\Delta\bar{q}/2}^{\infty} (a\Delta\bar{q} + 2s)^\alpha p(s)ds, \tag{4}$$

$$\overline{q'^2_c} = \int_{-a\Delta\bar{q}/2}^{\infty} (a\Delta\bar{q} + 2s - \overline{q_c})^2 p(s)ds, \tag{5}$$

with the cloud condensate specific humidity $q_c = q_l + q_i$, and the mean saturation deficit $\Delta\bar{q} = \overline{q_t} - q_s(\overline{T_l})$.

The probability distribution in Eqs (4) and (5) is given by

$$p(s) = \frac{1}{\sqrt{2\pi}\sigma_s} \exp\left(-\frac{s^2}{2\sigma_s^2}\right),$$

with the variance

$$\sigma_s^2 = \frac{1}{4}(a^2\overline{q'^2_t} - 2ab\overline{q'_t T'_l} + b^2\overline{T'^2_l}). \tag{6}$$

Solutions to Eqs (4) and (5) can be obtained either analytically (e.g., Mellor, 1977) or numerically, depending on the value of α. The assumption of Gaussian probability distributions in this approach leads to qualitatively good agreement with observations (Larson et al., 2001). However, in order to account for deviations of probability distributions from Gaussian distributions, fitted results based on cloud-resolving model (CRM) output (Chaboureau and Bechtold, 2002) are used to diagnose the condensate mixing ratios and cloud fraction,

$$C = \max\{0, \min[1, 0.5 + 0.36 \arctan(1.55Q_1)]\},$$

with

$$Q_1 = \frac{a\Delta\bar{q}}{\sigma_s}.$$

Unfortunately, results for other moments of the cloud condensate distribution are not presently available from CRM simulations.

The treatment of variance for s (Eq. (6)) is based on the approach by Chaboureau & Bechtold (2005). According to this approach, the total variance is decomposed into local turbulent (subscript t) and convective (subscript c) contributions, that is,

$$\sigma_s^2 = \sigma_t^2 + \sigma_c^2,$$

with

$$\sigma_t = c_\sigma l_c \left| a\frac{\partial \overline{q_t}}{\partial z} - \frac{b}{c_p}\frac{\partial \bar{h}}{\partial z}\right|,$$

with $c_\sigma = 0.2$. A mixing length $l_c = 600$ m in the cloudy part of the free troposphere is assumed to account for sub-grid-scale effects of radiative cooling at cloud top and buoyancy sorting on mixing, following Chaboureau and Bechtold (2005).

The diagnostic treatment of the convective contribution σ_c that was originally proposed by Chaboureau and Bechtold (2005) has been slightly modified to account for the aging of convectively generated perturbations caused by gravity waves. For typical horizontal grid sizes in global climate models (i.e., a few hundred kilometres), sources of variance from convection cannot be resolved. Gravity wave activity associated with convective events is expected to cause dissipation of the convectively injected variance. In order to account for this effect, the following approach is used:

$$\sigma_c(z, t) = \begin{cases} 3 \times 10^{-3}|M_c|a^{-1}, & \text{if } M_c \neq 0, \\ \sigma_c(z, t - \Delta t)\exp(-\Delta t/\tau_c), & \text{if } M_c = 0, \end{cases} \tag{7}$$

with convective mass flux M_c. As a modification to the original parameterization by Chaboureau and Bechtold (2005), the variance decays over time on a time scale of $\tau_c = 6$ h in Eq. (7), which is within the range of results from studies on the decay of deep convection (e.g., Khairoutdinov and Randall, 2002).

The statistical cloud scheme is also used to calculate humidity profiles in parameterizations of clear-sky radiative transfer. The mean specific humidity in the clear portion of the grid cells after adjustment to thermodynamic equilibrium is given by

$$q_{v,\text{clear}} = \frac{\overline{q_v} - C q_s}{1 - C}.$$

2 CLOUD MICROPHYSICS

A prognostic microphysics scheme is used for simulations of stratiform clouds. The governing equations for the mass mixing ratios of water vapour, cloud liquid water, and cloud ice are based on the approach by Lohmann and Roeckner (1996) and Lohmann (1996). Basic microphysical processes in the scheme are similar to those found in other state-of-the art global models.

Figure 1 and Table 3 provide an overview of microphysical processes in the cloud scheme.

Condensation and evaporation (Q_{cnd}) are treated as an instantaneous adjustment of the thermodynamic properties in

127

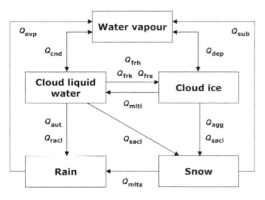

Fig. 1   Microphysical processes in layer clouds in CanAM4. See Table 3 for an explanation of terms.

the grid cells to equilibrium as given by the statistical cloud scheme.

The relative humidity in ice clouds may be either lower or higher than saturation over ice. Time scales for water vapour deposition (Q_{dep}) and sublimation (Q_{sub}) in ice clouds typically vary from minutes to hours. Consequently, relative humidities are affected by cloud dynamics and other cloud processes with finite time scales (e.g., Korolev and Isaac, 2006). For simplicity and numerical efficiency, non-equilibrium effects are omitted in CanAM4 for the deposition of water vapour onto ice crystals and sublimation of ice crystals.

Parameterizations of autoconversion of cloud droplets to rain (Q_{aut}) and accretion of cloud droplets by rain (Q_{racl}) are based on results of cloud-resolving model simulations (Khairoutdinov and Kogan, 2000). The mean autoconversion rates in the model grid cells are calculated using probability distributions of cloud water from the statistical cloud scheme by using a look-up table containing a numerical solution of Eq. (4), with $\alpha = 2.47$.

The original parameterization of autoconversion by Khairoutdinov and Kogan (2000) was modified for application in CanAM4. Firstly, the conversion rate is scaled up by a factor of 1.3, which provides better agreement of global mean model results and observations for cloud liquid water path and cloud radiative effects. The sensitivity of model

Table 3.   Summary of cloud microphysical processes.

Variable	Description
Q_{agg}	Aggregation of ice crystals
Q_{aut}	Autoconversion of cloud droplets to rain
Q_{cnd}	Condensation and evaporation of cloud liquid water
Q_{dep}	Deposition of water vapour and sublimation of cloud ice
Q_{evp}	Evaporation of rain
Q_{frh}	Homogeneous freezing of cloud droplets
Q_{frs}	Stochastic and heterogeneous freezing of cloud droplets
Q_{frk}	Contact freezing of cloud droplets
Q_{mlti}	Melting of cloud ice
Q_{mlts}	Melting of snow
Q_{racl}	Accretion of cloud droplets by rain
Q_{saci}	Accretion of ice crystals by snow
Q_{sacl}	Accretion of cloud droplets by snow
Q_{sub}	Sublimation of snow

results to this factor has been analyzed by Cole, Barker, Loeb, and von Salzen (2011). Secondly, the second indirect aerosol effect is not considered in the model given substantial uncertainties that are associated with the representation of this effect in models (e.g., Lohmann and Feichter, 2005). Hence, a global constant cloud droplet number concentration ($N_c = 50$ cm^{-3}) is used in the parameterization for autoconversion of cloud water into rain instead of Eq. (3). Although these changes may appear to be substantial, the approach is justifiable in the view of large overall uncertainties in parameterizations of autoconversion (Wood, 2005) and the uncertain role of cloud droplet concentrations in interactions between cloud microphysical and dynamical processes (Ackerman, Kirkpatrick, Stevens, and Toon, 2004).

The net conversion rate for autoconversion and accretion ($Q_{aut} + Q_{racl}$) is simulated by using an iterative semi-implicit method.

Yuan, Fu, and McFarlane (2006) identified inconsistencies in the treatments of accretion processes in the microphysics scheme by Lohmann and Roeckner (1996) which was used in an early developmental version of CanAM4. This was subsequently addressed in the development of CanAM4 by using the approach of Rotstayn (1997) for accretion of cloud droplets by snow (Q_{sacl}) and accretion of ice crystals by snow (Q_{saci}). For the latter, the collection efficiency that was proposed by Levkov, Rockel, Kapitza, and Raschke (1992) is used.

As a further modification to the original cloud scheme by Lohmann and Roeckner (1996), evaporation of rain (Q_{evp}) is also treated according to the approach of Rotstayn (1997) in CanAM4.

Rain and snow contents are diagnostically calculated from the corresponding precipitation fluxes $R_{r,s}$ for calculations of Q_{racl} and Q_{sub}, that is,

$$q_{r,s} = \frac{R_{r,s}}{\rho \overline{v}_{r,s}},$$

with the density of air, ρ, and terminal fall velocities $v_{r,s}$ for rain (subscript r) and snow (subscript s). The terminal fall velocities that are used in these calculations represent the mean fall velocities for spectra of droplet and crystal sizes (Rotstayn, 1997).

Sedimentation of ice crystals in CanAM4 is parameterized in terms of the mean ice crystal diameter, which is determined from the cloud ice content (Lohmann and Roeckner, 1996; Murakami, 1990). However, considerable uncertainty is associated with this process owing to the wide range of sizes and shapes of ice crystals in the atmosphere (Heymsfield, 2003). Additional uncertainties in ice crystal contents arise from complex interactions of sedimentation with other microphysical processes and mixing (Kay, Baker, and Hegg, 2006). In order to account for these uncertainties and to reduce systematic model biases for high clouds, the ice terminal fall velocity in CanAM4 is multiplied by a factor of $a_l = 6000$ instead of $a_l = 700$ in the original approach by Murakami (1990).

Another modification in CanAM4 is the scaling of the aggregation rate for ice crystals to snow (Q_{agg}). Lohmann (1996) used a formulation of the aggregation time scale that was originally proposed by Murakami (1990) but multiplied the time scale by a factor of $\gamma = 220$. In CanAM4, the same formulation is retained with $\gamma = 50$, which produces good agreement between simulation results and satellite observations for clouds.

The representation of the unresolved precipitation flux in microphysical calculations is described in Appendix B. The approach accounts for partial overlap of different cloud layers in the vertical.

e Moist convection

CanAM4 includes separate parameterizations for deep and shallow convection. Both parameterizations of convection use the same input profiles of temperature, moisture, and chemical tracer mixing ratios, which were output from the prognostic cloud scheme rendering them statically stable and at most fully saturated. Both schemes are permitted to be active in the same grid cells at any time within specific physical constraints for each scheme (von Salzen et al., 2005; Xie et al., 2002).

Similar to AGCM3, the cumulus parameterization of Zhang and McFarlane (1995) is used to represent the effects of deep convection (hereafter denoted by ZM) in the model. The ZM-parameterization is a bulk mass flux scheme which includes a representation of convective scale motions. It is designed to account for the effects of convective updrafts and downdrafts from evaporation of rain. As a modification of the original approach, the ZM-parameterization in CanAM4 is applied only to cumulus cloud ensembles with maximum cloud top heights above the ambient freezing level as predicted by the parameterization. This effectively limits the application of the parameterization to cumulonimbus and cumulus congestus types of clouds. A prognostic closure, based on convectively available potential energy (CAPE), is used (Scinocca and McFarlane, 2004).

Vertical momentum transfer in deep convection is included as a vertical flux term in the form:

$$\left.\frac{\partial(\rho\mathbf{V})}{\partial t}\right|_{conv} = -\frac{\partial[M_c(\mathbf{V}_c - \mathbf{V})]}{\partial z},$$

where $\mathbf{V}$ is the large-scale horizontal velocity, and $\mathbf{V}_c$ is the corresponding convective scale horizontal velocity. This quantity is determined by solving the convective scale momentum budget equation in the form

$$\frac{\partial(M_c\mathbf{V}_c)}{\partial z} + D\mathbf{V}_c - E\mathbf{V} = -f_c(\nabla_H p)_c, \tag{8}$$

where f_c is the fractional area covered by convective-scale updrafts, M_c is the deep convective updraft mass flux, D and E are, respectively, mass entrainment and detrainment rates

such that

$$\frac{\partial M_c}{\partial z} + D - E = 0.$$

These quantities are determined in the ZM-parameterization.

The pressure gradient force is parameterized following Gregory, Kershaw, and Inness (1997) as

$$-f_c(\nabla_H p')_c = \eta M_c \frac{\partial\mathbf{V}}{\partial z}. \tag{9}$$

With this parameterization it is easily shown that the convective scale horizontal velocity is a weighted average of the large-scale flow and its value in the absence of a convective scale pressure-gradient force,

$$\mathbf{V}_c = (1 - \eta)\mathbf{V}_0 + \eta\mathbf{V},$$

where $\mathbf{V}_0$ is the solution to Eq. (8) with the right-hand side set to zero.

This implies that the tendency associated with the cumulus momentum transfer is given by the value determined by ignoring the pressure-gradient force scaled by the factor $1 - \eta$. This relationship is used in implementing the parametrization. Gregory et al. (1997) suggested $\eta = 0.7$ based on cloud resolving model results and that is the default value used in CanAM4. However, Zhang and Wu (2003) have suggested a smaller value ($\eta = 0.55$) and more recently Romps (Unpublished manuscript) has argued that the formulation in Eq. (9) may produce non-physical results in some cases and that the choice $\eta = 0$ as in Schneider and Lindzen (1976) is preferable. Preliminary results show that CanAM4 simulations with $\eta = 0$ are no worse than with the default value. However, Romps (Unpublished manuscript) also advocated using a drag-law formulation for the convective scale momentum flux to account for situations with strong shear in the large-scale flow. Such a formulation has not yet been tested in CanAM4.

Effects of shallow convection are parameterized following von Salzen and McFarlane (2002) and von Salzen et al. (2005). In the parameterization, parcels of air are lifted from the planetary boundary layer (PBL) into the layer above the PBL. Shallow cumulus clouds are formed once the parcels reach the level of free convection (LFC), at which the parcels become positively buoyant. Above the LFC the parcels are modified by entrainment of environmental air into the ascending top of the cloud and also by organized entrainment at the lateral boundaries of the cloud. The cloud-top mixing produces horizontal inhomogeneities in cloud properties and vertical fluxes which are parameterized using joint probability density distributions of total water and moist static energy. The initial growth phase of the cumulus cloud is assumed to be terminated when its top reaches its maximum level. The growth phase is followed by instantaneous decay, with complete detrainment of cloudy air into the environment.

Tendencies of thermodynamical and chemical tracers in the shallow cumulus scheme are calculated based on continuity equations for mass, dry static energy, moisture, chemical tracer mixing ratios, and vertical momentum. The mixing parameters used in CanAM4 are consistent with the parameters used in the experiment DECORE_B, as described by von Salzen and McFarlane (2002). A parameterization of autoconversion has been included in the scheme in order to account for the effects of drizzle formation in shallow cumulus clouds following Lohmann and Roeckner (1996).

For implementation of the shallow convection scheme in CanAM4, the cloud base closure condition proposed by Grant (2001) is used. This approach is based on a simplified turbulent kinetic energy (TKE) budget for the convective boundary layer. It is based on the assumption of steady mean boundary layer flow by omitting effects of pressure perturbations and shear on the TKE. Under the assumption that the buoyancy flux can be approximated by a linear function of height and by further assuming that the vertical flux of TKE at cloud base is proportional to the cloud base mass flux, a simple expression for the cloud base mass is obtained,

$$M_b = \rho \left[\frac{1}{2}(1 - \alpha_g) - A_\varepsilon \right] w_*, \tag{10}$$

where, for $\overline{w'\theta_v'}^s > 0$,

$$w_* = \left(\frac{g \, \overline{w'\theta_v'}^s \, z_i}{\theta_v} \right)^{\frac{1}{3}}, \tag{11}$$

is the sub-cloud layer convective velocity scale, θ_v is the mean virtual potential temperature of the mixed layer, $\overline{w'\theta_v'}^s$ is the turbulent flux of virtual potential temperature at the surface, g is the gravitational acceleration, and ρ is the density of air. The variable z_i is the depth of the mixed layer, which is defined as that level in the model at which the gradient Richardson number exceeds Ri = 1. $\alpha_g = 0.2$ and $A_\varepsilon = 0.37$ in Eq. (10) are constants (Grant, 2001).

The parameterization of shallow convection is invoked only if the cloud tops, as predicted by this parameterization, are below the ambient freezing level.

f *Radiation*
1 OPTICAL PROPERTIES FOR GASES, CLOUDS, AEROSOLS, AND THE SURFACE
The absorption by gases in the atmosphere is parameterized using the Correlated-k Distribution (CKD) method (Li and Barker 2005) which replaces the band-mean absorption used in AGCM3 (Fouquart and Bonnel, 1980; Morcrette, 1984). The method upon which CKD is built, effectively sorts the gaseous absorption coefficients (k), which can vary greatly with wavenumber, over a particular wavenumber range into much smoother cumulative probability distributions. These sorted probability distributions form a cumulative probability space (CPS). Doing so greatly reduces the number of radiative transfer calculations while maintaining good accuracy relative to computationally expensive benchmark radiative transfer calculations over the rapidly varying k's. The CKD approach also improves the ability to model the overlapping absorption by more than one gas, especially relative to the method used in AGCM3. The gases accounted for in the CKD, which has four wavenumber intervals for the shortwave and nine intervals for the longwave, are listed in Table 4. Methods are described in Li and Barker (2005) to compute the combined absorption efficiently and accurately when there is more than one gas present. Further efficiencies are found by noting that for some integration points along the k distributions the gaseous absorption is very large relative to other absorbing and scattering constituents in the atmosphere (e.g., clouds and aerosols) and so the radiative transfer calculations can be greatly simplified by neglecting scattering and only considering absorption. In Table 4, two types of CPS intervals are listed: the *major* intervals are those for which the absorption by gases is small enough that scattering should be included in the radiative transfer calculations whereas for the *minor* intervals only absorption is included in the radiative transfer calculations.

In contrast to the optical properties of the gases, those for cloud particles and aerosols vary relatively slowly with wavenumber. Therefore, the parameterizations of these optical properties in CanAM4 are appropriately weighted mean values over each of the wavenumber bands listed in Table 4.

The cloud optical properties, specific extinction, single scattering albedo, and asymmetry parameter, are parameterized for each CKD band as a function of particle size and concentration. For liquid cloud particles the effective radius is computed assuming that the drop size is a gamma distribution (Ramaswamy and Li, 1996),

$$r_{e,l} = f \left(\frac{3\rho q_l}{4\pi\rho_l C(N_c \times 10^6)} \right)^{1/3}, \tag{12}$$

where q_l is the specific humidity of cloud liquid water, ρ_l is the density of water, C is the cloud fraction (Eq. (7)), and $f = 1.4$. For ice cloud particles their effective radius is computed using Lohmann and Roeckner (1996):

$$r_{e,i} = 83.8 \times 10^{-6} \left(\frac{10^3 \rho q_i}{C} \right)^{0.216}, \tag{13}$$

where q_i is the concentration of cloud ice. This effective radius is then related to a *generalized effective size* to account for non-spherical ice particles used to compute ice optical properties (Fu, 1996),

Table 4. The band spectrum ranges, absorbers, and the number of intervals for CPS. The intervals for CPS are divided into two categories, major and minor. See text for details.

Band	(cm^{-1})	Sub-band (cm^{-1})	Absorber	Number of interval in CPS	
				major	minor
Solar					
1	14500–50000	50000–42000: UVC 42000–37400: UVC 37400–35700: UVC 35700–34000: UVB 34000–32185: UVB, UVA 32185–30300: UVA 30300–25000: UVA 25000–20000: PAR 20000–14500: PAR	O$_2$, H$_2$O	6	3
2	8400–14500		H$_2$O, O$_3$, O$_2$	4	4
3	4200–8400		H$_2$O, CO$_2$, CH$_4$	6	4
4	2500–4200		H$_2$O, CO$_2$, CH$_4$	4	9
			Total	20	20
Infrared					
1	2200–2500		H$_2$O, CO$_2$, N$_2$O	1	5
2	1900–2200		H$_2$O, N$_2$O	1	1
3	1400–1900		H$_2$O	2	3
4	1100–1400		H$_2$O, N$_2$O, CH$_4$, CFC12	5	5
5	980–1100		H$_2$O, O$_3$, CO$_2$, CFC11, CFC12	2	4
6	800–980		H$_2$O, CO$_2$, CFC11, CFC12	3	0
7	540–800		H$_2$O, CO$_2$, N$_2$O, O$_3$	3	7
8	340–540		H$_2$O	6	3
9	0–340		H$_2$O	4	6
			Total	27	34

$$D_{ge} = \frac{8}{3\sqrt{3}} r_{e,i}. \qquad (14)$$

With the cloud water contents and effective particles sizes, the cloud optical properties for each band in the CKD are computed for liquid cloud particles at solar (Dobbie, Li, and Chýlek, 1999) and infrared (Lindner and Li, 2000) wavenumbers and for ice cloud particles at solar (Fu, 1996) and infrared (Fu, Yang, and Sun, 1998) wavenumbers.

Several species of aerosols are radiatively active in CanAM4. Under the assumption that ammonium sulphate ((NH$_4$)$_2$SO$_4$) dominates the chemical composition of sulphate aerosols, their optical properties are parameterized using Li, Wong, Dobbie, and Chýlek (2001). Sea salt and mineral dust optical properties are parameterized using Lesins, Chýlek, and Lohmann (2002) while the optical properties of black and organic carbon are parameterized using Bäumer, Lohmann, Lesins, Li, and Croft (2007). Volcanic aerosols in the stratosphere are assumed to be distributed between a climatological mean tropopause and 10 hPa and to be composed of 25% water (H$_2$O) and 75% sulphuric acid (H$_2$SO$_4$). A lognormal distribution with a geometric radius of 0.1336 µm (effective radius of 0.35 µm) and effective variance of 0.46978 µm^2 is assumed (Hansen and Travis, 1974). Their optical properties are computed using Mie computations for each CKD band.

The parameterizations of surface albedo and emissivity over land and sea ice are largely unmodified from AGCM3 (see Section 2i). For ocean solar surface albedo, CanAM4 uses the scheme of Jin, Charlock, and Rutledge (2002), which is dependent on solar zenith angle and surface wind speed, and accounts for the direct and diffuse components of the incident solar radiation. This scheme was modified to account for the effect of white-caps forming at relatively high wind speeds (>15 m s^{-1}; Monahan and MacNiocaill, 1986), which can enhance the surface albedo (Li et al., 2006). The ocean surface emissivity uses the broadband formulation of Hansen et al. (1983).

2 RADIATIVE TRANSFER

There are significant differences between the radiative transfer solver in AGCM3 and that used in CanAM4. The latter has the ability to use either the deterministic solvers described in Li (2002); Li, Dobbie, Räisänen, and Min (2005); Li and Barker (2005) or the Monte Carlo Independent Column Approximation (McICA; Barker et al., 2008; Pincus, Barker, and Morcrette, 2003). Although both schemes share the ability to account for radiative transfer through overlapping and horizontally inhomogeneous clouds, only the implementation of McICA will be described here.

The McICA solves the radiative transfer for each grid cell in CanAM4 by randomly sampling unresolved structure while systematically sampling each integration point in the CKD, thereby generating an unbiased estimate of grid cell–mean radiative fluxes. This requires, in addition to the optical properties described in the previous section, a radiative transfer solver and a method to model the unresolved structure, namely clouds. The radiative transfer solvers, one for solar and one for the infrared, are effectively those described in Li et al. (2005) and Li (2002) although they have been simplified to solve the radiative transfer only through overcast homogeneous clouds in multiple layers because this is all that is required for McICA. The unresolved cloud structure is provided by a stochastic cloud generator (Räisänen, Barker, Khairoutdinov, Li, and Randall, 2004).

For solar radiation, a two-stream solver, along with a delta-Eddington approximation, is used to calculate radiative transfer in the atmosphere, which leads to a numerically efficient approach that has a linear dependence on the number of vertical levels. The effect of atmospheric spherical curvature and refraction on the effective pathlength is accounted for by adjusting the solar zenith angle using

$$\frac{1}{\mu_e} = \frac{24.35}{2\mu_0 + \sqrt{498.5225\mu_0^2 + 1}},$$

where μ_0 is the cosine of solar zenith angle and μ_e is the cosine of effective solar zenith angle (Li and Shibata, 2006).

For infrared radiation, a two-stream solver is also used along with a methodology to account for the scattering by cloud and aerosol particles efficiently (Li, 2002). In the CKD model, wavenumber 2500 cm^{-1} delineates the solar and infrared regions for the application of each of the radiative transfer solvers. However, there is approximately 12 Wm^{-2} of solar radiation incident at the top of the atmosphere for wavenumbers less than 2500 cm^{-1}. This solar radiation is used as an upper boundary condition for the downward fluxes in the infrared radiative transfer calculations (Li, Curry, Sun, and Zhang, 2010).

Two aspects of the unresolved cloud structure are accounted for in the CanAM4 radiative transfer calculations: the horizontal variability of cloud condensate and the vertical overlap of cloud. The cloud condensate is assumed to follow a gamma distribution, based on previous studies (e.g., Barker, 1996):

$$p = \frac{1}{\Gamma(\nu)}\left(\frac{\nu}{\overline{q_c}}\right)^\nu q_c^{\nu-1} e^{-\nu q_c/\overline{q_c'^2}}, \tag{15}$$

where $\Gamma(\nu)$ is the gamma function, and $\nu = \overline{q_c}^2/\overline{q_c'^2}$, where $\overline{q_c}$ and $\overline{q_c'^2}$ are the mean and the variance of the cloud water content, respectively. The parameter ν in Eq. (15) is calculated using $\overline{q_c}$ and $\overline{q_c'^2}$ from the Gaussian probability distributions of quasi-conserved thermodynamic variables in the statistical cloud scheme (Section 2d1). No variability is assumed in cloud particle size (i.e., it is horizontally uniform in a model layer) although this assumption can be relaxed.

Cloud vertical overlap is modelled using Hogan and Illingworth (2000) and Räisänen et al. (2004),

$$C_{k,l} = \alpha_{k,l}C_{k,l}^{\max} + (1 - \alpha_{k,l})C_{k,l}^{\mathrm{ran}}, \tag{16}$$

where $C_{k,l}$ is the vertically projected cloud fraction for two layers k and l, $C_{k,l}^{\max} = \max(C_k, C_l)$ (maximum overlap), and

$C_{k,l}^{\mathrm{ran}} = C_k + C_l - C_kC_l$ (random overlap). The variable $\alpha_{k,l}$ is the overlap parameter which is given as:

$$\alpha_{k,l} = \exp\left(-\int_{z_k}^{z_l}\frac{dz}{L(z)}\right), \tag{17}$$

where $L(z)$ is the decorrelation length and z is the height. Because the layer cloud fractions C are computed in CanAM4, it is necessary to set $L(z)$ to specify the vertical overlap of clouds. Based on analysis of observations and output from cloud resolving models (Barker et al., 2008; Hogan and Illingworth, 2003; Pincus, Hannay, Klein, Xu, and Hemler, 2005), $L(z)$ was set to 2 km for cloud fraction and 1 km for cloud water content.

g Turbulence

Turbulent transfer of scalar quantities in the boundary layer involves both local, down-gradient transfer processes and non-local processes that are often associated with the occurrence of upward heat and moisture fluxes at the surface.

Non-local mixing of heat, moisture, and chemical tracers occurs in simulations with CanAM4 if the direction of the buoyancy flux at the surface is upward. The approach is based on the assumption that non-local effects can be represented by relaxing local values of liquid/ice water static energy, total water, and other quasi-conserved scalar quantities in the boundary layer, $\bar{\chi}$, to a vertically homogeneous reference state, $\overline{\chi_r}$, over a specified time period, τ, that is,

$$\left.\frac{\partial\bar{\chi}}{\partial t}\right|_{\mathrm{nl}} = \frac{1}{\tau}(\chi_r - \bar{\chi}).$$

Similar to AGCM3, the reference state is determined by assuming that the vertical flux must vanish at the top of the mixing region, that is,

$$\chi_r = \frac{1}{1-\sigma_t}\int_{\sigma_t}^1 \bar{\chi}d\sigma + \frac{g\tau\rho_s\overline{w'\chi'}^s}{p_s(1-\sigma_t)},$$

where $\sigma_t = p_t/p_s$ is the sigma coordinate for the top of the mixing region. The variables p_s and ρ_s are, respectively, the pressure and density of the air at the surface, and $\overline{w'\chi'}^s$ is the surface flux for χ.

The top of the mixing region is determined iteratively by strapping layers together, beginning with the bottom layer alone and adding additional layers as long as the computed value of the virtual potential temperature in the topmost layer of the mixing region is larger than the ambient value in the layer directly above the top one in the mixing region. It can be shown that the treatment of non-local mixing in CanAM4 corresponds to the standard encroachment formula for growth of a convectively driven mixed layer if the temporal variation of the surface flux is negligible (McFarlane et al., 2006).

The mixing adjustment time scale τ is based on the eddy turnover time scale (Abdella and McFarlane, 1997), that is,

$$\tau = \frac{\alpha_t z_t}{\max[(0.6w_*^2 + u_*^2)^{1/2}, 0.03]},$$

with the convective velocity scale w_*, friction velocity u_*, $\alpha_t = 1$, and

$$z_t = \frac{p_s}{g} \int_{\sigma_t}^{1} \rho^{-1} \, d\sigma.$$

Once vertical profiles for liquid/ice water static energy and total water are determined, the statistical cloud scheme is used to determine profiles of temperature, specific humidity, and cloud condensate.

The local down-gradient transfer of momentum, liquid/ice water static energy, total water, and prognostic trace constituents associated with turbulent transfer are accounted for by using diffusivities which are functions of the vertical wind shear and the local gradient Richardson number. The formulation used is qualitatively similar to that of AGCM2 (McFarlane et al., 1992, 2006). Eddy diffusivities for momentum, heat, and tracers are of the form

$$K = l^2 \left|\frac{\partial \mathbf{V}}{\partial z}\right| f(\mathrm{Ri}),$$

where $\mathbf{V}$ is the horizontal wind vector and z the distance above the terrain. The factor $f(\mathrm{Ri})$ depends on the gradient Richardson number Ri (McFarlane et al., 1992).

A unified formulation for the mixing length is used in CanAM4 for consistent treatment of local mixing under clear- and cloudy-sky conditions. It combines features of the approach that is used in AGCM3 with the statistical cloud scheme in CanAM4.

In the absence of clouds, the mixing length in the convectively mixed layer is given by

$$l = \max\left(\frac{l_u l_d}{l_u + l_d}, l_{\min}\right),$$

where l_u and l_d are length scales for upward and downward turbulent transfer, respectively (Lenderink and Holtslag, 2004). The minimum mixing length is given by

$$l_{\min} = \max\left(\frac{l_m l_w}{l_m + l_w}, l_0\right),$$

where $l_m = 75$ m, $l_w = 0.5 l_u$, and $l_0 = 10$ m. In the atmosphere, l_u and l_d will generally depend on the stability of the atmosphere and the efficiency of non-local mixing. However, the following simple approximations are used in CanAM4:

$$l_u = \kappa z,$$
$$l_d = \kappa(z_t - z),$$

with the von Kármán constant κ and the height above the surface, z.

Above the convectively mixed layer, it is assumed that the mixing length decreases monotonically with height according to

$$l = l_0 + (l_{\min} - l_0)\left(\frac{\sigma}{\sigma_t}\right)^2,$$

if no clouds are present.

Under cloudy conditions, a mixing length of $l_c = 600$ m in the free troposphere is assumed, as described in Section 2d1. For all-sky conditions, the mixing length is calculated from a linear combination of l and l_c,

$$l_a = (1 - C)l + Cl_c,$$

with cloud fraction C.

The treatment of mixing in CanAM4 yields realistic profiles of water and heat in CanAM4 (e.g., Zhu et al., 2005).

h *Surface fluxes*

Surface exchanges of heat, moisture, and momentum follow the treatment of Abdella and McFarlane (1996). The approach is based on the similarity theory of Monin and Obukhov (1954) according to which the surface wind stress, sensible heat flux, and moisture flux, are related to the wind, temperature, and specific humidity near the top of the surface layer through friction velocity, temperature, and humidity scales.

CanAM4 accounts for the effects of wind gustiness on surface fluxes of momentum, heat, moisture, and tracers based on an approach that was proposed by Redelsperger, Guichard, and Mondon (2000). For this, an effective absolute wind speed V_{eff} is used in the calculations of surface fluxes, with

$$V_{\mathrm{eff}} = \sqrt{U^2 + V^2 + V_1^2 + V_2^2},$$

where (U, V) is the wind speed in the first model layer and

$$V_1 = \beta w_*$$
$$V_2 = \log(1 + 6.69 P_c - 0.476 P_c^2),$$

with w_* from Eq. (11), $\beta = 0.55$, and the convective surface precipitation rate P_c (in cm d^{-1}). V_1 and V_2 refer to contributions of gustiness from boundary layer free convection and deep convection, respectively.

i *Land and ocean surface*

Similar to AGCM3, the calculation of energy and moisture fluxes at the land surface is carried out within the Canadian Land Surface Scheme (CLASS) module that was first introduced in AGCM3 (McFarlane et al., 2006). Development of the CLASS model started in the late 1980s. The module has undergone certain modifications since then. The basic physics underlying the scheme is outlined in two papers, Verseghy (1991) and Verseghy et al. (1993), and the changes implemented over the following decade are described in Verseghy (2000). The version of CLASS currently used in CanAM4 is referred to as version 2.7. A brief outline of its structure is provided below.

Each land surface grid cell treated by CLASS can have up to four subareas: bare soil, vegetation-covered soil, snow-covered soil, and soil covered by both vegetation and snow. The incoming shortwave and longwave radiation, the ambient air temperature, humidity, and wind speed, and the precipitation rate at the current time step are supplied by the atmospheric model. The energy and moisture budgets of each subarea are calculated separately within CLASS, and the surface fluxes are averaged over the grid cell and passed back to the atmospheric model.

The soil profile is divided into three horizontal layers, of thicknesses 0.10, 0.25 and 3.75 m. The texture of each layer and the overall depth to bedrock are derived from the global dataset assembled by Webb, Bartlein, Harrison, and Anderson (1993). The hydraulic properties of the soil layers are obtained from the soil texture using relationships developed by Cosby, Hornberger, Glapp, and Ginn (1984). The layer temperatures and liquid and frozen moisture contents are carried as prognostic variables and are stepped forward in time using the fluxes calculated at the top and bottom of each layer. Energy fluxes are obtained from the solution of the surface energy balance, expressed as a function of the surface temperature and solved by iteration. The soil albedo and thermal properties vary with texture and moisture content. Moisture fluxes are determined using classic Darcy theory in the case of drainage and capillary rise and after the method of Mein and Larson (1973) in the case of infiltration. If the surface infiltration capacity is exceeded, water is allowed to pond on the surface up to a maximum depth which varies by land cover. Continental ice sheets are modelled in the same way as bare soil, using the thermal properties of ice instead of soil minerals. Snow is modelled as a fourth, variable-depth soil layer with its own separate layer temperature, carried prognostically. Density and albedo vary exponentially with time, from fresh-snow values to specified background values, according to relationships derived from field data. Melting occurs if either the surface temperature or the snow pack layer temperature is projected to rise above 0°C In this case, the excess energy is used to melt part of the snow pack and the temperature is set back to 0°C. Meltwater percolates into the pack and refreezes until the entire layer reaches 0°C, at which point any further melt is allowed to reach the soil surface. Snowmelt decreases the thickness of the pack until a limiting depth of 0.10 m is reached; after this, the snow pack is assumed to become discontinuous, and a fractional snow cover is calculated by setting the depth back to 0.10 m and employing conservation of mass.

Vegetation properties such as height, leaf area index, albedo and rooting depth are assigned to each vegetation type on the basis of measurements gleaned from the literature. Derived properties such as the shortwave radiation extinction coefficient, the canopy gap fraction, the roughness lengths for heat and momentum, and the annual cycle of leaf area index are determined separately for coniferous trees, deciduous trees, crops, and grass, and are then averaged over the grid cell to define the bulk canopy characteristics. The canopy temperature and the liquid and frozen intercepted water are carried as prognostic variables. The interception capacity is calculated as a function of leaf area index. Stomatal resistance to transpiration is parameterized as a function of incoming shortwave radiation, air vapour pressure deficit, canopy temperature and soil moisture, using functional relationships similar to those presented by Stewart (1988).

3 Experimental design

Simulations using CanAM4 were performed following the specifications outlined by Taylor, Stouffer, and Meehl (2011) for the Atmospheric Model Intercomparison Project (AMIP), which is a subset of CMIP5. A five-member ensemble of CanAM4 simulations was generated by starting each simulation from slightly perturbed initial conditions. Each simulation started on 1 January 1949, spinning up CanAM4 for 1 year, with the analysis period starting on 1 January 1950 and ending 31 December 2009. Monthly mean AMIP sea surface temperature (SST) and sea-ice boundary condition data were used for the lower boundary. These are the same boundary condition data required for the CMIP5 project (Hurrell, Hack, Shea, Caron, and Rosinski, 2008). See Taylor, Williamson, and Zwires (2008) for a description of this dataset. Time varying monthly averaged concentrations for CO_2, CH_4, O_3, N_2O, effective CFC11 and CFC12 were those of the historical period used in the CMIP5 project and extended to 2010 using the Representative Concentration Pathway (RCP4.5) scenario (Moss et al., 2010). Aerosols were interactive in the simulations, with decadal emissions from anthropogenic and biomass burning sources interpolated to monthly time increments for SO_2 and black and organic carbon (Lamarque et al., 2010). Stratospheric volcanic effects are modelled via the use of specified time- and latitude-varying (four latitude bands) stratospheric aerosol optical depth. The data used were those of Sato, Hansen, McCormick, and Pollack (1993), as extended by the Hadley Centre using an exponential decay out to a minimum value of 0.00001 in 1999 and remaining constant thereafter. Therefore, there were effectively no stratospheric volcanic effects after 1999.

Annual mean total solar irradiance variations are specified according to the CMIP5 protocol. These were determined by direct multiple regression of the sunspot and facular time series with a time series of total solar irradiance, as described in Fröhlich and Lean (2004). See http://sparcsolaris.gfz-potsdam.de/cmip5.php for details.

Ozone is specified for radiative transfer calculations using the transient, three-dimensional ozone fields recommended for CMIP5 and described in Cionni et al. (2011). For the period of time covered by the historical segment of the ozone database, 1850–2009, a zonally averaged version of the CMIP5 ozone was used in CanAM4 without further modification.

In CanAM4, the vegetation types present over each grid cell are obtained from the Global Land Cover 2000 (GLC2000) global dataset (Bartholomé and Belward, 2005).

4 Comparison between CanAM4 and AGCM3

Given the substantial number of new or improved parameterizations in CanAM4 that were described in previous sections, CanAM4 provides a more physically complete representation of the atmosphere than does AGCM3. For instance, radiation schemes based on correlated-k distribution and McICA are increasingly replacing other, less accurate radiation schemes in climate models (Barker et al., 2008; Oreopoulos et al., 2012). As another example, the parameterization for transient shallow convection by von Salzen and McFarlane (2002) has led to improved representations of low clouds and convective mixing in CanAM4 (von Salzen et al., 2005) and the global climate model ECHAM5-HAM (Isotta, Spichtinger, Lohmann, and von Salzen, 2011). Furthermore, detection and attribution of climate change and climate projections benefit substantially from the introduction of a prognostic aerosol scheme (e.g., Gillett et al., 2012).

Improvements in parameterizations have led to more skilful seasonal predictions. A coupled forecast system based on CanAM4 produces a markedly higher skill than an earlier system that is based on AGCM3 (Merryfield et al., Unpublished manuscript). Compared to AGCM3, simulations with the CanAM4-based forecast system produce a more vigorous El Niño-Southern Oscillation (ENSO) when employing an identical ocean model, which is generally in better agreement with observations.

For a comparison of broad climatological features, CanAM4 and AGCM3 were configured to run AMIP-type simulations similar to those described in Section 3. Given the formulation of AGCM3, mixing ratios of greenhouse gases and the forcings could not be made identical in these simulations. For example, background aerosols in AGCM3 are specified as constants, assuming present-day conditions. Also, variations in solar irradiance and volcanic aerosol are not accounted for in AGCM3. Overall, simulated and observed climatological mean results agree slightly better with observations in CanAM4 than AGCM3 for a range of different atmospheric quantities (Fig. 2). The most apparent difference between CanAM4 and AGCM3 is an increase in

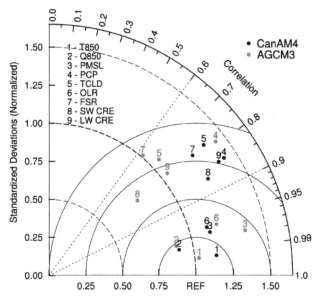

Fig. 2   Taylor diagram (Taylor, 2001) for CanAM4 (black) and AGCM3 (red). The radial coordinate gives the magnitude of the total standard deviation, normalized by the observed value, and the angular coordinate gives the correlation with observations. Numbers indicate model-based datasets compared with observations for global temperature (1) and specific humidity (2) at 850 hPa, mean sea level pressure (3), precipitation (4), total cloud amount (5), outgoing longwave (6) and shortwave (7) radiation at the top of the atmosphere, and shortwave (8) and longwave (9) cloud radiative effects. Mean model results and observations during the time period 2003–08 are used. Observations are from the European Centre for Medium-range Weather Forecasts Re-Analysis (ERA) Interim reanalysis (Dee et al., 2011) for 1–3, the Global Precipitation Climatology Project (GPCP; Adler et al., 2003) for 4, ISCCP D2 (Rossow, Walker, Beuschel, and Roiter, 1996) for 5, and CERES EBAF (Loeb et al., 2009) for 6–9.

spatial variability for many of the simulated quantities, in particular for quantities related to clouds (e.g., for datasets 5, 7, 8, and 9 in Fig. 2). On a global scale, the general increase in spatial variability in the results of CanAM4 tends to counteract improvements in spatial correlation between model results and observations, which leads to similar errors in patterns for both models.

In the following, the analysis of model results will only be based on results from CanAM4 in order to provide a benchmark for studies of clouds and precipitation in the second part of this paper and for other model applications. A more detailed analysis of AGCM3 simulation results is not pursued further given complications associated with differences in forcings and diagnostic capabilities relative to CanAM4. The general performance of this model has been documented in a number of earlier publications.

5 Mean distributions of clouds and precipitation

Clouds cover a wide range of spatial and temporal scales in the atmosphere. The diverse nature of clouds and associated radiative properties poses a major challenge for those seeking to

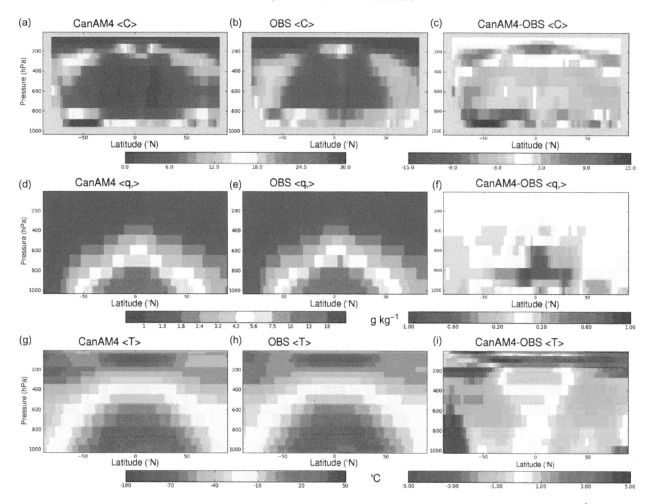

Fig. 3   Simulated and observed zonally and temporally averaged cloud amounts (panels a to c), specific humidity (panels d to f; units: g kg⁻¹), and temperature (panels g to i; units: °C) for CanAM4 (left column), CALIPSO-GOCCP satellite observations, and the ERA-interim reanalysis. Differences between model results and observations are displayed in the last column. A logarithmic scale is used for panels d and e. The corresponding time periods are June 2006 to June 2009 for cloud fractions and January 1989 to December 2009 for specific humidities and temperatures. Zonal mean cloud amounts from CALIPSO-GOCCP and COSP were interpolated to pressure levels for the comparison using the simulated geopotential height. Grid points with missing data appear as grey areas. Large temperature anomalies over Antarctica are caused by differences in extrapolation below topography in the different datasets.

understand the role of clouds in climate. Satellite simulators have recently become available for detailed and accurate comparisons between cloud-related quantities from models and satellite-based datasets. For instance, systematic biases in the representation of cloud amounts and optical thickness of different types of clouds in global climate models were identified through application of the International Satellite Cloud Climatology Project (ISCCP; Rossow and Schiffer, 1999) simulator (Klein and Jakob, 1999; Webb, Senior, Bony, and Morcrette, 2001).

Clouds simulated by CanAM4 can be compared with cloud properties from ISCCP, and other satellite platforms, through the use of the CFMIP Observational Simulator Package (COSP; Bodas-Salcedo et al., 2011). To ensure consistency between the diagnosed cloud properties and radiative fluxes, the simulators in COSP were modified so that each satellite simulator used the same subgrid-scale clouds as the CanAM4 radiation (Section 2f).

A comparison of retrieved cloud amount from the GCM-Oriented Cloud-Aerosol Lidar and Infrared Pathfinder Satellite Observations (CALIPSO) Cloud Product (CALIPSO-GOCCP; Chepfer et al., 2010) against that simulated by CanAM4 and the COSP is shown in Fig. 3. Zonal mean results for cloud amounts are compared to results for specific humidity and temperature. Broad features in the observations are reasonably well captured by the model. However, amounts of low and mid-level clouds tend to be underestimated in CanAM4 at levels above 925 hPa (Fig. 3c). Similar to CanAM4, most models tend to underestimate amounts of low and mid-level clouds (Zhang et al., 2005), without a known single cause. For CanAM4, underestimates for mid-level clouds in the tropics are related to different causes. Firstly, effects of cumulus congestus clouds are insufficiently parameterized in CanAM4, as is evident from an underestimate of humidity in the tropical free troposphere at mid- and low levels (Fig. 3f). In addition, a more detailed analysis of the

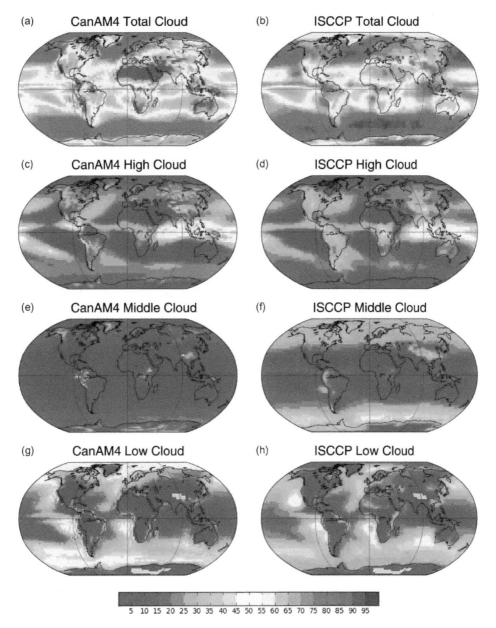

Fig. 4   Mean total cloud fraction for CanAM4 (a) and ISCCP D2 during the time period January 1996 to December 2005. Results are broken down into contributions from high (panels c and d), middle (panels e and f), and low (panels g and h) top clouds using the ISCCP/COSP cloud simulator tool in CanAM4. The corresponding pressure intervals are from 50 to 440 hPa (high), 440 to 680 hPa (middle) and 680 hPa to surface (low). Only the contributions of stratiform clouds and shallow cumulus to total cloud amounts are considered for CanAM4.

cloud data gives evidence that cloud top heights and humidity are underestimated in CanAM4 for marine regions that are mainly affected by stratocumulus clouds. Finally, free-tropospheric extratropical clouds in CanAM4 occur at levels that are too high as is evident from Fig. 3c. The formation of anomalous high-level free-tropospheric extratropical clouds is related to unrealistically cold conditions in the upper free troposphere (Fig. 3i), which points to the general circulation as a potential cause of this bias.

Cloud fractions simulated by the ISCCP/COSP simulator in CanAM4 agree reasonably well with results from ISCCP D2 for broad climatological features (Fig. 4). Over the subtropical ocean, total cloud fractions are slightly smaller in CanAM4, which can be largely attributed to an overall lack of low-level clouds and humidity. Small-scale spatial inhomogeneities for simulated amounts of clouds (e.g., near the west coast of South America) are caused by numerical truncation which is associated with the spectral transform in regions with sharp gradients in advected quantities (the so-called Gibbs effect). A lack of mid-level clouds largely explains smaller total cloud fractions in the extratropics. Note that differences in polar regions may not necessarily imply model biases because passive satellite retrievals tend to be less robust over snow and ice found at high latitudes.

Results for cloud fractions in Fig. 4 cannot be directly compared to zonal mean cloud amounts in Fig. 3. Results in Fig. 4 represent vertically projected cloud fractions for cloud tops within different pressure ranges whereas results in Fig. 3 refer to horizontally overlapping cloud amounts at different latitudes and heights. Furthermore, owing to relatively wide pressure ranges that are used for the plots in Fig. 4, vertical shifts in amounts of low and high clouds that are apparent in Fig. 3 do not notably affect results in Fig. 4. In fact, good agreement is found between mean cloud amounts from ISCCP and CALIPSO-GOCCP when both datasets are averaged over the same pressure ranges (not shown).

Amounts and location of clouds in the vertical have considerable implications for radiative transfer in the atmosphere. Combined effects of scattering and absorption of radiation by different types of clouds lead to a large net sink of energy in the atmosphere in the global and annual mean (Trenberth, Fasullo, and Kiehl, 2009). The Clouds and Earth's Radiant Energy System (CERES) Energy Balanced and Filled (EBAF) dataset provides detailed and accurate satellite-based estimates of cloud contributions to atmospheric energy budgets (Loeb et al., 2009). Monthly mean results from CERES EBAF (edition 2) for cloud radiative effects (CRE) were used here. The CRE is calculated as the difference between net all-sky and net clear-sky radiative fluxes at the top of the atmosphere and is sometimes referred to as cloud radiative forcing. Typically the CRE is calculated in models by rerunning the radiative transfer calculations at each time step with the clouds removed and by applying the same large-scale humidity profiles in the calculations. However, in CanAM4 rather than use the large-scale humidity profiles in the radiative transfer calculations we use the clear-sky humidity profiles which are provided by the statistical cloud scheme. This results in CanAM4 computing clear-sky radiative fluxes that are more consistent with satellite-based clear-sky fluxes which are determined from cloud-free footprints (e.g., Sohn, Nakajima, Satoh, and Jang, 2010) and by extension more consistent CREs.

The global and annual mean averaged net CRE for March 2000 to December 2007 is -20.6 W m^{-2} for CERES and -23.1 W m^{-2} for CanAM4. This can be compared with other estimates for the present-day net CRE from ISCCP FD (-23.6 W m^{-2}; Zhang, Rossow, Lacis, Oinas, and Mishchenko, 2004) and other global climate models (-23.3 W m^{-2}; Meehl et al., 2007). As mentioned in Section 2, the magnitude of the simulated global mean net CRE depends on the values of several parameters in parameterizations in CanAM4, which are subject to considerable, often unknown, uncertainty. Consequences of parameter choices for simulated climate, clouds, and precipitation have been addressed in numerous studies (e.g., Cole et al., 2011; Sanderson, 2011; Scinocca and McFarlane, 2004; von Salzen et al., 2005).

There is good overall agreement between observed and simulated patterns of the net CRE from CERES EBAF and CanAM4, respectively (Figs 5a and 5b). Similar to CERES EBAF, CanAM4 produces a negative net CRE in the relatively cloudy extratropics and coastal stratus regions. On the other hand, slightly positive forcings are found for regions that are covered by bare ground, ice, or snow.

A breakdown of the net CRE into shortwave (SW) and longwave (LW) components gives evidence for characteristic contributions of low and high clouds to the net CRE (Fig. 5). Observed and simulated SW CREs are dominated by contributions from highly reflective low and mid-level clouds in extratropical and coastal stratus regions (see Fig. 4). Absorption and re-emission of infrared radiation by high-level clouds largely explains the results for the LW CRE. Substantial compensation of negative SW CREs by positive LW CREs is found in regions with deep cloud layers in the convectively active tropics. There is good agreement for global mean results (SW CRE -48.4 W m^{-2} for CanAM4, -47.2 W m^{-2} for CERES EBAF; LW CRE 25.3 W m^{-2} for CanAM4, 26.6 W m^{-2} for CERES EBAF).

Regional differences between results from CanAM4 and observations are similar to biases in several other models (e.g., Williams and Webb, 2009, suppl. material). For instance, the SW CRE tends to be underestimated for stratus clouds in coastal areas in eastern portions of ocean basins, consistent with biases for cloud fractions (Fig. 4). Maxima in LW CRE are unrealistically shifted from the Amazon region to central America and from the eastern to the western tropical Pacific Ocean in CanAM4. The consistency of these differences with biases in the fractions of low and high clouds points to model shortcomings related to parameterizations for clouds and convection or their interactions with other processes in the model.

It is worthwhile to note that the overall good agreement between simulated and observed SW CRE is partly a consequence of compensating biases in the model. Cole et al. (2011) found that cloud-mean albedo for clouds located at low and mid-levels in CanAM4 are larger than those observed by CERES. This is attributable to CanAM4 simulating cloud optical depths, via large liquid water paths, that are too large for these cloud types. These large optical depths are then partly compensated for by cloud fractions that are too small. Similar biases have been found in other models (Zhang et al., 2005) although Klein et al. (2012) have shown that biases are less pronounced in CMIP5 models compared with earlier versions of the same models, including CanAM4.

There is good agreement for mean precipitation from CanAM4 and observations from GPCP during the time period January 1979 to December 2007. Given that precipitation tends to be associated with deep cloud layers with high cloud tops, results for precipitation and LW CRE consistently point to common biases in the representation of clouds in CanAM4, including the shift of clouds from the Amazon

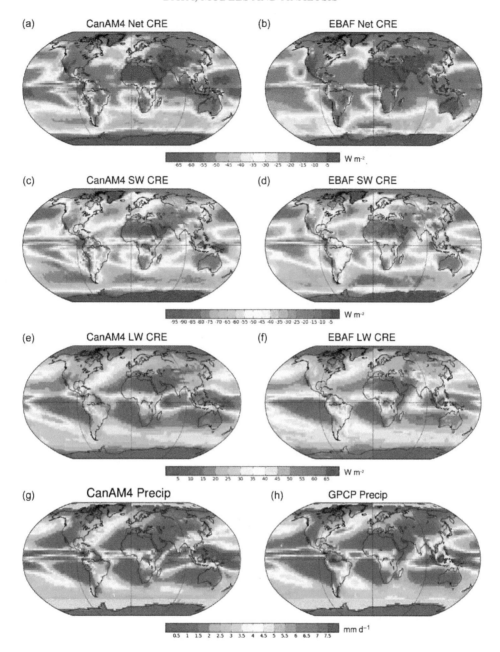

Fig. 5 Mean net, SW, and LW cloud radiative effects from simulations with CanAM4 (left column, panels a, c and e) and observations from CERES EBAF (right column, panels b, d and f) for March 2000 to December 2007. Mean precipitation for CanAM4 (g) and GPCP (h) for January 1979 to December 2007. Grid points with missing data appear as grey areas. For the comparisons, monthly mean observations were first regridded to match the grid used in CanAM4 and accounting for missing data points, as appropriate. Subsequently, results were averaged over the time period for which datasets overlap in time.

region to central America (see Fig. 4). Some of the biases in mean distributions for CRE and precipitation are also associated with biases in the responses of clouds to variations in temperatures, as addressed in the second part of this paper.

6 Summary and conclusions

Parameterizations for clouds, radiation, and other physical processes in the fourth generation Canadian Atmospheric Global Climate Model (CanAM4) have been described.

Numerous changes to parameterizations have resulted in a substantial improvement in model functionality between CanAM4 and a previous version of the model (AGCM3). For instance, previously missing effects of shallow convection, cloud microphysical processes, and aerosol life cycles were added in CanAM4. A noticeable difference between these versions of the model is a larger variability of cloud-related results in CanAM4 which is associated with an increased number and complexity of prognostic parameterizations for clouds and aerosols in CanAM4.

Comparisons of the results of CanAM4 with satellite-based observations for clouds and precipitation give evidence for overall realistic climatological mean results. There is a high consistency between biases in cloud radiative effects and biases in cloud amounts. Biases in the vertical distribution of the clouds apparently play a more subtle role in biases in cloud radiative effects. Biases for clouds and humidity in the tropical free troposphere are likely related to insufficient mixing from cumulus congestus clouds in CanAM4. Low biases in high cloud amounts over the Amazon region in CanAM4 are associated with underestimates in precipitation and biases in cloud radiative effects. Results in other regions also yield consistent biases in mean cloud amount, precipitation, and cloud radiative effects (e.g., the eastern North Atlantic and Indian Ocean).

Improvements to parameterizations for convection and the introduction of a scheme for turbulent kinetic energy are planned in the future to address shortcomings in the simulations of clouds.

Results for clouds are further analyzed in the second part of this paper, which addresses responses of clouds and precipitation to short-term variations in temperature and atmospheric stability.

Acknowledgements

We thank three anonymous reviewers for helpful comments. We further thank Ulrike Lohmann, Phil Austin, Howard Barker, and co-workers for contributions to the development of cloud and aerosol parameterizations, and Steven Lambert and Slava Kharin for providing diagnostic tools and datasets. Helpful comments on the manuscript by Greg Flato and Phil Austin are acknowledged. CERES data were obtained from the NASA Langley Research Center CERES ordering tool at http://ceres.larc.nasa.gov/. ISCCP data were obtained from http://eosweb.larc.nasa.gov/PRODOCS/isccp/table_isccp. html. GPCP precipitation data were provided by the World Meteorological Organization's World Data Centre at NOAA's National Climatic Data Centre, from http://ncdc. noaa.gov/oa/wmo/wdcamet-ncdc.html. Major funding was provided by Environment Canada, the Canadian Foundation for Climate and Atmospheric Sciences (CFCAS), the Climate Change Action Fund (CCAF), and the National Sciences and Engineering Research Council of Canada (NSERC).

Appendix A: Mass fixer

CanAM4 includes a method that locally corrects for any residual changes in globally integrated tracer amounts which result from transport of the hybrid variable in the model. As described in Section 2b, advection of a hybridized instead of a physical variable necessarily leads to sources or sinks of global tracer mass. Different methods for tracer mass correction have been tested. According to the method currently used in CanAM4, the corrected tracer mass mixing ratio,

after the application of the mass fixer, is given by

$$q_{corr}(t) = q(t)[1 + (c - 1)f], \qquad (A1)$$

where $q(t)$ is the initial tracer mass mixing ratio before the correction. The variable c is the ratio of the predicted (i.e., correct) to the initially simulated globally integrated tracer mass,

$$c = \frac{\iiint \left[q(t - \Delta t) + \Delta t \left. \frac{dq}{dt} \right|_{phys} \right] dV}{\iiint q(t)dV},$$

where t is the current time and Δt is the model time step. $dq/dt|_{phys}$ is the time rate of change for the tracer mass mixing ratio between the current and previous time step owing to non-transport processes (i.e., from calculations in the physics part of the model). Non-conservation of global mass from transport calculations implies that $c \neq 1$ in general.

In Eq. (A1), f is used to modulate the magnitude of the mass correction. In AGCM3, $f = 1$ was chosen so the same scaling factor is applied in all grid cells (McFarlane et al., 2006). However, the approach in CanAM4 is to vary the magnitude of the mass correction locally in such a way that the correction tends to be stronger in grid cells that experience large net physical sources or sinks of tracer mass compared to grid cells with weaker sources or sinks. Therefore, corrections for tracer mass are typically much smaller in the stratosphere than in the troposphere for tracer fields that are dominated by emission and removal in the troposphere. In practice, this approach ensures that the impact of the mass correction on local tracer budgets is small relative to the impact of physical processes. In detail,

$$f = (1 - g)k \left| \frac{dq}{dt} \right|_{phys} + g,$$

with

$$k = \frac{\iiint q(t)dV}{\iiint q(t) \left| \frac{dq}{dt} \right|_{phys} dV}$$

$$g = \begin{cases} 0, & \text{if } c \geq 1, \\ \max\left[\min\left(\frac{\frac{1 - \varepsilon}{1 - c} - k \left| \frac{dq}{dt} \right|_{phys,max}}{1 - k \left| \frac{dq}{dt} \right|_{phys,max}}, 1 \right), 0 \right], & \text{if } c < 1, \end{cases}$$

where $|dq/dt|_{phys,max}$ is the maximum of the absolute tendency that occurs for all grid cells. With this approach, $q_{corr} \geq q\varepsilon$, with $\varepsilon = 10^{-9}$. The lower bound for q_{corr} ensures that the tracer mass mixing ratio is large enough in the variable transformation according to Eq. (1). Furthermore, $f = 1$ for $c = \varepsilon$.

Appendix B: Overlap of precipitation with clouds

Precipitation that is produced within a layer of the atmosphere may partly overlap with clouds in grid cells below that layer. The overlap between the precipitation flux and the clouds is variable in the vertical and is parameterized according to the maximum-random overlap rule (Geleyn and Hollingsworth, 1979). The fraction of a grid cell which is affected by precipitation is given by

$$C_{k \to l}^{p} = 1 - (1 - C_k^p) \prod_{m=k+1}^{l} \frac{1 - \max(C_{m-1}^p, C_m^p)}{1 - C_{m-1}^p}. \quad \text{(B1)}$$

Here, k refers to the topmost grid cell and l the lowest grid cell at the bottom of the layer $k \to l$, with $k < l$; C_m^p is the local fraction of the cloud which is affected by precipitation,

$$C_m^p = \begin{cases} C_m, & if \quad P_{m-1} > 0 \\ 0, & \text{otherwise} \end{cases},$$

where P_{m-1} is the local precipitation flux between grid cells $m-1$ and m. Therefore, it is usually true that $C_m^p < C_m$; if not the entire atmospheric column is affected by precipitation. This approach ensures that only clouds that produce rain will actually contribute to the rain flux.

It is further assumed that $C_{k \to l}^p$ can be decomposed into a cloudy-sky and a clear-sky component,

$$C_{k \to l}^p = C_{k \to l}^{p,\text{cld}} + C_{k \to l}^{p,\text{clr}}.$$

$C_{k \to l}^{p,\text{clr}}$ is determined under the assumption of maximum overlap between the precipitation flux and local cloud fraction in each grid cell, so that

$$C_{k \to l}^{p,\text{clr}} = C_{k \to l}^p - C_l^p.$$

For simplicity, precipitation fluxes are assumed to be horizontally homogeneous on all scales within the cloudy- and clear-sky components of grid cells.

Separate equations according to the approach in Eq. (B1) are solved for rain and snow.

References

Abdella, K., & McFarlane, N. (1997). Parameterization of the surface-layer exchange coefficients for atmospheric models. *Journal of the Atmospheric Sciences, 54*, 1850–1867.

Abdella, K., & McFarlane, N. A. (1996). Parameterization of the surface-layer exchange coefficients for atmospheric models. *Boundary-Layer Meteorology, 80*, 223–248.

Ackerman, A. S., Kirkpatrick, M. P., Stevens, D. E., & Toon, O. B. (2004). The impact of humidity above stratiform clouds on indirect aerosol climate forcing. *Nature, 432*, 1014–1017.

Adler, R. F., Huffman, G. J., Chang, A., Ferraro, R., Xie, P.-P., Janowiak, J., ... Nelkin, E. (2003). The version 2.1 global precipitation climatology project (GPCP) monthly precipitation analysis (1979-present). *Journal of Hydrometeorology, 4*, 1147–1167.

Arora, V. K., Scinocca, J. F., Boer, G. J., Christian, J. R., Denman, K. L., Flato, G. M., ... Merryfield, W. J. (2011). Carbon emission limits required to satisfy future representative concentration pathways of greenhouse gases. *Geophysical Research Letters, 38*. doi:10.1029/2010GL046270

Atkinson, R., Baulch, D. L., Cox, R. A., Hampson, J.R. F., & Troe, J. (1989). Evaluated kinetics and photochemical data for atmospheric chemistry: Supplement III. *Journal of Physical and Chemical Reference Data, 88*, 881–1097.

Barker, H. W. (1996). A parameterization for computing grid-averaged solar fluxes for inhomogeneous marine boundary layer clouds. Part I: Methodology and homogeneous biases. *Journal of the Atmospheric Sciences, 53*, 2298–2303.

Barker, H. W., Cole, J. N. S., Morcrette, J.-J., Pincus, R., Räisänen, P., von Salzen, K., & Vaillancourt, P. A. (2008). The Monte Carlo independent column approximation: An assessment using several global atmospheric models. *Quarterly Journal of the Royal Meteorological Society, 134*, 1463–1478.

Bartholomé, E., & Belward, A. S. (2005). GLC2000: A new approach to global land cover mapping from earth observation data. *International Journal of Remote Sensing, 26*, 1959–1977.

Bäumer, D., Lohmann, U., Lesins, G., Li, J., & Croft, B. (2007). Parameterizing the optical properties of carbonaceous aerosols in the Canadian Centre for Climate modeling and analysis atmospheric general circulation model with impacts on global radiation and energy fluxes. *Journal of Geophysical Research, 112*. doi:10.1029/2006JD007319

Bodas-Salcedo, A., Webb, M. J., Bony, S., Chepfer, H., Dufresne, J.-L., Klein, S. A., ... John, V. O. (2011). COSP: Satellite simulation software for model assessment. *Bulletin of the American Meteorological Society, 92*, 1023–1043.

Boer, G. J. (1995). *A hybrid moisture variable suitable for spectral GCMs.* Research Activities in Atmospheric and Oceanic Modelling, report No. 21, WMO/TD 665, World Meteorological Organization, Geneva.

Brasseur, G. P., Hauglustaine, D. A., Walters, S., Rasch, P. J., Müller, J.-F., Granier, C., & Tie, X. X. (1998). MOZART, a global chemical transport model for ozone and related chemical tracers 1. Model description. *Journal of Geophysical Research, 103*, 28265–28290.

Chaboureau, J.-P., & Bechtold, P. (2002). A simple cloud parameterization derived from cloud resolving model data: Diagnostic and prognostic applications. *Journal of the Atmospheric Sciences, 59*, 2362–2372.

Chaboureau, J.-P., & Bechtold, P. (2005). Statistical representation of clouds in a regional model and the impact on the diurnal cycle of convection during tropical convection, cirrus and nitrogen oxides (TROCCINOX). *Journal of Geophysical Research, 110*. doi:10.1029/2004JD005645

Chameides, W. L. (1984). The photochemistry of a remote marine stratiform cloud. *Journal of Geophysical Research, 89*, 4739–4755.

Chepfer, H., Bony, S., Winker, D., Cesana, G., Dufresne, J. L., Minnis, P., ... Zeng, S. (2010). The GCM-oriented CALIPSO cloud product (CALIPSO-GOCCP). *Journal of Geophysical Research, 115*. doi:10.1029/2009JD01225

Cionni, I., Eyring, V., Lamarque, J. F., Randel, W. J., Stevenson, D. S., Wu, F., ... Waugh, D. W. (2011). Ozone database in support of CMIP5 simulations: Results and corresponding radiative forcing. *Atmospheric Chemistry and Physics, 11*, 11267–11292.

Cole, J., Barker, H. W., Loeb, N. G., & von Salzen, K. (2011). Assessing simulated clouds and radiative fluxes using properties of clouds whose tops are exposed to space. *Journal of Climate, 24*, 2715–2727.

Cosby, B. J., Hornberger, G. M., Glapp, R. B., & Ginn, T. R. (1984). A statistical exploration of the relationships of soil moisture characteristics to the physical properties of soils. *Water Resources Research, 20*, 662–690.

Croft, B., Lohmann, U., & von Salzen, K. (2005). Black carbon ageing in the Canadian Centre for Climate modelling and analysis atmospheric general circulation model. *Atmospheric Chemistry and Physics, 5,* 1931–1949.

Dee, D. P., Uppala, S. M., Simmons, A. J., Berrisford, P., Poli, P., Kobayashi, S., … Vitart, F. (2011). The ERA-interim reanalysis: Configuration and performance of the data assimilation system. *Quarterly Journal of the Royal Meteorological Society, 137,* 553–597.

DeMoore, W. B., Sander, S. P., Golden, D. M., Hampson, R. F., Howard, C. J., Ravishankara, A. R., Kolb, C. E., & Molina, M. H. (1992). *Chemical kinetics and photochemical data for use in stratospheric modelling.* JPL Publication 92–20, Jet Propulsion Laboratory, Pasadena, CA.

Dentener, F., & Crutzen, P. J. (1994). A 3-dimensional model of the global ammonia cycle. *Journal of Atmospheric Chemistry, 19,* 331–369.

Dentener, F., Kinne, S., Bond, T., Boucher, O., Cofala, J., Generoso, S., … Wilson, J. (2006). Emissions of primary aerosol and precursor gases in the years 2000 and 1750 prescribed data-sets for AeroCom. *Atmospheric Chemistry and Physics, 6,* 4321–4344.

Dobbie, J. S., Li, J., & Chýlek, P. (1999). Two and four stream optical properties for water clouds and solar wavelengths. *Journal of Geophysical Research, 104,* 2067–2079.

Dufresne, J.-L., Quaas, J., Boucher, O., Denvil, S., & Fairhead, L. (2005). Contrasts in the effects on climate of anthropogenic sulfate aerosols between the 20th and the 21st century. *Geophysical Research Letters, 32,* 1–4.

Fairall, C. W., Davidson, K. L., & Schacher, C. E. (1983). An analysis of the surface production of sea-salt aerosols. *Tellus, Series B, 35,* 31–39.

Fécan, F., Marticorena, B., & Bergametti, G. (1999). Parametrization of the increase of the aeolian erosion threshold wind friction velocity due to soil moisture for arid and semi-arid areas. *Annales Geophysicae, 17,* 149–157.

Forster, P., Ramaswamy, V., Artaxo, P., Berntsen, T., Betts, R., Fahey, D. W., … van Dorland, R. (2007). Changes in atmospheric constituents and in radiative forcing. In S. Solomon, D. Qin, M. Manning, Z. Chen, M. Marquis, K. B. Averyt, M. Tignor, & H. L. Miller (Eds.), *Climate Change 2007: The Physical Science Basis. Contribution of Working Group I to the Fourth Assessment Report of the Intergovernmental Panel on Climate Change* (pp. 129–234). Cambridge, United Kingdom and New York, NY, USA: Cambridge University Press.

Fouquart, Y., & Bonnel, B. (1980). Computation of solar heating of the earth's atmosphere: A new parameterization. *Contributions to Atmospheric Physics* (German title: Beitrage Physique Atmosphere), *53,* 35–62.

Fröhlich, C., & Lean, J. (2004). Solar radiative output and its variability: Evidence and mechanisms. *Astronomy and Astrophysical Reviews, 12,* 273–320.

Fu, Q. (1996). An accurate parameterization of the solar radiative properties of cirrus clouds for climate models. *Journal of Climate, 9,* 2058–2082.

Fu, Q., Yang, P., & Sun, W. B. (1998). An accurate parameterization of the infrared radiative properties of cirrus clouds for climate models. *Journal of Climate, 11,* 2223–2237.

Fyfe, J. C., Merryfield, W. J., Kharin, V., Boer, G. J., Lee, W.-S., & von Salzen, K. (2011). Skillful predictions of decadal trends in global mean surface temperatures. *Geophysical Research Letters, 38.* doi:10.1029/2011GL049508

Geleyn, J., & Hollingsworth, A. (1979). An economical analytical method for computation of the interaction between scattering and line absorption of radiation. *Contributions to Atmospheric Physics, 52,* 1–9.

Gillett, N. P., Arora, V. K., Flato, G. M., Scinocca, J. F., & von Salzen, K. (2012). Improved constraints on 21st-century warming derived using 160 years of temperature observations. *Geophysical Research Letters, 39.* doi:10.1029/2011GL050226

Grant, A. L. M. (2001). Cloud-base fluxes in the cumulus-capped boundary layer. *Quarterly Journal of the Royal Meteorological Society, 127,* 407–421.

Gregory, D., Kershaw, R., & Inness, P. M. (1997). Parametrization of momentum transport by convection. II: Tests in single-column and general circulation models. *Quarterly Journal of the Royal Meteorological Society, 123,* 1153–1183.

Hansen, J., Russell, G., Rind, D., Stone, P., Lacis, A., Lebedeff, S., … Travis, L. (1983). Efficient three-dimensional global models for climate studies: Models I and II. *Monthly Weather Review, 111,* 609–662.

Hansen, J. E., & Travis, L. (1974). Light scattering in planetary atmospheres. *Space Science Reviews, 16,* 527–610.

Heymsfield, A. J. (2003). Properties of tropical and midlatitude ice cloud particle ensembles. Part II: Applications for mesoscale and climate models. *Journal of the Atmospheric Sciences, 60,* 2592–2611.

Hogan, R. J., & Illingworth, A. J. (2000). Deriving cloud overlap statistics from radar. *Quarterly Journal of the Royal Meteorological Society, 126,* 2903–2909.

Hogan, R. J., & Illingworth, A. J. (2003). Parameterizing ice cloud inhomogeneity and the overlap of inhomogeneities using cloud radar data. *Journal of the Atmospheric Sciences, 60,* 756–767.

Holzer, M. (1996). Optimal spectral topography and its effect on model climate. *Journal of Climate, 9,* 2443–2463.

Hurrell, J. W., Hack, J. J., Shea, D., Caron, J. M., & Rosinski, J. (2008). A new sea surface temperature and sea ice boundary dataset for the community atmosphere model. *Journal of Climate, 21,* 5145–5153.

Isotta, F. A., Spichtinger, P., Lohmann, U., & von Salzen, K. (2011). Improvement and implementation of a parameterization for shallow cumulus in the global climate model ECHAM5-HAM. *Journal of the Atmospheric Sciences, 68,* 515–532.

Jin, Z., Charlock, T. P., & Rutledge, K. (2002). Analysis of broadband solar radiation and albedo over the ocean surface at COVE. *Journal of Atmospheric and Oceanic Technology, 19,* 158–601.

Kay, J. E., Baker, M., & Hegg, D. (2006). Microphysical and dynamical controls on cirrus cloud optical depth distributions. *Journal of Geophysical Research, 111.* doi:10.1029/2005JD006916

Kettle, A. J., Andreae, M. O., Amouroux, D., Andreae, T. W., Bates, T. S., Berresheim, H., … Uher, G. (1999). A global database of sea surface dimethylsulfide (DMS) measurements and a procedure to predict sea surface DMS as a function of latitude, longitude, and month. *Global Biogeochemical Cycles, 13,* 399–444.

Khairoutdinov, M., & Kogan, Y. (2000). A new cloud physics parameterization in a large-eddy simulation model of marine stratocumulus. *Monthly Weather Review, 128,* 229–243.

Khairoutdinov, M. F., & Randall, D. A. (2002). Similarity of deep continental cumulus convection as revealed by a three-dimensional cloud-resolving model. *Journal of the Atmospheric Sciences, 59,* 2550–2566.

Klein, S. A., & Jakob, C. (1999). Validation and sensitivities of frontal clouds simulated by the ECMWF model. *Monthly Weather Review, 127,* 2514–2531.

Kohfeld, K. E., & Harrison, S. P. (1999). DIRTMAP: The geological record of dust. *Earth-Science Reviews, 54,* 81–114.

Korolev, A., & Isaac, G. A. (2006). Relative humidity in liquid, mixed-phase, and ice clouds. *Journal of the Atmospheric Sciences, 63,* 2865–2880.

Lamarque, J.-F., Bond, T., Eyring, V., Granier, C., Heil, A., Klimont, Z., … van Vuuren, D. (2010). Historical (1850–2000) gridded anthropogenic and biomass burning emissions of reactive gases and aerosols: Methodology and application. *Atmospheric Chemistry and Physics, 10,* 7017–7039.

Lander, J., & Hoskins, B. J. (1997). Believable scales and parameterizations in a spectral transform model. *Monthly Weather Review, 125,* 292–303.

Laprise, R., & Girard, E. (1990). A spectral general circulation model using a piecewise-constant finite element representation on a hybrid vertical coordinate system. *Journal of Climate, 3,* 32–52.

Larson, V. E., Wood, R., Field, P. R., Golaz, J.-C., Vonder Haar, T. H., & Cotton, W. R. (2001). Small-scale and mesoscale variability of scalars in cloudy boundary layers: One-dimensional probability density functions. *Journal of the Atmospheric Sciences, 58,* 1978–1994.

Lenderink, G., & Holtslag, A. A. M. (2004). An updated length-scale formulation for turbulent mixing in clear and cloudy boundary layers. *Quarterly Journal of the Royal Meteorological Society, 130,* 3405–3427.

Lesins, G., Chýlek, P., & Lohmann, U. (2002). A study of internal and external mixing scenarios and its effect on aerosol optical properties and direct radiative forcing. *Journal of Geophysical Research, 107.* doi:10.1029/2001JD000973

Levkov, L., Rockel, B., Kapitza, H., & Raschke, E. (1992). 3D mesoscale numerical studies of cirrus and stratus clouds by their time and space evolution. *Beitrage Physique Atmosphere, 65,* 35–58.

Li, J. (2002). Accounting for unresolved clouds in a 1D infrared radiative transfer model. Part I: Solution for radiative transfer, including cloud scattering and overlap. *Journal of the Atmospheric Sciences, 59,* 3302–3320.

Li, J., & Barker, H. W. (2005). A radiation algorigthm with correlated-k distribution. Part I: Local thermal equilibrium. *Journal of the Atmospheric Sciences, 62,* 286–309.

Li, J., Curry, C. L., Sun, Z., & Zhang, F. (2010). Overlap of solar and infrared spectra and the shortwave radiative effect of methane. *Journal of the Atmospheric Sciences, 67,* 2372–2389.

Li, J., Dobbie, S., Räisänen, P., & Min, Q. (2005). Accounting for unresolved clouds in a 1-D solar radiative-transfer model. *Quarterly Journal of the Royal Meteorological Society, 131,* 1607–1629.

Li, J., Scinocca, J., Lazare, M., McFarlane, N., von Salzen, K., & Solheim, L. (2006). Ocean surface albedo and its impact on radiation balance in climate models. *Journal of Climate, 19,* 6314–6333.

Li, J., & Shibata, K. (2006). On the effective solar pathlength. *Journal of the Atmospheric Sciences, 63,* 1365–1373.

Li, J., Wong, J. G. D., Dobbie, J. S., & Chýlek, P. (2001). Parameterization of the optical properties and growth of sulfate aerosols. *Journal of the Atmospheric Sciences, 58,* 193–209.

Lindner, T. H., & Li, J. (2000). Parameterization of the optical properties for water clouds in the infrared. *Journal of Climate, 13,* 1797–1805.

Loeb, N. G., Wielicki, B. A., Doelling, D. R., Smith, G. L., Keyes, D. F., Kato, S., … Wong, T. (2009). Toward optimal closure of the earth's TOA radiation budget. *Journal of Climate, 22,* 748–766.

Lohmann, U. (1996). Sensitivität des Modellklimas eines globalen Zirkulationsmodells der Atmosphäre gegenüber Änderungen der Wolkenmikrophysik. *Examensarbeit Nr. 41,* Max-Planck-Institut für Meteorologie, Hamburg, Germany.

Lohmann, U., & Feichter, J. (2005). Global indirect aerosol effects: A review. *Atmospheric Chemistry and Physics, 5,* 715–737.

Lohmann, U., & Roeckner, E. (1996). Design and performance of a new cloud microphysics scheme developed for the ECHAM general circulation model. *Climate Dynamics, 12,* 557–572.

Lohmann, U., von Salzen, K., McFarlane, N., Leighton, H. G., & Feichter, J. (1999). Tropospheric sulphur cycle in the Canadian general circulation model. *Journal of Geophysical Research, 104,* 26833–26858.

Ma, X., von Salzen, K., & Cole, J. (2010). Constraints on interactions between aerosols and clouds on a global scale from a combination of MODIS-CERES satellite data and climate simulations. *Atmospheric Chemistry and Physics, 10,* 9851–9861.

Maahs, H. G. (1983). Kinetics and mechanism of the oxidation of S(IV) by ozone in aqueous solution with particular reference to SO_2 conversion in nonurban tropospheric clouds. *Journal of Geophysical Research, 88,* 10721–10732.

Marticorena, B., & Bergametti, G. (1995). Modeling the atmospheric dust cycle: 1. Design of a soil-derived dust emission scheme. *Journal of Geophysical Research, 100,* 16415–16430.

Martin, L. R. (1984). Kinetic studies of sulfite oxidation in aqueous solution. In J. G. Calvert (Ed.), *Acid precipitation Series Volume 3: SO_2, NO and NO_2 Oxidation Mechanisms: Atmospheric Considerations* (pp. 63–100). Boston: Butterworth Publishers.

McFarlane, N. A., Boer, G. J., Blanchet, J.-P., & Lazare, M. (1992). The Canadian Climate Centre second-generation general circulation model and its equilibrium climate. *Journal of Climate, 5,* 1013–1044.

McFarlane, N. A., Scinocca, J. F., Lazare, M., Harvey, R., Verseghy, D., & Li, J. (2006). *The CCCma third generation Atmospheric General Circulation Model (AGCM3).* Technical report, Canadian Centre for Climate Modelling and Analysis, Victoria, BC, Canada.

Meehl, G. A., Stocker, T. F., Collins, W. D., Friedlingstein, P., Gaye, A. T., Gregory, J. M., … Zhao, Z.-C. (2007). Global climate projections. In S. Solomon, D. Qin, M. Manning, Z. Chen, M. Marquis, K. B. Averyt, M. Tignor, & H. L. Miller (Eds.), *Climate Change 2007: The Physical Science Basis. Contribution of Working Group I to the Fourth Assessment Report of the Intergovernmental Panel on Climate Change* (pp. 747–845). Cambridge, United Kingdom and New York, NY, USA: Cambridge University Press.

Mein, R. G., & Larson, C. L. (1973). Modelling infiltration during steady rain. *Water Resources Research, 9,* 384–394.

Mellor, G. L. (1977). The Gaussian cloud model relations. *Journal of the Atmospheric Sciences, 34,* 356–358.

Merryfield, W. J., McFarlane, N., & Lazare, M. (2003). *A generalized hybrid transformation for tracer advection.* Research activities in atmospheric and oceanic modelling, CAS/JSC WGNE Blue Book, report No. 33, WMO/TD 1161, World Meteorological Organization, Geneva, 13–14.

Monahan, E. C., & MacNiocaill, G. (1986). *Oceanic whitecaps and their role in air-sea exchange processes.* Dordrecht, Holland: D. Reidel Publishing Company.

Monin, A. S., & Obukhov, A. M. (1954). Basic regularity in turbulent mixing in the surface layer of atmosphere. *Trudy Geofizicheskogo Instituta, Akademiya Nauk S.S.S.R., 24,* 163–187.

Morcrette, J.-J. (1984). Sur la paramétrisation du rayonnement dans les modèles de la circulation générale atmosphérique. *Thèse de docotorat,* Université des sciences et techniques de Lille, Lille.

Moss, R. H., Edmonds, J. A., Hibbard, K. A., Manning, M. R., Rose, S. K., van Vuuren, D. P., … Wilbanks, T. J. (2010). The next generation of scenarios for climate change research and assessment. *Nature, 463.* doi:10.1038/nature08823

Murakami, M. (1990). Numerical modeling of dynamical and microphysical evolution of an isolated convective cloud - The 19 July 1981 CCOPE cloud. *Journal of the Meteorological Society of Japan, 68,* 107–128.

Oreopoulos, L., Mlawer, E., Delamere, J., Shippert, T., Cole, J., Fomin, B., … Rossow, W. B. (2012). The continual intercomparison of radiation codes: Results from phase I. *Journal of Geophysical Research, 117.* doi:10.1029/2011JD016821

Pham, M., Müller, J. F., Brasseur, G. P., Granier, C., & Mégie, G. (1989). A three-dimensional study of the tropospheric sulfur cycle. *Journal of Geophysical Research, 100,* 26061–26092.

Pincus, R., Barker, H. W., & Morcrette, J.-J. (2003). A fast, flexible, approximate technique for computing radiative transfer in inhomogeneous cloud fields. *Journal of Geophysical Research, 108.* doi:10.1029/2002JD003322

Pincus, R., Hannay, C., Klein, S. A., Xu, K.-M., & Hemler, R. (2005). Overlap assumptions for assumed probability distribution function cloud schemes in large-scale models. *Journal of Geophysical Research, 110.* doi:10.1029/2004JD005100

Räisänen, P., Barker, H. W., Khairoutdinov, M. F., Li, J., & Randall, D. A. (2004). Stochastic generation of subgrid-scale cloudy columns for large-scale models. *Quarterly Journal of the Royal Meteorological Society, 130,* 2047–2068.

Ramaswamy, V., & Li, J. (1996). A line-by-line investigation of solar radiative effects in vertically inhomogeneous low clouds. *Quarterly Journal of the Royal Meteorological Society, 122,* 1873–1890.

Redelsperger, J. L., Guichard, F., & Mondon, S. (2000). A parameterization of mesoscale enhancement of surface fluxes for large-scale models. *Journal of Climate, 13,* 402–421.

Rockel, B., Raschke, E., & Weyres, B. (1991). A parameterization of broad band radiative transfer properties of water, ice, and mixed clouds. *Beitrage Physique Atmosphere, 64,* 1 12.

Rossow, W. B., & Schiffer, R. A. (1999). Advances in understanding clouds from ISCCP. *Bulletin of the American Meteorological Society, 80,* 2261–2288.

Rossow, W. B., Walker, A. W., Beuschel, D. E., & Roiter, M. D. (1996). *International Satellite Cloud Climatology Project (ISCCP). Documentation of new cloud datasets.* WMO/TD-No. 737, World Meteorological Organization, Geneva.

Rotstayn, L. D. (1997). A physically based scheme for the treatment of stratiform clouds and precipitation in large-scale models. I: Description and evaluation of the microphysical processes. *Quarterly Journal of the Royal Meteorological Society, 123*, 1227–1282.

von Salzen, K., Leighton, H. G., Ariya, P. A., Barrie, L. A., Gong, S. L., Blanchet, J.-P., … Kleinman, L. I. (2000). Sensitivity of sulphate aerosol size distributions and CCN concentrations over North America to SO_x emissions and H_2O_2 concentrations. *Journal of Geophysical Research, 105*, 9741–9765.

von Salzen, K., & McFarlane, N. A. (2002). Parameterization of the bulk effects of lateral and cloud-top entrainment in transient shallow cumulus clouds. *Journal of the Atmospheric Sciences, 59*, 1405–1429.

von Salzen, K., McFarlane, N. A., & Lazare, M. (2005). The role of shallow convection in the water and energy cycles of the atmosphere. *Climate Dynamics, 25*, 671–688.

Sanderson, B. M. (2011). A multimodel study of parametric uncertainty in predictions of climate response to rising greenhouse gas concentrations. *Journal of Climate, 24*, 1362–1377.

Sato, M., Hansen, J. E., McCormick, M. P., & Pollack, J. B. (1993). Stratospheric aerosol optical depths, 1850–1990. *Journal of Geophysical Research, 98*, 22987–22994.

Schneider, E. K., & Lindzen, R. S. (1976). Influence of stable stratification on the thermally driven tropical boundary layer. *Journal of the Atmospheric Sciences, 33*, 1301–1307.

Scinocca, J. F., & McFarlane, N. A. (2000). Anisotrophy in the parameterization of drag due to orographically forced flows. *Quarterly Journal of the Royal Meteorological Society, 126*, 2353–2393.

Scinocca, J. F., & McFarlane, N. A. (2004). The variability of modeled tropical precipitation. *Journal of the Atmospheric Sciences, 61*, 1993–2015.

Scinocca, J. F., McFarlane, N. A., Lazare, M., Li, J., & Plummer, D. (2008). The CCCma third generation AGCM and its extension into the middle atmosphere. *Atmospheric Chemistry and Physics, 8*, 7055–7074.

Slingo, J. M. (1987). The development and verification of a cloud prediction scheme for the ECMWF model. *Quarterly Journal of the Royal Meteorological Society, 113*, 899–927.

Sohn, B. J., Nakajima, T., Satoh, M., & Jang, H.-S. (2010). Impact of different definitions of clear-sky flux on the determination of longwave cloud radiative forcing: NICAM simulation results. *Atmospheric Chemistry and Physics, 10*, 11641–11646.

Spiro, P. A., Jacob, D. J., & Logan, J. A. (1992). Global inventory of sulfur emissions with 1°×1° resolution. *Journal of Geophysical Research, 97*, 6023–6036.

Stewart, J. B. (1988). Modelling surface conductance of pine forest. *Agricultural and Forest Meteorology, 43*, 19–35.

Taylor, K. E. (2001). Summarizing multiple aspects of model performance in a single diagram. *Journal of Geophysical Research, 106*, 7183–7192.

Taylor, K. E., Stouffer, R. J., & Meehl, G. A. (2011). A summary of the CMIP5 experiment design. *CMIP5 internal document*, Retrieved from http://cmip-pcmdi.llnl.gov/cmip5/docs/Taylor_CMIP5_design.pdf, pp. 33.

Taylor, K. E., Stouffer, R. J., & Meehl, G. A. (2012). An overview of CMIP5 and the experiment design. *Bulletin of the American Meteorological Society, 93*, 485–498.

Taylor, K. E., Williamson, D., & Zwiers, F. (2008). AMIP II sea surface temperature and sea ice concentration boundary conditions. Retrieved from PCMDI website: http://www-pcmdi.llnl.gov/projects/amip/AMIP2EXPO SN/BCS/amip2bcs.php

Trenberth, K. E., Fasullo, J. T., & Kiehl, J. (2009). Earth's global energy budget. *Bulletin of the American Meteorological Society, 90*, 311–323.

Verseghy, D. (1991). CLASS—A Canadian land surface scheme for GCMS. 1. soil model. *International Journal of Climatology, 11*, 111–133.

Verseghy, D. L. (2000). The Canadian Land Surface Scheme (CLASS): Its history and future. *Atmosphere-Ocean, 38*, 1–13.

Verseghy, D. L., McFarlane, N. A., & Lazare, M. (1993). A Canadian land surface scheme for GCMs: II. Vegetation model and coupled runs. *International Journal of Climatology, 13*, 347–370.

Webb, M., Senior, C., Bony, S., & Morcrette, J.-J. (2001). Combining ERBE and ISCCP data to assess clouds in the Hadley Centre, ECMWF and LMD atmospheric climate models. *Climate Dynamics, 17*, 905–922.

Webb, III, T., Bartlein, P. J., Harrison, S. P., & Anderson, K. H. (1993). Vegetation, lake levels, and climate in eastern North America for the past 18,000 years. In H. E. Wright, Jr., J. E. Kutzbach, T. Webb, III, W. F. Ruddiman, F. A. Street-Perrot, & P. J. Bartlein (Eds.), *Global Climates since the Last Glacial Maximum* (pp. 415–467). Minneapolis, MN: University of Minnesota Press.

Williams, K. D., & Webb, M. J. (2009). A quantitative performance assessment of cloud regimes in climate models. *Climate Dynamics, 33*, 141–157.

Wood, R. (2005). Drizzle in stratiform boundary layer clouds. Part II: Microphysical aspects. *Journal of the Atmospheric Sciences, 62*, 3034–3050.

Xie, S., Xu, K.-M., Cederwall, R. T., Bechtold, P., Del Genio, A. D., Klein, S. A., … Zhang, M. (2002). Intercomparison and evaluation of cumulus parametrizations under summertime midlatitude continental conditions. *Quarterly Journal of the Royal Meteorological Society, 128*, 1095–1136.

Yuan, J., Fu, Q., & McFarlane, N. (2006). Tests and improvements of GCM cloud parameterizations using the CCCma SCM with the SHEBA data set. *Atmospheric Research, 82*, 222–238.

Zhang, G. J., & McFarlane, N. A. (1995). Sensitivity of climate simulations to the parameterization of cumulus convection in the CCC-GCM. *Atmosphere-Ocean, 3*, 407–446.

Zhang, G. J., & Wu, X. (2003). Convective momentum transport and perturbation pressure field from a cloud-resolving model simulation. *Journal of the Atmospheric Sciences, 60*, 1120–1139.

Zhang, M. H., Lin, W. Y., Klein, S. A., Bacmeister, J. T., Bony, S., Cederwall, R. T., … Zhang, J. H. (2005). Comparing clouds and their seasonal variations in 10 atmospheric general circulation models with satellite measurements. *Journal of Geophysical Research, 110*. doi:10.1029/2004JD005021

Zhang, Y., Rossow, W. B., Lacis, A. A., Oinas, V., & Mishchenko, M. I. (2004). Overlap assumptions for assumed probability distribution function cloud schemes in large-scale models. *Journal of Geophysical Research, 109*. doi:10.1029/2003JD004457

Zhu, P., Bretherton, C., Köhler, M., Cheng, A., Chlond, A., Geng, Q., … Stevens, B. (2005). Intercomparison and interpretation of single-column model simulations of a nocturnal stratocumulus-topped marine boundary layer. *Monthly Weather Review, 133*, 2741–2758.

Sensitivity of Climate Simulations to the Parameterization of Cumulus Convection in the Canadian Climate Centre General Circulation Model

G.J. Zhang

California Space Institute Scripps Institution of Oceanography University of California at San Diego, La Jolla, California

and

Norman A. McFarlane

Canadian Centre for Climate Modelling and Analysis University of Victoria, Victoria

ABSTRACT *A simplified cumulus parameterization scheme, suitable for use in GCMs, is presented. This parameterization is based on a plume ensemble concept similar to that originally proposed by Arakawa and Schubert (1974). However, it employs three assumptions which significantly simplify the formulation and implementation of the scheme. It is assumed that an ensemble of convective-scale updrafts with associated saturated downdrafts may exist when the atmosphere is locally conditionally unstable in the lower troposphere. However, the updraft ensemble is comprised only of those plumes which are sufficiently buoyant to penetrate through this unstable layer. It is assumed that all such plumes have the same upward mass flux at the base of the convective layer. The third assumption is that moist convection, which occurs only when there is convective available potential energy (CAPE) for reversible ascent of an undiluted parcel from the sub-cloud layer, acts to remove CAPE at an exponential rate with a specified adjustment time scale.*

The performance of the scheme and its sensitivity to choices of disposable parameters is illustrated by presenting results from a series of idealized single-column model tests. These tests demonstrate that the scheme permits establishment of a quasi-equilibrium between large-scale forcing and convective response. However, it is also shown that the strength of convective downdrafts is an important factor in determining the nature of the equilibrium state. Relatively strong down-drafts give rise to an unsteady irregularly fluctuating state characterized by alternate periods of deep and shallow convection.

The effect of using the scheme for GCM climate simulations is illustrated by presenting selected results of a multi-year simulation carried out with the Canadian Climate Centre GCM using the new parameterization (the CONV simulation). Comparison of these results with those for a climate simulation made with the standard model (the CONTROL simulation, as documented by McFarlane et al., 1992) reveals the importance of other parameterized processes in determining the ultimate effect of introducing the new convective scheme. The radiative response to changes in the cloudiness regime is particularly important in this regard.

RÉSUMÉ *On présente un schéma simplifié de paramétrisation des cumulus, utilisable dans les MCG. La paramétrisation est basée sur un concept de panache semblable à celui proposé par Arakawa et Schubert (1974). Toutefois, il utilise trois hypothèses qui simplifient significativement la formulation et la mise en oeuvre du schéma. On suppose qu'un ensemble de courants ascendants à l'échelle convective et les courants ascendants saturés associés sont possibles lorsque l'atmosphère est localement conditionnellement instable dans la basse troposphère. L'ensemble des courants ascendants ne comprend cependant que les panaches qui ont suffisamment de flottabilité pour pénétrer la couche instable. On suppose que tout panache semblable a le même flux de masse ascendant à la base de la couche convective. La troisième supposition est que la convection humide, présente seulement lorsqu'il y a de l'énergie convective potentielle disponible (CAPE) pour une montée réversible d'une particule non diluée de la couche de bas nuages, enlève la CAPE à un taux exponentiel avec une échelle de temps spécifiquement ajustable.*

On illustre la performance du schéma et sa sensibilité aux choix de paramètres disponibles en présentant les résultats d'une série de tests idéalisés de modèle à simple colonne. Ces tests montrent que le schéma permet d'établir un quasi-équilibre entre le forçage à grande échelle et la réponse convective. Cependant, on montre aussi que la force des courants descendants convectifs est un facteur important pour déterminer la nature de l'état d'équilibre. Les courants descendants relativement forts entraînent un état fluctuant irrégulièrement instable caractérisé par des périodes alternées de convection profonde et mince.

On montre l'effet de l'utilisation du schéma dans les simulations climatiques du MCG en présentant des résultats sélectionnés d'une simulation de plusieurs années à l'aide du MCG du Centre climatologique canadien qui utilise les nouvelles paramétrisations (CONV). La comparaison de ces résultats avec ceux d'une simulation climatique produite par le modèle standard (simulation CONTROL documentée par McFarlane et al., 1992), montre l'importance d'autres processus paramétrés pour déterminer l'effet ultime de l'introduction du nouveau schéma convectif. A ce sujet, la réponse radiative aux changements dans le régime de nébulosité est particulièrement importante.

1 Introduction

It has been recognized for decades that the effects of cumulus clouds must be parameterized in numerical weather prediction and general circulation models. However, there is not a consensus regarding the representation of these effects in large-scale models. The parameterization schemes currently in use range in complexity from simple moist convective adjustment schemes that are similar to that proposed by Manabe et al. (1965) almost three decades ago to complicated mass flux schemes utilizing and elaborating the basic concepts set forth by Arakawa and Schubert (1974). These also include adaptations of the parameterization concepts of Kuo (1965, 1974) and in some cases (e.g., Tiedtke, 1989) the chosen parameterization scheme employs the mass flux approach while using moisture convergence closure conditions as suggested by Kuo (1974). The validity of this closure approach has recently been questioned (Emanuel, 1991) and, in any event, is not required to ensure a close relationship between large-scale convergence and latent heat release due to moist convection. For example, Gregory and Rowntree (1990) show that such is the case for simulations made with their bulk cumulus parameterization scheme in which the initial upward mass flux is specified in terms of the parcel buoyancy in the lower part of the convective layer.

Recently Betts (1986) proposed an adjustment procedure based on the concept (supported to some extent by observations) that moist convection acts to relax the atmosphere toward a state in which the large-scale virtual temperature structure in the lower troposphere (below the freezing level) is close to that for a parcel undergoing reversible moist ascent. Since this reference state is not an unstable one, persistent convection requires that the tendency for convective processes to stabilize the atmosphere be balanced by the counteracting effects of other processes acting on larger scales. Such a quasi-equilibrium assumption is also a fundamental part of the Arakawa and Schubert (1974) parameterization scheme and is also utilized in a distinctly different manner in the buoyancy sorting scheme recently proposed by Emanuel (1991).

A simple moist convective adjustment scheme is used in the Canadian Climate Centre general circulation model (CCC GCM). This model simulates the current climate reasonably well (McFarlane et al., 1992 hereafter denoted as MBBL; Boer et al., 1992). Nevertheless, there are deficiencies in the simulated climate. One of the more pronounced of these is the one that motivated the work presented in this paper, namely that the tropical troposphere is systematically colder than the observed climatological state. Although there are several other factors that may contribute to this deficiency, the use of the moist convective adjustment to parameterize convective effects is at least partly responsible. As in other such schemes, unstable air at each layer is mixed with that in the adjacent layer above it, so that heat and moisture are transported upward while latent heat is released from the lower layer when it is found to have a relative humidity in excess of some specified threshold value. One consequence of such localized exchange is that heat and moisture are not effectively transported from the planetary boundary layer into the conditionally stable regions of the upper troposphere.

The purpose of this paper is to present a parameterization of penetrative cumulus convection that alleviates the above deficiencies in climate simulations made with the CCC GCM. This parameterization scheme utilizes some of the basic concepts set forth by Arakawa and Schubert (1974) to construct a bulk representation of the effects of an ensemble of cumulus clouds. In particular, it is assumed that these effects can be represented in terms of an ensemble of entraining updrafts with an associated evaporatively driven ensemble of convective-scale downdrafts.

A parameterization scheme based on such an ensemble concept can not be constructed in closed form without invoking a set of assumptions that enable determination of the contributions of each member of the ensemble to the total convectivescale vertical mass flux. A fundamental part of the original formulation of Arakawa and Schubert is the quasi-equilibrium assumption which achieves this by requiring that the consumption of potential energy by the convective ensemble exactly balances the rate at which it is produced by larger scale processes. This assumption is implemented in practice by requiring that the action of the ensemble be such as to restore the work function (defined ignoring liquid water loading) for each member of the ensemble to a prescribed equilibrium value. These equilibrium values, specified as a function of the depths of the individual updrafts, are believed to be characteristic (Lord et al., 1982).

Moorthi and Suarez (1992) have shown that, in some circumstances, the standard implementation of the Arakawa-Schubert scheme as described by Lord et al. (1982) gives rise to an artificial elimination of cloud types from the ensemble. They have proposed an alternative, more economical, formulation in which relaxation to a reference state is accomplished over a period of several dynamical time steps by invoking single members of the ensemble at each of those steps and requiring that their work functions be only partially adjusted toward the specified characteristic values.

The parameterization scheme presented here was also motivated by a desire for simplicity and economy. These features are obtained by invoking two assumptions that significantly simplify implementation of the scheme. The first of these specifies the distribution of the individual drafts that are used to construct the bulk updrafts and downdrafts utilized in the scheme. It is assumed that these all have the same initial mass flux (cloud-base mass flux for updrafts, draft-top mass flux for downdrafts) and characteristic fractional entrainment rates limited to a range of values that depends on the large-scale thermodynamical structure of the atmosphere.

The second assumption is that, in the absence of other effects, cumulus convection acts to relax the atmosphere to a state that is neutrally buoyant for undiluted reversible ascent of a parcel whose initial equivalent potential temperature is equal to the large-scale mean value in the sub-cloud layer (assumed to be contained within the planetary boundary layer). A quasi-equilibrium can be established when production of convective available energy by other processes balances consumption of this quantity by moist convection.

In this paper the mathematical formulation of the new convective parameterization is presented and some features of its response to larger scale forcing conditions are illustrated by presentation of results from a simple idealized column model in which convective instability is generated and maintained by imposing a surface energy flux in association with idealized radiative cooling and large-scale ascent/descent profiles. The rest of the paper is devoted to illustration and discussion of the effects of the new convection scheme on simulations of the general circulation made with the Canadian Climate Centre GCM.

2 Convective effects on large-scale temperature and moisture fields

Cumulus convection affects the large-scale temperature and moisture fields through subgrid-scale transport and condensation. As indicated in the introductory remarks, it is assumed here that these processes can be parameterized in terms of a quasisteady ensemble of updrafts and downdrafts which, in accordance with the well known representation of Arakawa and Schubert (1974), gives the following formulation for the effects of cumulus convection on the larger scale temperature and moisture fields:

$$C_p \left(\frac{\partial T}{\partial t} \right)_{cu} = \frac{1}{\rho} \frac{\partial}{\partial z} (M_u S_u + M_d S_d - M_c S) + L(c-e) \quad (1a)$$

$$\left(\frac{\partial q}{\partial t} \right)_{cu} = -\frac{1}{\rho} \frac{\partial}{\partial z} (M_u q_u + M_d q_d - M_c q) + e - c \quad (1b)$$

where the net vertical mass flux within the convective region, M_c, is made up of upward (M_u) and downward (M_d) components. Here, c and ε are respectively the large-scale mean rates of condensation and of evaporation; q, q_u and q_d are

respectively the large-scale, and the convective scale updraft and downdraft components of the specific humidity field; and S, S_u, S_d are respectively the corresponding values of dry static energy (defined in the usual way as $S_{u,d} = C_p T_{u,d} + gz$).

In cumulus convection, high moist static energy air is transported out of the sub-cloud layer in updrafts and lower moist static energy air is imported into this layer by downdrafts. In the absence of evaporation of precipitation below cloud base, these processes give rise to temperature and moisture tendencies in the sub-cloud layer that are represented as

$$C_p \left(\rho \frac{\partial T}{\partial t} \right)_m = \frac{1}{\Delta z_b} (M_b[S(z_b) - S_u(z_b)] + M_d(z_b)[S(z_b) - S_d(z_b)]) \quad (2a)$$

$$\left(\rho \frac{\partial q}{\partial t} \right)_m = \frac{1}{\Delta z_b} (M_b[q(z_b) - q_u(z_b)] + M_d(z_b)[q(z_b) - q_d(z_b)]) \quad (2b)$$

where Δz_b is the depth of the sub-cloud layer, quantities with subscript m are vertical mean values for the sub-cloud layer, z_b is the height of the cloud base and M_b is the cloud updraft mass flux at this level. The first terms on the right-hand sides of equations (2) represent the net gain or loss of heat and moisture caused by updrafts exiting from the sub-cloud layer, and the second terms are those caused by downdrafts detraining into the sub-cloud layer.

Equations (1) and (2) are derived using the assumption, common to all such parameterization schemes, that quantities characterizing the mean environment of cumulus clouds are closely approximated by the mean values for an area, typically several thousand kilometers in magnitude, that include both the relatively small one occupied by cumulus clouds and the significantly larger one surrounding them. In the following sections, no distinction is made between such large-scale mean values and those typical of the mean environment of cumulus clouds.

3 The cloud model

a Updraft Ensemble

The bulk mass flux and the associated cloud properties in equations (1) and (2) represent the effects of an ensemble of cumulus clouds that may exist in a larger scale circulation regime. As in Arakawa and Schubert (1974), these effects are represented in terms of an ensemble of cloud updrafts and downdrafts. Each updraft is represented as an entraining plume with a characteristic fractional entrainment rate. Detrainment is confined to a thin layer near the plume top where the mass carried upward is expelled into the environment. The air detrained from updrafts is assumed to be saturated and have the same temperature as the environmental air. In the absence of liquid water, this air would be nearly neutral or slightly positively buoyant with respect to the local environmental air. However, since the detrained cloudy air may also contain liquid water, it may in fact be slightly negatively buoyant with respect to environmental air, or a mixture of cloudy and environmental air may in fact be slightly negatively buoyant.

This simple updraft formulation has been used extensively in observational studies of the coupling between convective systems and the larger scale flow and thermodynamic fields. The pioneering study of Ogura and Cho (1973) showed that in the tropics, a bimodal distribution of updrafts, comprised of deep and shallow populations, is typically needed to account for the observed large scale response to moist convection. This finding has been confirmed by later studies although it has also been found (Johnson, 1976) that allowing for convective and meso-scale downdrafts reduces somewhat the requirement for a shallow updraft population. Moreover Johnson (1976, 1980) presents evidence indicating an out-of-phase relationship between maxima in deep and shallow convective activity when the effects of convective-scale downdrafts are taken into account.

While fewer studies have been done for extra-tropical systems, the available literature suggests that they may differ somewhat from tropical ones. For example, the studies of Lewis (1975) and Zhang and McFarlane (1991) suggest that CAPE values are often larger and shallow convection is less common.

In constructing the parameterization scheme outlined below, we have chosen to focus attention on representing the effects of deep convective clouds. In the tropics, the deep convective cloud population, if represented in terms of the simple one-dimensional entraining plume updraft model, can be regarded as being comprised of those plumes which penetrate through the conditionally unstable region which typically extends up to the middle of the troposphere. A characteristic feature of this region is that the large-scale mean ambient and saturated moist static energies (h, h^*) decrease with height. Thus we assume that the top of the shallowest of the convective updrafts is no lower than the minimum in h^*.

Mainly for the sake of simplicity, we assume that the updraft population is comprised of a set of plumes that have a common value for the cloud-base mass flux. Although the studies of Ogura and Cho (1973) and Yanai et al. (1976) suggest that there may be a bias in favor of the deeper clouds, they do support the contention that there is a reasonably broad distribution of deep updrafts. The study of Johnson (1980) suggests a significant variation in the deep cloud population throughout easterly waves that were commonly observed in the GATE region. In the trough region of the composite wave, where deep convective activity is most pronounced, detrainment occurs over a broad region from just above the level of minimum h^* to the top of the convective layer, suggesting a relatively broad spectrum of deep clouds. Thus the assumption of a uniform distribution of deep updrafts is not inconsistent with many of the observational results. It also provides useful analytical simplifications which enable us to construct a simple bulk cloud model based on the updraft ensemble concept. In particular, the ensemble cloud updraft mass flux has a simple analytical form:

$$M_u = M_b \int_0^{\lambda_D} \frac{1}{\lambda_0} e^{\lambda(z-z_b)} d\lambda = \left(\frac{M_b}{\lambda_0 (z-z_b)} \right) (\exp(\lambda_D(z)(z-z_b))-1) \quad (3)$$

where M_b is the ensemble cloud base updraft mass flux. Interpreting this as an ensemble mass flux allows us to identify $\lambda_D(z)$ as the fractional entrainment rate of the updraft plume that detrains at height z, while λ_0 is the maximum entrainment rate allowed. The mass flux at the base of the convective layer is $(M_b d\lambda/\lambda_0)$ for that plume with fractional entrainment rate in the interval $(\lambda, \lambda + d\lambda)$.

The quantity λ_D is defined by the requirement that the temperature of the plume that detrains at height z is equal to the environment value, which is ensured by requiring that

$$h_b - h^* = \lambda_D \int_{z_b}^{z} [h_b - h(z')] \exp[\lambda_D(z'-z)] dz' \quad (4)$$

where h^* is the moist static energy for a saturated state having the environmental temperature and pressure.

In equation (4) the moist static energy profile, $h(z)$, is presumed to be known. As this equation is non-linear in λ_D, it must be solved by an appropriate numerical method. In practice, a Newton-Raphson procedure generally converges rapidly with a reasonable first guess. We have also found that a quite accurate non-iterative approximate solution can, in most circumstances, be obtained using the reversion of series procedure outlined in the Appendix.

For a typical conditionally unstable atmospheric column, $\lambda_D(z)$ has a maximum in the lower troposphere and decreases with height. The maximum entrainment rate λ_0, which is associated with the shallowest plume of the ensemble, must be set. This is done by making use of the fact that h^* typically has a minimum in the middle troposphere in a conditionally unstable atmosphere. The ensemble is limited to updrafts that detrain at or above the height of minimum h^*. Thus if z_0 is this height, $\lambda_0 = \lambda_D(z_0)$ is the maximum entrainment rate in the cloud ensemble. This choice insures that detrainment is confined to the conditionally stable region of the atmospheric column.

In practice, $\lambda_D(z)$ is evaluated numerically for each layer in the model by proceeding downward from the top of the convective region in the upper troposphere and in typical cases, it is single-valued above z_0 and decreases with height. However, if h^* has secondary minima between the top of the convective region and z_0, $\lambda_D(z)$ will not vary monotonically with height. In such cases, $\lambda_D(z)$ is set to the larger of the values obtained for the ambient layer and the layer directly above it.

Detrainment is confined to regions where $\lambda_D(z)$ decreases with height so that the total detrainment $D_u(z)$ is zero below z_0 and

$$D_u(z) = -\frac{M_b}{\lambda_0} \frac{\partial \lambda_D}{\partial z} \exp(\lambda_D(z)(z-z_b)) \quad (5a)$$

above z_0. The corresponding total entrainment rate is given by

$$E_u = \frac{M_b}{\lambda_0} \int_0^{\lambda_D} \lambda e^{\lambda(z-z_b)} d\lambda = \frac{\partial M_u}{\partial z} - D_u. \quad (5b)$$

It is assumed that the air that detrains from the convective ensemble is saturated at the environmental temperature and pressure. All condensation occurs within updrafts (i.e., $c = C_u$).

148

Condensed water detrained from updrafts evaporates locally into the environment. With these assumptions, the equations governing the budgets of ensemble mean dry static energy S_u, water vapor mixing ratio q_u and cloud water content l for cloud updrafts are

$$\frac{\partial}{\partial z}(M_u S_u) = (E_u - D_u)S + LC_u \qquad (6a)$$

$$\frac{\partial}{\partial z}(M_u q_u) = E_u q - D_u q^* - C_u \qquad (6b)$$

$$\frac{\partial}{\partial z}(M_u l) = -D_u l_d + C_u - R_r \qquad (6c)$$

where C_u and R_r are respectively the net condensation and conversion from cloud liquid water to rainwater in the updrafts and l_d is the liquid water content of detrained cloudy air, for simplicity taken here to be identical to the ensemble mean cloud water, l.

Conversion of cloud water to rainwater is represented using the empirical formulation discussed by Lord (1982). In this formulation, precipitation production is taken to be proportional to the vertical flux of cloud water in the updraft as

$$R_r = C_0 M_u l \qquad (7)$$

where $C_0 = 2 \times 10^{-3}$ m^{-1}. Lord (1982) notes that this simple representation of precipitation production has been found to give good agreement with observed liquid water contents in observational studies of tropical convective systems.

While the focus of our parameterization is on deep clouds, there are circumstances in which the cumulus cloud population is predominantly shallow. This is typically true in the trade-wind convective regime where large-scale subsidence acts to suppress deep convection. The shallow cloud population which develops in this regime acts predominantly to moisten and cool the upper part of the convective layer and to warm and dry the lower part. Precipitation rates are relatively small so that the net convective heating is also small. In an overall sense, cumulus clouds act predominantly to re-distribute heat and moisture within the convective layer. Studies of the heat and moisture budgets of the Trade Wind boundary layer (e.g. Betts, 1975, 1982; Cho, 1977) raise doubts about the adequacy of the simple one-dimensional steady-state updraft model as a means of representing the effects of cumulus clouds in this regime. Cloud life-cycle effects, cloud top entrainment and non-penetrative downdrafts may all be important. However, the broad features of cumulus cloud effects can be captured at least qualitatively by limiting the production of precipitation in shallow updrafts. Cooling and moistening in the upper part of the cloud layer is then represented as *in-situ* evaporation of detrained cloud water while heating and drying in the lower part is associated with compensating subsidence. In reality, of course, most of the evaporation probably occurs in shallow downdrafts originating near cloud tops.

The practice of inhibiting rain water production within a region directly above cloud base is also used in other parameterization schemes (Tiedtke, 1989; Emanuel, 1991; Slingo et al., 1994) but the depth of this region is usually specified and invariant with location. Here we suppress conversion of cloud water to rain water below the freezing level so that shallow convection does not generate precipitation. In practice, the pressure at the freezing level is usually near 600 mb in the tropics and the sub-cloud layer is typically of the order of 500 m deep so that the depth of the region in which precipitation production is inhibited is usually in the range of 3 to 4 km in the tropics. These values are comparable to those specified in other schemes. However, we do recognize that this admittedly *ad-hoc* limitation of precipitation production to clouds which extend above the freezing level may have shortcomings. It does not account for circumstances (e.g. over the tropical oceans) where weak environmental winds and high liquid water contents may promote substantial precipitation from shallower clouds. Some authors (e.g. Slingo et al., 1994) attempt to account for this in a simple and practical way by assigning fixed but different depths for non-precipitating cumulus clouds over land and ocean surfaces.

b Downdraft Ensemble

Downdrafts have been found important in the interaction between cumulus convection and larger scale processes. Indeed, the marked cooling and drying of the sub-cloud layer that often follows organized convection in both the tropics and the mid-latitudes (e.g. Frank and McBride, 1989; Zhang and McFarlane, 1991) is likely the result of detrainment of low energy air imported from mid-tropospheric levels by downdrafts.

In this study, downdrafts are assumed to exist when there is precipitation production in the updraft ensemble. The downdrafts start at or below the bottom of the updraft detrainment layer, which as indicated above is also the layer in which the minimum in h^* occurs and penetrates down to the sub-cloud layer. Knupp (1987) provides some observational evidence to support such a choice for the level at which penetrative downdrafts are established. There are also studies (Betts, 1976) which lend some support to the idea that penetrative downdrafts originate closer to the base of the convective layer. In the present study, the downdraft initiation level is chosen to be coincident with the minimum of h if lower than the base of the detrainment layer. This assumption is used by Grell et al. (1991) in a version of the Arakawa-Schubert scheme which includes saturated downdrafts. Their study suggests, on the basis of semi-prognostic tests, that initiation of saturated downdrafts at this level, so that they can transport lower-energy air more efficiently to the sub-cloud layer, may be preferable to initiating them higher up in the convective region. Detrainment of downdrafts is confined to the sub-cloud layer. As for updrafts, it is assumed that the bulk effect of downdrafts can be represented in terms of an ensemble of plumes that have the same mass flux at the top of the downdraft region. The initial downward mass flux for this ensemble is taken to be proportional to the cloud base mass flux for the updraft ensemble. With these assumptions, the ensemble downdraft mass flux is given by

$$M_d = \frac{-\alpha M_b}{\lambda_m (z_D - z)} \{\exp(\lambda_m (z_D - z)) - 1\} \qquad (8)$$

where λ_m is the maximum downdraft entrainment rate.

The proportionality factor, α, is chosen so as to ensure that the strength of the downdraft ensemble is constrained both by the availability of precipitation and by the requirement that the net mass flux at cloud base be positive (i.e. upward). The total precipitation, produced in updrafts between cloud base (z_b) and the top of the convective layer (z_T) is given by

$$PCP = \int_{z_b}^{z_T} R_r \, dz. \qquad (9)$$

It is assumed that the downdrafts are maintained in a saturated state by evaporation of rain water. The evaporation, per unit of α, in downdrafts is given by

$$EVP = \left(\frac{1}{\alpha}\right) \int_{z_b}^{z_D} \left(\left(\frac{\partial}{\partial z} M_b q_d - E_d q \right) dz \right). \qquad (10)$$

The proportionality factor is chosen to be of the form

$$\alpha = \mu \left[\frac{PCP}{PCP + EVP} \right]. \qquad (11)$$

This form for α ensures that the downdraft mass flux vanishes in the absence of precipitation. It also ensures that evaporation in the downdraft can not exceed a fraction μ of the precipitation. We have also chosen to impose the constraint that the net mass flux at cloud base must be positive. This constraint is ensured by requiring that

$$\mu \le \lambda_m (z_D - z_b)/(\exp[\lambda_m (z_D - z_b)] - 1). \qquad (12)$$

For most of the work presented below, μ has the fixed value of 0.2 and, consistent with the idea that downdrafts have substantially higher entrainment rates than the corresponding updrafts, $\lambda_m = 2\lambda_0$ (as in Zhang and Cho, 1991) but is also constrained to be no larger than

$$\lambda_{max} = 2/(z_D - z_b). \qquad (13)$$

This constraint on λ_m ensures that the magnitude of the downdraft mass flux at cloud base is no larger than 65% of the corresponding updraft mass flux. As demonstrated below, single column tests show that the factor $PCP/(PCP + EVP)$ in equation (11) is in general sufficiently small to ensure that the actual magnitude of the downdraft mass flux at cloud base is substantially less than that of the associated updraft.

Johnson (1978) finds considerable variation in the optimal value of α within tropical waves. In the southern portion of the composite wave that he analyses, the values of this quantity are as large as 0.67 and the corresponding ensemble downdraft mass flux at cloud base has a magnitude which is about one-half that of the corresponding updraft mass flux. In contrast, values of α in the central portion of the wave do not exceed 0.5.

Some interesting responses were found in single-column model tests when μ is given by the equality condition in equation (12). As illustrated below, the typical magnitude of the downdraft mass flux at cloud base for this choice is close to onehalf that of the updraft mass flux, typical of the larger values shown in Johnson's (1978) paper. In some circumstances, this "strong" downdraft formulation gives rise to a temporally variable convective regime in which both deep and shallow updrafts are present. Selected results from these tests are presented in section 5 below.

4 Closure condition

As discussed in section 1 above, the closure condition used here is based on the assumption that cumulus clouds consume convective available potential energy at a specified rate. We define this quantity as

$$A = g \int_{IL}^{EL} \frac{(\Theta_{vp} - \Theta_v)}{\Theta_v} dz \qquad (14)$$

where $\Theta_{vp} = \Theta_p (1 + 1.608 q_p - q_m)$ is the virtual potential temperature of undiluted air parcels following a reversible moist adiabat; and $\Theta_v = \Theta(1 + 0.608 q)$ is the large-scale virtual potential temperature. Subscript p is for parcel properties and IL and EL are, respectively, the parcel-originating level and the equilibrium level in the upper troposphere where air parcels become non-buoyant, and q_m is the water vapor mixing ratio at IL.

It is shown in section 2 that the large-scale temperature and moisture changes in both the cloud layer and sub-cloud layer are linearly proportional to the cloud base updraft mass flux M_b. Therefore, the CAPE change due to convection can be symbolically written as:

$$\left(\frac{\partial A}{\partial t} \right)_c = -M_b F \qquad (15)$$

where F is the CAPE consumption rate per unit cloud base updraft mass flux. This quantity, which is required to be positive for cumulus convection to exist, is evaluated by formally differentiating the integrand in (14) with respect to t and substituting from equations (1) and (2), evaluated for unit mass flux using the updraft/downdraft properties determined as discussed in the foregoing section. The contributions associated with variations of Θ_{vp} are due to the effects of convection on the sub-cloud layer as expressed in equations (2) while those due to variation of Θ_v are due to the large-scale effects of moist convection in the cloud layer as expressed in equation (1).

The closure condition in this study is that CAPE is consumed at an exponential rate by cumulus convection with characteristic time scale τ. This assumption implies that

$$M_b = \frac{A}{\tau F}. \qquad (16)$$

Sensitivity to the choice of the time scale is discussed in the following section in the context of a simple single-column model.

5 Results from single column tests

This section is devoted to presentation of results from tests of the convective scheme using a simple single-column model in which convective available energy is generated by the combined action of surface fluxes of heat and moisture, radiative cooling, and large-scale ascent. These processes are specified in simple idealized ways. Pressure is used as a vertical coordinate with equally spaced layers, 50 mb in depth, between the 1000 and 50 mb levels. The large-scale variables (T and q) are specified at the mid-layer pressure levels 975, 925, ..., 75 mb. The cloud properties M_u, M_d, M_c, S_u, S_d, q_u, q_d, l are at the levels of the layer interfaces and the rest of the cloud properties (D_u, E_u, E_d, C_u) are at the mid-layer levels.

A time step of 20 minutes is used for all of the model experiments in this study. Initial values of the large-scale variables (T, q) are specified using a mean sounding from the GATE data.

a "Convective-Radiative" Equilibrium

In the "convective-radiative" equilibrium test, the atmosphere is forced by a specified cooling rate in the troposphere ("radiative cooling") and sensible and latent heat fluxes at the surface. Dry adjustment is activated when the atmosphere is statically unstable. Large-scale condensation occurs when the atmosphere reaches saturation. The radiative cooling rate is constant from 150 mb to the surface, with a magnitude of 2×10^{-5} (°C/s), and decreases linearly to zero at and above the 100 mb level.

Sensible and latent heat fluxes at the surface (assumed to have a fixed pressure of 1000 mb) are specified using the following bulk aerodynamic formulae:

$$F_s = C_D V (Q_s - Q_1) \rho_s \qquad (17a)$$

$$F_q = C_D V (q_s - q_1) \rho_s \qquad (17b)$$

where $C_D = 2 \times 10^{-3}$ is the drag coefficient, V is a typical surface wind speed, Q_s, and Q_1 are respectively the potential temperature at the surface and the lowest model layer, q_1 and q_s are respectively the lowest model-level water vapor mixing ratio and the surface saturation value. Here we use $Q_s = 300$ K, $V = 5$ m s^{-1}.

Numerous tests were made with this simple column model. Selected results from these tests are summarized in Table 1 and some of the experiments are discussed more fully below. The results of the first three experiments, summarized in Table 1, reveal that a nearly steady equilibrium state, in which there is a finite amount of CAPE, is reached when the simple radiative convective configuration of the model is used. The time required to achieve equilibrium and the magnitude of the final value of CAPE both depend on the choice of the adjustment time, τ. A consequence of the proportionality between the equilibrium value of CAPE and τ is that the cloud-base mass flux and the convective precipitation rate are almost independent of τ at equilibrium.

Figure 1 displays the vertical profiles of the cloud mass flux, entrainment and detrainment averaged over the last 500 hours of the time integration for the first experiment in Table 1. The convective heating and drying profiles are shown in Figure 2 and the corresponding profiles of h, h^*, specific humidity and dry static energy (normalized by C_p) are shown in Figure 3. The weak shallow downdraft, relatively weak entrainment and strong detrainment in a narrow region at the top of the convective layer are related to the vertical structure of h and h^*. The marked difference between the initial and equilibrium structures of these quantities, especially with regard to the moist static energy, is a consequence of a pronounced drying and a moderate warming in the convective layer. Since surface evaporation is the only imposed moisture source in this experiment, the equilibrium state is one in which there is no net moisture tendency in the convective layer. The humidity structure required to ensure this is one in which the vertical gradient of q vanishes within the convective layer. Within the detrainment layer (which is confined to a single model layer), the environment is saturated and detrainment of cloud water is balanced by large-scale condensation. Thus the specific humidity is constant everywhere within the convective layer and equal to the saturation value at the mid-point of the detrainment layer.

The shallowness of downdrafts in the experiment discussed above is a direct consequence of the assumption that they start from the level of minimum h. It is noteworthy that, despite being rather shallow and weak, they do account for most of the convective cooling and drying in the sub-cloud layer, as shown in Figure 2.

The quasi-steady equilibrium illustrated in Figure 2 is associated exclusively with deep convective clouds. Although we have not attempted to devise a distinct parameterization for shallow cumulus clouds, we have found that using the

TABLE 1. Summary of column model experiments.

| Exp. No. | τ (hours) | Downdraft strength (μ) | Max. Top (Mb) | Large-scale vertical vel. | CAPE(J/Kg) | | | |
					Char.	max.	mean	min.
1	2	0.2	200	No	Steady		800	
2	4	0.2	200	No	Steady		1400	
3	6	0.2	200	No	Steady		1800	
4	2	Str.	150	No	Fluct.	1000	650	400
5	2	Str.	150	Periodic	Periodic and Fluct.	1100		100

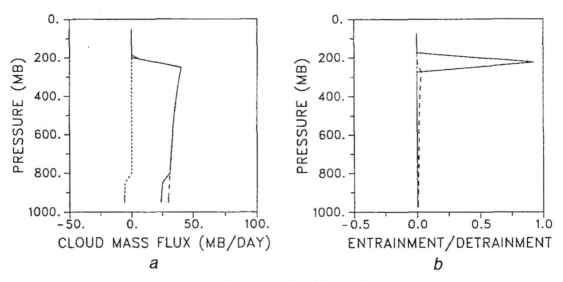

Fig. 1 Vertical profiles of (a) net cloud mass flux (solid line), updraft mass flux (dotted line) and downdraft mass flux (dashed line), units: mb day; (b) updraft mass detrainment (solid line), updraft mass entrainment (dashed line) and downdraft mass entrainment (dotted line), units: $10^{-5} \, s^{-1}$.

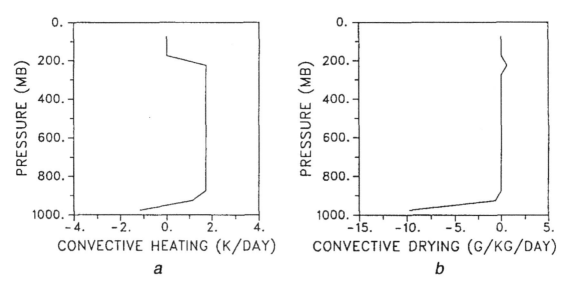

Fig. 2 Vertical profiles of (a) convective heating (units: °K/day) and (b) convective drying (units: g/kg/day).

"strong downdraft" formulation, as defined in section 4 above, results in a much less regular temporal behavior, in which shallow clouds occur spontaneously. This is illustrated in Figure 4a which shows the time evolution of CAPE for the "strong downdraft" case (exp. 4 in Table 1). The nearly steady equilibrium found in the previous experiment is replaced by one in which CAPE varies irregularly from about 400 J/kg to 1000 J/kg. The mass flux and the entrainment/detrainment profiles, averaged over the last 500 hours (Figures 4b and 4c), are distinctly different as well. The updraft mass flux has a lower-level maximum at 800 mb and a second upper-level maximum at 200 mb, the lower one being the result of a greater abundance of shallower convection events in this experiment. This is also reflected in the mean vertical

structure of the updraft detrainment which has two distinct maxima, the lower one (at 700 mb) being associated with shallower convective events. The mean downdraft is associated exclusively with the deeper convective events and is relatively stronger than in the previous experiments.

The lower half of the troposphere is moister (Figure 5b) due to moistening by the liquid water and moist air detrained from shallow convection. The cooling associated with evaporation of detrained liquid water is reflected in the mean vertical structure of the moist static energy (Figure 5a), which has an inversion near the top of the region occupied by shallow convection.

The results shown in the above experiments clearly demonstrate the potential importance of downdrafts in the convective stabilization process. The cooling and drying of the

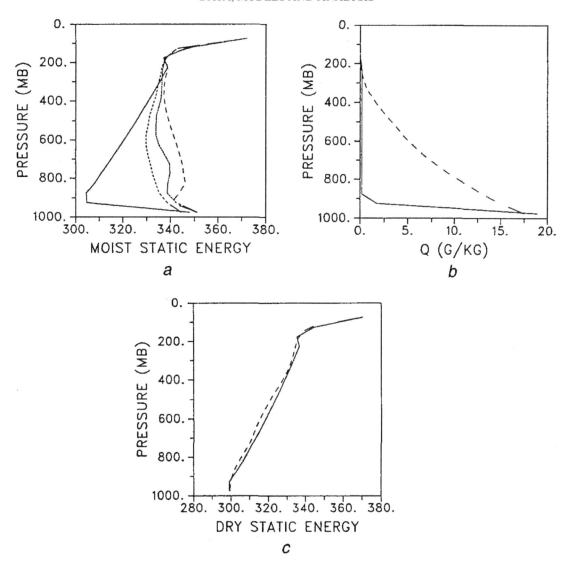

Fig. 3   Vertical profiles of (*a*) the moist static energy and its saturation value; the solid line and the long dashed line are the averaged actual and saturation moist static energy, the short-dashed and the dotted lines are those at the initial time, respectively, units: J/kg; (*b*) averaged water vapor mixing ratio (solid line) and initial state (dashed line), units: g/kg; (*c*) averaged dry static energy (normalized by the heat capacity of the air $C_p = 1004.6$ J/kg/K) (solid line) and initial state (dashed line), units: °K.

sub-cloud layer due to downdrafts acts to stabilize the atmosphere column. However, for the simple column model used here, this effect is relatively small for weaker downdrafts and is easily offset by surface heat and moisture fluxes and the imposed radiative cooling. It is possible in these circumstances to achieve a quasi-steady equilibrium between the convective stabilization and the forced large-scale destabilization. On the other hand, strong downdrafts provide a powerful mechanism for stabilizing the atmospheric column. When these are present, deep convection may occur intermittently because it consumes CAPE more rapidly than CAPE is built up by the large-scale processes. In the time periods between deep convective events, shallow convection develops and prevails until the convectively unstable layer becomes sufficiently deep that a deep convection episode can again occur.

The presence of both deep and shallow convection produces the double peak structure in time-averaged cloud updraft mass flux and detrainment profiles as shown in Figure 4. In this respect, our results are qualitatively similar to those obtained by Satoh and Hayashi (1992) using a simple radiative-convective model in which cumulus convection is parameterized in terms of a bulk updraft cloud model in which the mass flux is assumed to be proportional to the amount of convective instability. They found that, for large proportionality constants (i.e. larger mass fluxes for a given amount of convective instability), the temporal evolution is irregular and is characterized by the intermittent occurrence of both shallow and deep convection. However, for a sufficiently small proportionality constant, only deep convection exists and a steady equilibrium is obtained.

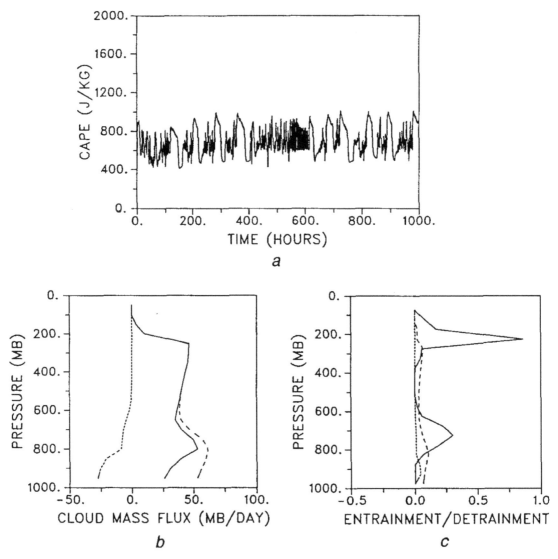

Fig. 4 (*a*) Time evolution of CAPE for the stronger downdraft experiment, (*b*) and (*c*) are the same as Figs 1*a* and 1*b* except with stronger downdrafts.

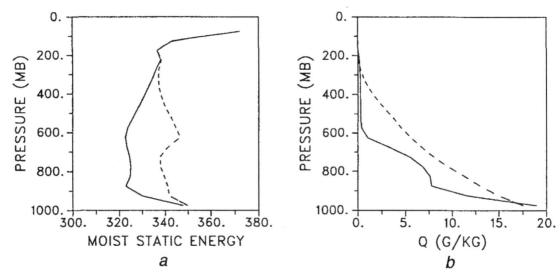

Fig. 5 Vertical profiles for the strong downdraft experiment, (*a*) Moist static energy (solid line) and its saturation value (dashed line), (*b*) water vapor mixing ratio (solid line) and its initial value (dashed line). Units are as in Fig. 3.

b *Effects of Imposed Large-scale Vertical Motion*

It is well known that deep convection in the atmosphere is enhanced in regions of large-scale ascent and suppressed in regions of subsidence. In this section, results are presented from experiments in which, in addition to the prescribed "radiative cooling", a large-scale vertical motion profile is also imposed. This is specified in terms of the total time derivative of pressure ($\omega = dp/dt$) which is specified in the following form:

$$\omega = 4\omega_m(p_0 - p)(p - p_1)/(p_0 - p_1)^2 \qquad (18)$$

where $p_0 = 1000$ mb and $p_1 = 150$ mb. The maximum vertical motion, ω_m, occurs at 575 mb.

The maximum ascent typical of mean conditions in convergence zones of the tropics is of the order of a few tens of millibars per day. Wavelike synoptic-scale disturbances in the tropics impose a fluctuating component on this mean state such that large-scale ascent is enhanced in the trough regions of the waves while vertical ascent is suppressed in the ridges. To simulate the convective response in such circumstances, we impose the following temporally periodic amplitude factor,

$$\omega_m = -75[1 + \cos(2\pi t/\tau_s)] \qquad (19)$$

where the period, τ_s, of the imposed sinusoidal variation is specified to be 5 days and the units of ω_m are (mb)/(day).

In general we have found that the convective response is controlled by the imposed vertical motion even when strong downdrafts are allowed in the convection scheme. Figure 6 shows the temporal evolution of CAPE for such a case

(exp. 5). The rather irregularly fluctuating character associated with strong downdrafts is strongly modulated by a periodic component of temporal variation which is the response to that in the imposed large-scale ascent profile. A similar behavior is found for convective precipitation (not shown) which has a periodic component that varies between 5 mm/day in the undisturbed sector and 18 mm/day in the disturbed part. It also has irregular fluctuations imposed on this periodic component. The effect of these is to make the precipitation range somewhat larger (between 0 and 22 mm/day). The predominance of the periodic component of the response, evidence of a quasi-equilibrium between the imposed large-scale forcing and the convective response, is established quickly.

Other features of the quasi-equilibrium nature of the convective response to the imposed large-scale forcing are apparent in Figure 7, which depicts the ensemble mean profiles of the convective mass flux and associated updraft detrainment rate for disturbed and undisturbed periods. These profiles were obtained by separately averaging the relevant quantities for 8 periods of enhanced ascent (disturbed periods) and the associated 8 periods of reduced ascent (undisturbed periods).

The vertical structure of ensemble mean mass flux (Figure 7a) during disturbed periods is similar to that of the imposed vertical motion field, reflecting the fact that large-scale forcing during these periods is predominantly due to the imposed vertical ascent. However, during undisturbed periods, the mean vertical ascent is much weaker so that the imposed radiative cooling makes a relatively larger contribution to the large-scale forcing. The time-averaged vertical profile of the

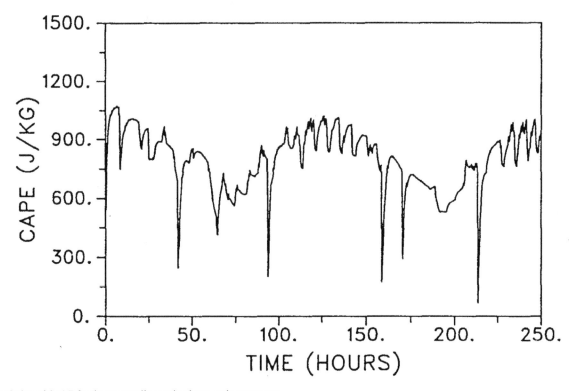

Fig. 6   Evolution of CAPE for the temporally varying large-scale ascent case.

mass flux in this regime has two maxima, indicating a bi-modal character for the convection.

The detrainment profiles also clearly reflect the different characteristics of the two regimes. The double maxima present in the mean detrainment profile for the undisturbed periods indicates that in this regime the convective response consists of contributions from both deep and shallow elements. In contrast, the broad single maximum of mean detrainment that is found in the upper troposphere during the disturbed periods is characteristically associated with a predominance of deep convection. The presence of significantly non-zero detrainment values throughout the region between the 500 and 800 mb levels confirms, however, that shallower convective elements are also sufficiently abundant to make a significant contribution. The broader spectrum of convective elements excited during disturbed periods is the convective response needed to offset the enhanced destabilization in the middle part of the troposphere associated with the imposed large-scale vertical motion field.

The overall behavior of the convective activity in this simple simulation of the response to tropical wave forcing is broadly similar to that shown by Johnson (1980), although unlike his scheme, ours does not include mesoscale downdrafts. Although shallower clouds are present in our simulation, the shallow cloud population differs from that found by Johnson (1980). In his results, shallow clouds are predominant in the undisturbed portion of the wave and detrain mainly in the lower troposphere. This type of response is suppressed in our simulation largely as a result of our assumption that no detrainment occurs in the region below the minimum in h^*.

We have also carried out an experiment designed to simulate the effects of large-scale subsidence in suppressing deep convection. In this case, only non-precipitating shallow convection occurs. The effects of convection are entirely confined to the region below the 700 mb level. The heating and drying profiles (not shown) for this case show that cooling and moistening occur in the upper part of the convective layer in association with detrainment of cloud water while warming and drying occur in the lower part. These processes maintain a well-defined convective boundary layer capped by a sharp inversion, qualitatively similar to what is found in the trade-wind regime.

This is of course an expected result which is consistent with those demonstrated in other studies (e.g. Tiedtke, 1989; Gregory and Rowntree, 1990). We also find, as did those authors, a tendency of the scheme to produce near-saturated conditions at the base of the inversion, possibly indicative of insufficient vertical mixing at the top of the convective layer. We have not, however, followed their lead in attempting to correct this in the simple column model by forcing the scheme to overshoot into the inversion. Gregory and Rowntree (1990) note that such split final detrainment schemes can, with quite reasonable assumptions, lead to excessive moistening and cooling in the inversion layer. Moreover, the simple column model used here does not include the vertical mixing from background diffusion and radiative feed-back processes that are part of the full GCM.

c Moist Convective Adjustment

The results of the simple column model experiments discussed in the foregoing sections have illustrated the effect of the penetrative cumulus convection scheme in a variety of situations. It is natural to ask how different these results are from those that are obtained with the simple column model using the moist convective adjustment scheme currently employed in the CCC GCM (MBBL). Figure 8 shows the thermodynamic structure averaged over the last 500 hours for the "convectiveradiative" equilibrium obtained with the simple column model using this moist convective adjustment scheme. The moist static energy profiles (Figure 8a) indicate that the equilibrium atmospheric state is conditionally unstable throughout the troposphere. The minimum of the moist static energy is located in the upper troposphere near the 175 mb level, whereas in the corresponding deep convection simulation it is located below the 800 mb level.

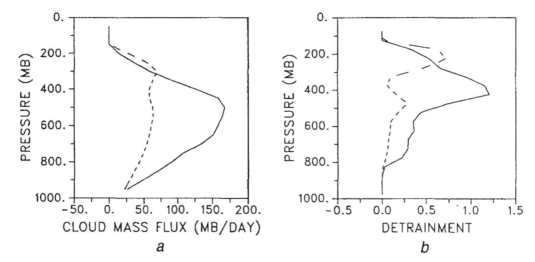

Fig. 7 Vertical profiles of net mass flux and detrainment for the large-scale ascent case. Solid, disturbed cases, dashed, undisturbed periods.

In contrast to the penetrative convection scheme, the convective adjustment scheme is local in nature. It checks the stability of an air column, layer by layer, starting from the lowest pair of layers. Air is exchanged between adjacent layers where the lapse rate at the interface between them exceeds an imposed threshold value. This threshold value is chosen to be a linear combination of the wet and dry adiabatic values with weighting such that the wet adiabatic value is used when the relative humidity of the lower layer is 100%. Any supersaturation which remains after, or results from, this exchange process is condensed and precipitated. Clearly this process is much less efficient than the penetrative convective parameterization at transporting moist static energy from the lower levels to the upper troposphere. Comparison between the moist static energy and its saturation value indicates that the atmosphere is very close to saturation whereas the air in the deep convection runs is far from saturated.

Figure 8b displays the time-averaged moisture structure in comparison with the initial state. The averaged state is moister below the 500 mb level and drier above. This is qualitatively different from what is found in the penetrative convection runs, where the air is much drier in the middle and lower troposphere (except the sub-cloud layer) and moister in the upper troposphere. The dry static energy profiles (Figure 8c) show that the averaged atmosphere is less dry-statically stable and colder than the initial state in the upper troposphere. It is slightly warmer just above the sub-cloud layer. This is again in qualitative contrast to the results for the deep convection runs where the air is warmer almost everywhere, especially in the upper half of the troposphere. Due to the inefficiency in the vertical transport through convective adjustment, the moisture provided through surface evaporation tends to accumulate in the lower troposphere.

6 Effect on GCM climate simulations

The new cumulus parameterization has also been tested in the second generation version of the CCC GCM (McFarlane et al., 1992, hereafter referred to as MBBL). This version of the GCM also contains a simple thermodynamic oceanic mixed-layer module coupled to a thermodynamic sea-ice module. The internal heat transports in the oceanic mixed layer and the vertical heat transport at the lower boundary of the sea ice are specified as outlined in MBBL. These specified internal heat

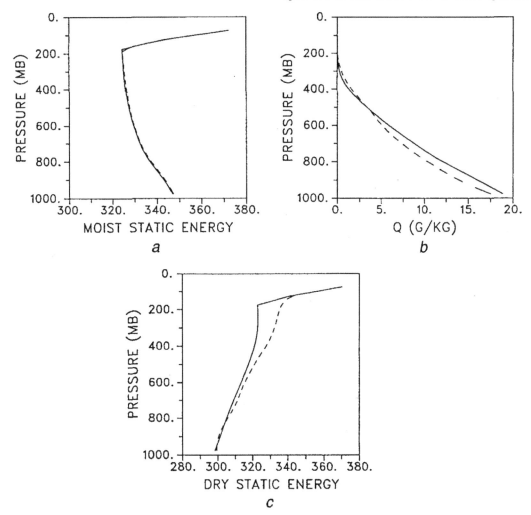

Fig. 8   Vertical profiles of (*a*) moist static energy (solid line) and its saturation value (dashed line); (*b*) water vapor mixing ratio (solid line) and its initial value (dashed line); and (*c*) dry static energy (solid line) and its initial value (dashed line) for the moist convective adjustment experiment.

transports are retained in unmodified form in the version of the GCM which contains the new cumulus parameterization. This allows direct comparison to the multi-year climate simulation (hereafter referred to as the CONTROL simulation) that has been documented by MBBL.

For the experiment in which the cumulus parameterization is used for the climate simulation, the moist convective adjustment scheme in the operational version of the CCC GCM is modified so that it acts only to remove dry static instabilities. Apart from these modifications, the version of the GCM used is as described by MBBL.

The climate simulation using the penetrative convection scheme was initiated from a May 1 restart file of the multi-year simulation made with this standard version of the GCM. The experimental simulation (hereinafter referred to as the CONV simulation) was terminated at the end of February of the third simulated year. A significant amount of "climate drift" is apparent in the last two years of the CONV simulation. Reasons for this are discussed below. The results shown in the following sub-sections are ensemble means for the December–February and June–August periods unless otherwise indicated.

a Zonal Mean Temperature and Moisture Fields

The CONTROL climate simulation has a cold bias throughout the tropical troposphere in all seasons. Implementing the penetrative cumulus parameterization removes this bias and, in fact, replaces it with a warm bias in the upper troposphere. This is illustrated in Figure 9 which shows the zonally averaged differences between simulated temperature fields and 9-year ensemble means of objectively analysed fields produced at the ECMWF.

Figure 10, which depicts the corresponding specific humidity anomalies, shows that implementation of the penetrative convection scheme also results in a drier lower troposphere in the tropics. This effect is quite consistent with the results of the column tests discussed above. However, in this case it enhances the dry anomaly which is already present in the CONTROL simulation. In part, this feature may be associated with an inadequate representation of the effects of shallow convection. Results of other shorter simulations (not shown) also indicate that this dry anomaly is reduced, but not removed, by implementing the "strong downdrafts" version of the convection scheme.

Other experiments also reveal, not surprisingly, that the vertical structure of the humidity field in the lower troposphere

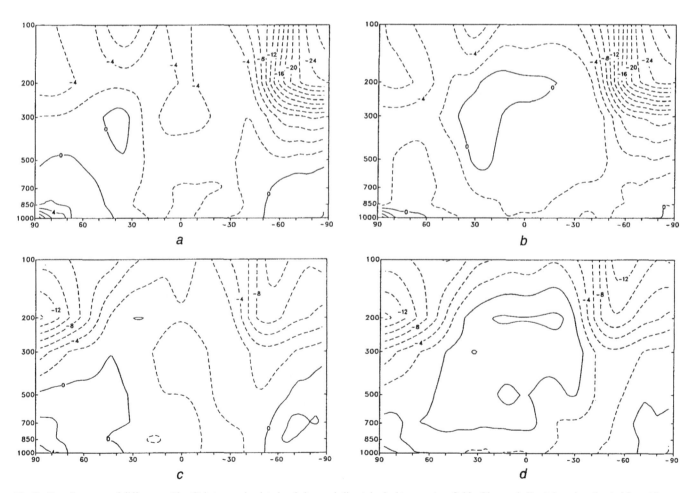

Fig. 9   Zonally averaged differences (deg C) between simulated and observed climatological temperature fields. Observed climatology is estimated from 10 years of ECMWF objectively analyzed fields, (a) Control, Dec.–Feb.; (b) CONV, Dec.–Feb.; (c) CONTROL, June–Aug.; (d) CONV, June–Aug.

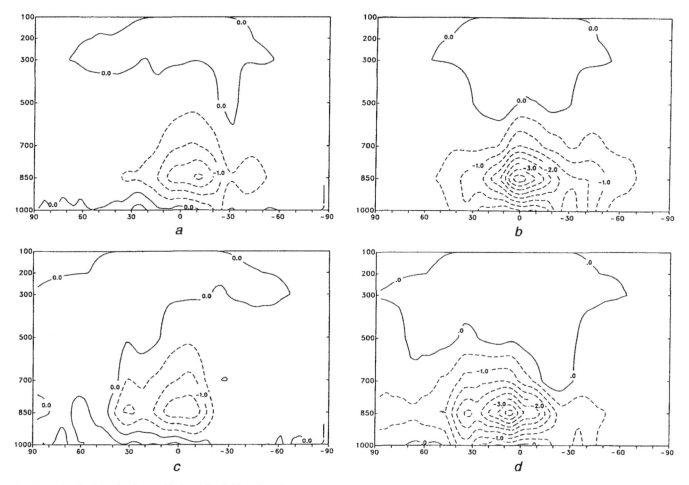

Fig. 10 As in Fig. 9 but for the specific humidity field. Units: g/kg.

is sensitive to vertical resolution and the treatment of vertical transfer of moisture due to turbulent mixing in the boundary layer. The zonally averaged relative humidity structure for both simulations shows a tendency for trapping moisture near the surface, suggesting that in both experiments there may be insufficient vertical transfer of moisture due to a combination of boundary-layer processes and shallow cumulus convection.

b Precipitation and Tropical Circulation Patterns

The main features of the climatological precipitation patterns are captured in both simulations (Figure 11). There are, however, pronounced local differences between the two simulations of this field. A particularly interesting difference is in the simulation of the Indian and Southeast Asian monsoon regime during the boreal summer. This regime is characterized by large amounts of precipitation over India and South-East Asia associated with the low-level jet that crosses the equator off the east coast of Africa and flows northeastward over the Arabian sea and eastward over India. Although both simulations reproduce the main features of this circulation regime, the CONTROL simulation has a deficit

of precipitation over India and an excessive eastward and northward extension of the precipitation pattern over the southwestern Pacific ocean. These features are alleviated by introduction of the penetrative convection parameterization, although comparison with the observed climatological patterns suggests that there is still a deficit of precipitation over western India in the CONV simulation. This may in part be the result of inadequately resolving important topographic features such as the Western Ghat mountains.

The changes to summer monsoon simulations in these experiments are very similar to those of the single-season experiments discussed in greater detail by Zhang (1994). The model used in that study is a version of the CCC model with higher vertical resolution, a different land surface scheme and with sea surface temperatures specified from climatological values. The monsoon circulation regime for a simulation made without any explicit parameterization of moist convection is compared to one made with the same penetrative convection scheme as was used in the present study. An interesting conclusion of that study is that simulations made with no explicit moist convection scheme are qualitatively similar to those made with the moist convective adjustment scheme of the CONTROL run in the present study. In both cases, the

159

eastward extension of the monsoon precipitation pattern is associated with excessive westerly flow (not shown here for the CONTROL run) in the lower troposphere and excessive moisture convergence. These features are not present in the simulations made with the penetrative convection scheme.

c Clouds, Radiation and the Surface Energy Budget
An important part of the simulated climate change resulting from introduction of the penetrative convection scheme is associated with the radiative response to changes in the spatial distribution of clouds and their optical properties. Although the

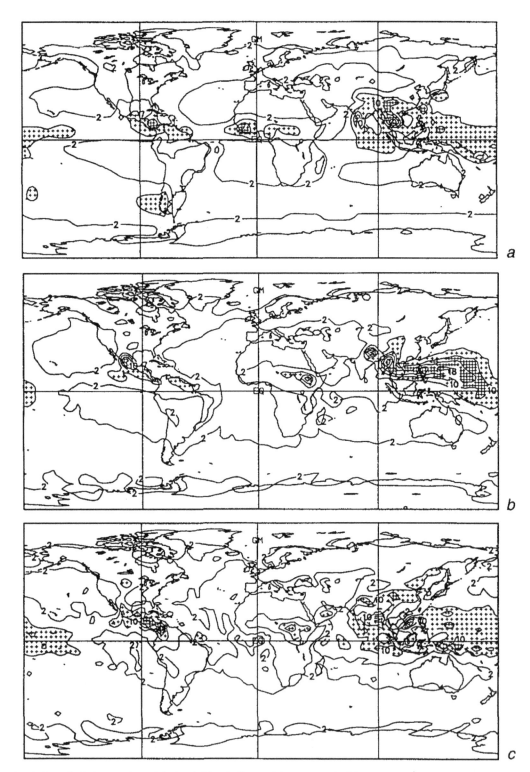

Fig. 11   Simulated and climatological (Jeager) precipitation fields, (*a*) Observed June–Aug.; (*b*) CONTROL June–Aug.; (*c*) CONV June–Aug.; (*d*) Observed Dec.–Feb.; (*e*) CONTROL Dec.–Feb.; (*f*) CONV Dec.–Feb. Units: mm/day.

globally averaged total fractional cloud cover is almost the same in both experiments (near 51% in all seasons), there are significant differences in the spatial distribution of cloudiness. Figure 12 shows the zonally averaged vertical distribution of the change in cloudiness for the summer and winter seasons. In general,

introduction of the penetrative convection scheme results in substantial reduction of low cloud amounts, particularly in the tropics, and an associated increase in high cloud amounts.

As discussed by McFarlane et al. (1992), the optical properties of clouds are determined in the model as a function of

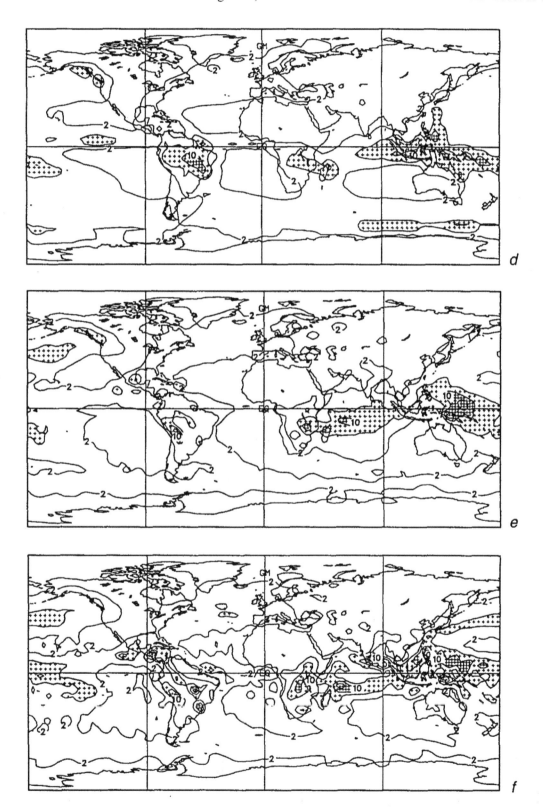

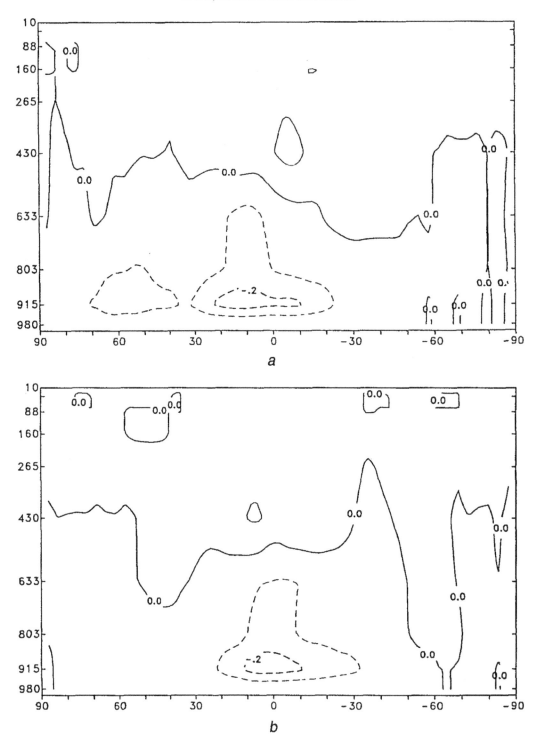

Fig. 12   Zonally averaged change in simulated cloudiness (CONV-CONTROL). (*a*) June–Aug.; (*b*) Dec.–Feb.

the cloud liquid water content which, in turn, is parameterized in terms of the adiabatic water content that may occur with a specified amount of local lifting. Thus it varies with local conditions and allows changes in cloudiness to be accompanied by changes in the optical properties of the clouds. This is reflected, for example, in the changes in the zonally averaged representative cloud optical depth depicted in Figure 13. This quantity is defined as

$$[\tau] = [C\tau]/[C]$$

where τ is the local optical depth of a 100 mb thick cloudy layer and C is the local fractional cloud amount.

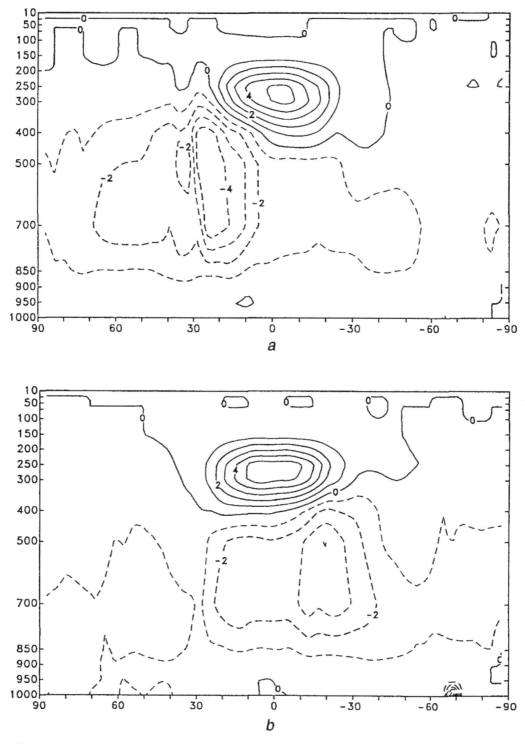

Fig. 13   As in Fig. 12 but for the representative optical depth.

The changes in cloud amount and optical properties have important consequences for the radiation budget. This is illustrated for the boreal summer in Figure 14. The out-going long-wave radiation (OLR) is reduced as a consequence of the more abundant high clouds (Figure 14a). These clouds also reflect more solar radiation resulting in a general increase in the planetary albedo (Figure 14b). Although these changes have opposing effects on the net radiation at the top of the atmosphere, their combined effect is to reduce both the net incoming radiation in the summer hemisphere and the net outgoing radiation in the winter hemisphere (Figure 14c). Qualitatively similar effects are also found for the winter (December–February) season.

At the surface of the earth, the net radiative energy flux is generally downward at all but the high latitudes of the winter hemispheres. This downward net radiative flux is partially offset by the energy loss from the surface due to upward fluxes of sensible and latent heat associated with evaporation at the surface. The combined effect of these processes is to give a net downward flux of energy at the surface in the summer hemisphere and a net upward flux in the winter hemisphere.

Figure 15, which depicts the changes (CONV-CONTROL), shows that introduction of the convection scheme results in generally increased magnitudes for the zonally averaged values of the individual flux terms. An exception is the sharply decreased net radiative flux at high latitudes in the summer hemispheres. This is due to the increased surface albedo associated with increased sea ice extent in the CONV simulation. Elsewhere the changes in sensible and latent heat fluxes tend to compensate those for the net radiative flux. The combined effect of these changes is reduced net downward flux at extra-tropical latitudes in the summer hemispheres and reduced net upward flux at high latitudes in the winter hemisphere.

The heat capacity of the land surface layer is small and there is no net flux of heat within the layer. Thus the seasonally averaged net flux at the surface is nearly zero over land surfaces. An exception may be found in high latitudes and/or over high terrain where a small net heat flux may be required to balance the latent heat exchange associated with freezing and thawing of soil moisture and melting of snow. These considerations apply also to the CONV simulation. Thus differences in the net surface flux are also generally small over land surfaces.

In contrast, in regions occupied by open ocean and sea ice, the net surface flux is of significant magnitude and spatially variable. In these regions, the local net energy gain/loss due to surface fluxes is balanced by changes of the storage of thermal

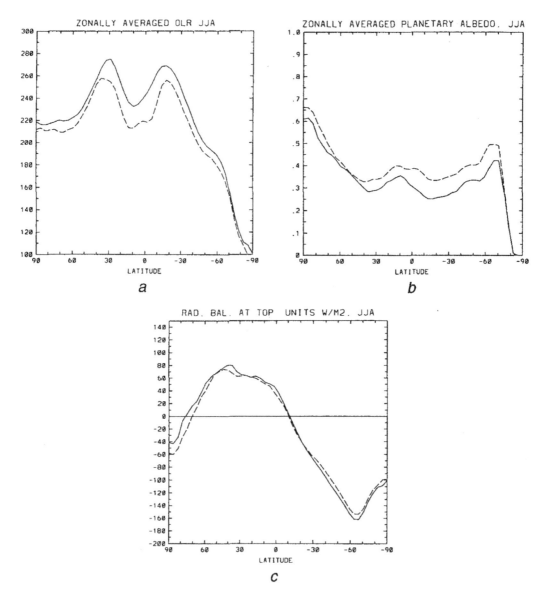

Fig. 14   Zonally averaged values of (a) Out-going long-wave radiation W/m², (b) Planetary albedo, (c) Net radiative flux at the top of the atmosphere W/m². Solid lines are the CONTROL simulation, dashed lines CONV.

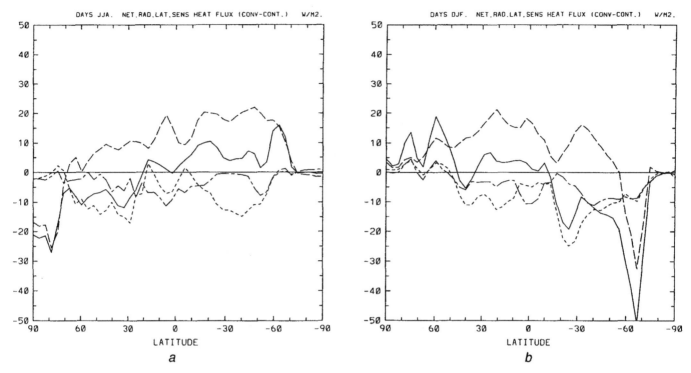

Fig. 15 Zonally averaged differences for terms in the surface energy balance for the CONTROL simulation. (*a*) June–Aug. (*b*) Dec.–Feb. Solid: net (positive downward) energy flux at the surface. Long dashed: net radiative flux. Short dashed: Latent heat flux due to evaporation from the surface. Long-short dashed: Sensible heat flux. Units: W/m².

energy in the sub-surface layers, sub-surface heat transport and, where present, freezing or melting of sea ice and snow. The sub-surface heat flux is specified under the sea ice and within the 50 m thick oceanic mixed layer in both simulations. As discussed in MBBL, these vary on a monthly time scale in such a way as to ensure that the evolution of sea surface temperatures and sea ice in the CONTROL simulation is acceptably close to observed climatology throughout the annual cycle. This sub-surface flux specification is unchanged for the CONV simulation. Thus the differences in net surface flux imply that for this simulation, there is an energy flux imbalance for the surface layer. Maintenance of a climatological equilibrium in which sea surface temperatures and sea ice extent remain close to observed values throughout the annual cycle is not possible in these circumstances.

Figure 16, which depicts the ensemble mean surface temperature differences between the two simulations, shows that while the CONV simulation is warmer over some parts of the land surface in the tropics, it is generally colder elsewhere. Although the largest cooling is in higher latitudes over land and sea ice areas in the winter hemispheres, there is also significant cooling elsewhere, particularly at high latitudes in the summer hemisphere. The substantial heat capacity of the oceanic surface layer ensures that surface temperatures in those regions respond relatively slowly to the energy flux imbalance implied by the difference fields shown in Figure 16. Thus the CONV simulation was not in equilibrium when it was terminated. The globally averaged monthly mean values

of the surface air temperature for the last two years are less than the corresponding ensemble mean values for the CONTROL simulation, suggesting that it was drifting toward a generally cooler climatic state.

7 Discussion and conclusions

The bulk cumulus parameterization presented in this paper is simple enough to be relatively easily implemented in a GCM. It is designed to represent mainly the effects of deep convection and does retain the important features of penetrative convection schemes of this type. The simple column tests presented show that the convection scheme comes into equilibrium with imposed forcing. Although the nature of the equilibrated state is sensitive to the specification of the various disposable parameters, particularly the adjustment time scale and the initial downdraft mass flux, the temporally averaged features of the equilibrium states obtained with the new cumulus scheme are broadly similar. In general, the effect of the new parameterization scheme is significantly different from that of the moist convective adjustment scheme currently used in the second generation version of the CCC GCM. In particular, the radiative–convective equilibrium obtained using the convective adjustment scheme yields substantially colder temperatures in the upper troposphere. The regime is also somewhat less moist in that region but more so in the lower troposphere.

A limitation of these simple column tests is that they do not allow some of the more important components of the

165

large-scale forcing to respond to convective heating. However, the full GCM climate simulations corroborate many of the results of the column tests. In particular, it is shown that the cold anomaly in the tropics for the CONTROL simulation is removed by replacing the moist convective adjustment scheme with the new cumulus ensemble scheme. This change is accompanied by drying in the lower troposphere and moistening in the upper troposphere.

An interim specification of the disposable parameters is made for the version of the cumulus parameterization scheme used in the CONV simulation. The choice of 2 hours for the adjustment time used in the GCM experiment is based on results from a few monthly simulation tests which tend to suggest that it is somewhat optimal in the sense that it ensures that the precipitation regime in the tropics is predominantly convective while preventing excessive stabilization due to the choice of a value which is too small. Similarly short tests of the sensitivity to specification of the initial downdraft mass flux suggest that the GCM simulations are somewhat less sensitive to variations in this parameter than are the column model simulations. However, use of the "strong downdraft" version does reduce somewhat the lower troposphere drying (relative to the CONTROL simulation). This is consistent with the results of the column model simulations.

A significant reduction in lower tropospheric cloudiness and a slight increase of cloud cover in the upper troposphere

JJA. (CONV-CONTROL) FOR GRND TEMPERATURE. UNITS DEG C.

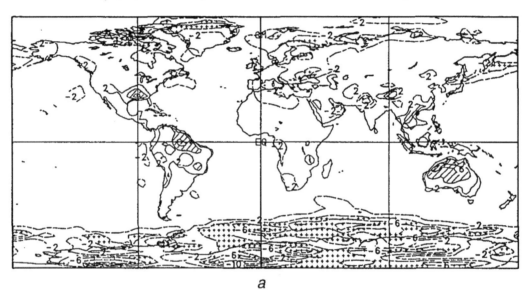

a

DJF. (CONV-CONTROL) FOR GRND TEMPERATURE. UNITS DEG C.

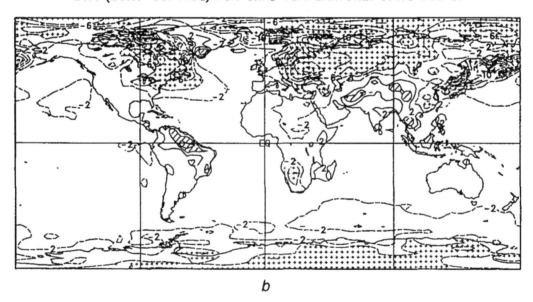

b

Fig. 16   Spatial distribution of the difference (CONV-CONTROL) of the surface temperature for (*a*) June-Aug.; (*b*) Dec.-Feb. Hatched regions have values larger that 4C, shaded less than –4C.

accompany the changes in thermodynamic structure. These changes are also accompanied by important changes in cloud optical properties and consequently of the radiation budget. The planetary albedo is increased and out-going longwave radiation is reduced in the CONV simulation. Although these changes have opposite effects on the energy balance, there is a general reduction of the magnitude of the net radiative flux at the top of the atmosphere.

These are also significant changes in the energy balance at the surface. The net downward radiative flux is increased in tropical and middle latitudes but sharply decreased in high latitudes of the summer hemisphere. This reduction is associated with the albedo change that accompanies increased sea-ice cover. The increased downward radiative flux in the tropics and middle latitudes is partially compensated by increased upward fluxes of sensible and latent heat. Both the net surface flux and changes to it are small over land. However, over oceanic regions, the changes are more complicated and include areas of substantial net increases and decreases. Since these changes are not accompanied by changes in the sub-surface flux in the mixed layer or undersea ice, sea surface temperatures and sea-ice mass and extent change as well.

It is of interest to compare these results to those presented by Washington and Meehl (1993). They show that addition of a penetrative convection scheme to a version of the NCAR CCM2 that includes a 50 m thick oceanic mixed layer results in an increase in upper tropospheric cloud with an associated enhancement of the planetary albedo. This was also accompanied by an increase in the global mean surface temperature even though there was a net increase in the rate of heat loss from the surface due to a substantial increase in surface evaporation. Thus the reduction in solar radiation at the surface due to increased reflectivity of cirrus clouds was not sufficient to inhibit enhancement of the greenhouse effect. The authors show, however, that increasing the albedo of cirrus clouds that occur over warm pools of the ocean results in smaller global mean increases in surface temperature and evaporation. It is also found that introduction of penetrative convection alone enhances the global mean surface warming due to doubling of CO_2 but allowing the additional feedback associated with the enhancement of cirrus cloud albedos over the warm ocean surface nearly offsets this. This is associated with a reduced radiative input at the surface, which is balanced mainly by a reduction in the heat loss due to evaporation.

These results are consistent with those of Boer (1993), who showed that the negative feedback associated with increased reflectivity of cirrus clouds plays a significant role in determining the equilibrium response to doubling of CO_2 as simulated with the CCC GCM. In particular, it results in a significantly smaller increase in surface evaporation than is typically found in earlier simulations of this type. Although cloud cover decreases in the tropics in the doubled CO_2 simulation, the optical thickness of upper tropospheric clouds increases in a manner similar to that depicted in Figure 15. It should be noted that the version of the model used for the doubled CO_2 experiment is identical in its physical parameterization to that used here for the CONTROL run (Boer et al., 1992).

Although the responses of individual components in the CONV experiment are qualitatively similar to that found by Washington and Meehl (1993), and somewhat different from what is found for a doubling of CO_2, the net effect is slight cooling rather than warming of the surface. This is predominantly associated with enhanced evaporation rather than reduced solar input. This effect is, not surprisingly, sensitive to features of the convection scheme and the way that it couples to other parameterized processes in the boundary layer. The version of the penetrative convection scheme used in this simulation tends to be biased toward deep convection, which acts to dry and cool the boundary layer. Although these effects are partially balanced by increased surface evaporation and sensible heat flux, a slight reduction of the low level moisture also results, in contrast to the finding of Washington and Meehl (1993).

Although introduction of the penetrative convection scheme corrects some significant biases in climate simulations, the changes in the radiation budget that give rise to the drift toward a cooler surface are an undesirable feature of the response. Correcting these defects requires modifications of other parameterizations also and is the subject of current work. Apart from this, the scheme as used here is not completely satisfactory in its representation of shallow convection. This may be reflected to some extent in the enhanced dryness of the lower troposphere in the CONV simulation, although it is also somewhat reduced when stronger downdrafts are used, and is also sensitive to vertical resolution and boundary-layer mixing. Nevertheless, it may be desirable to incorporate an explicit shallow convection scheme to complement and/or replace, the penetrative scheme in circumstances where shallow convection is expected to occur predominantly, as is done, for example, in the Betts-Miller scheme as used by Slingo et al. (1994).

A representation of the vertical momentum transfer by cumulus clouds, similar in form to that proposed by Zhang and Cho (1991), has also been designed and tested in seasonal GCM simulations. That work is reported separately (Zhang and McFarlane, 1995).

Appendix: Determining entrainment rates

To determine $\lambda_D(z)$, we consider a cloud type with entrainment rate λ. If clouds are assumed in a steady state, the moist static energy h_u of the updraft satisfies

$$\frac{\partial h_u}{\partial z} + \lambda(h_u - h) = 0. \qquad (A1)$$

Assuming that the updraft moist static energy is equal to the environmental (approximately the large-scale mean) value, h_b, at the detrainment level, the updraft air is saturated and has the same temperature as the environmental air ($h_u \cong h^*$), if follows from the above equation that

$$h_b - h^*(z) = \lambda_D(z) \int_{z_b}^{z} (h_u - h) \, dz'. \qquad (A2)$$

In principle, eqs. (A1) and (A2) can be used to determine $\lambda_D(z)$ by iteration. A Newton-Raphson procedure can be used effectively provided a reasonable first guess is supplied for h_u, or alternatively of $\lambda_D(z)$. The procedure outlined here utilizes reversion of series to derive a high-order Newton-Raphson scheme that provides a first-guess estimate that is often sufficiently accurate that further refinement by iteration is not required.

The updraft moist static energy can be expanded in a Taylor series in λ as

$$h_u(\lambda, z) = h_u(0, z) + \sum_n \left(\frac{1}{n!}\right) \left.\frac{\partial^n h_u}{\partial \lambda^n}\right|_{\lambda=0} \lambda^n. \qquad (A3)$$

The first term on the right-hand side of (A3) is h_b and the derivatives with respect to λ can be obtained by noting that h_b is independent of λ and using (A1) to show that

$$\left.\frac{\partial^n h_u}{\partial \lambda^n}\right|_{\lambda=0} = (-1)^n n! I_n \qquad (A4)$$

where I_n is the nth order integral

$$I_n = \int_{z_b}^{z} \cdots (h_b - h) \, dz^{(n)}. \qquad (A5)$$

Substitution of (A3), (A4) and (A5) into (A2) and applying reversion of series yields:

$$\lambda = \Delta + \frac{I_2}{I_1} \Delta^2 + \frac{(2I_2^2 - I_1 I_3)}{I_1^2} \Delta^3 + \cdots \qquad (A6)$$

where

$$\Delta = \frac{h_b - h^*}{I_1}.$$

The first three terms of (A6) are often sufficient to provide an accurate estimate of $\lambda_D(z)$.

References

ARAKAWA, A. and W.H. SCHUBERT. 1974. Interaction of cumulus cloud ensemble with the large-scale environment, Part I. *J. Atmos. Sci.* **31**: 674–701.

BETTS, A.K. 1976. The thermodynamic transformation of the tropical subcloud layer by precipitation and downdrafts. *J. Atmos. Sci.* **33**: 1008–1020.

——. 1986. A new convective adjustment scheme. Part I: Observational and theoretical basis. *Quart. J. Roy. Meteorol. Soc.* **112**: 677–691.

BOER, G.J.; N.A. MCFARLANE and M. LAZARE. 1992. Greenhouse gas-induced climate change simulated with the CCC second-generation general circulation model. *J. Climate* **5**: 1045–1077.

——. 1993. Climate change and regulation of the surface moisture and energy budgets. *Clim. Dyn.* **8**: 225–239.

BOLTON, D. 1990. The computation of equivalent potential temperature. *Mon. Wea. Rev.* **108**: 1046–1053.

CHO, H.R. 1977. Contribution of cumulus life-cycle effects to the large-scale heat and moisture budget equations. *J. Atmos. Sci.* **34**: 87–97.

EMANUEL, K.A. 1991. A scheme for representing cumulus convection in large-scale models. *J. Atmos. Sci.* **48**: 2313–2335.

FRANK, W.M. and J.L. MCBRIDE. 1989. The vertical distribution of heating in AMEX and GATE cloud clusters. *J. Atmos. Sci.* **46**: 3464–3478.

GREGORY, D.R. and P.R. ROWNTREE. 1990. A mass flux convective scheme with representation of cloud ensemble characteristics and stability dependant closure. *Mon. Wea. Rev.* **118**: 1483–1506.

GRELL, G.; Y.-H. KUO and J.R. PASCH. 1991. Semi-prognostic tests of cumulus parameterization schemes in the middle latitudes. *Mon. Wea. Rev.* **119**: 5–31.

JOHNSON, R.H. 1976. The role of convective-scale precipitation downdrafts in cumulus and synoptic-scale interactions. *J. Atmos. Sci.* **33**: 1890–1910.

——. 1978. Cumulus transports in a tropical wave composite for phase III of GATE. *J. Atmos. Sci.* **35**: 484–494.

——. 1980. Diagnosis of mesoscale motions during phase III of GATE. *J. Atmos. Sci.* **37**: 733–753.

KNUPP, K.R. 1987. Downdrafts within High Plains cumulonimbi, part I: General kinematic structure. *J. Atmos. Sci.* **44**: 987–1008.

KUO, H.-L. 1965. On formation and intensification of tropical cyclones through latent heat release by cumulus convection. *J. Atmos. Sci.* **22**: 40–63.

——. 1974. Further studies of the properties of cumulus convection on the large scale flow. *J. Atmos. Sci.* **31**: 1232–1240.

LEWIS, J.M. 1975. Tests of the Ogura-Cho model on a prefrontal squall line case. *Mon. Wea. Rev.* **103**: 764–778.

LORD, S.J. 1982. Interaction of a cumulus cloud ensemble with large-scale environment. Part III: Semi-prognostic test of the Arakawa-Schubert cumulus parameterization. *J. Atmos. Sci.* **39**: 88–103.

MANABE, S.; J. SMAGORINSKY and R.F. STRICKLER. 1965. Simulated climatology of a general circulation model with a hydrologic cycle. *Mon. Wea. Rev.* **93**: 769–798.

MCFARLANE, N.A.; G.J. BOER, J.-P. BLANCHET and M. LAZARE. 1992. The Canadian Climate Centre second generation circulation model and its equilibrium climate. *J. Climate*, **5**: 1013–1044.

MOORTHI, S. and M.J. SUAREZ. 1992. Relaxed Arakawa-Schubert: A parameterization of moist convection for general circulation models. *Mon. Wea. Rev.* **120**: 978–1002.

OGURA, Y. and H.-R.CHO. 1973. Diagnostic determination of cumulus cloud populations from observed large-scale variables. *J. Atmos. Sci.* **30**: 1276–1286.

SLINGO, J.N.; M. BLACKBURN, A. BETTS, R. BRUGGE, K. HODGES, M. MILLER, L. STEENMAN-CLARK and J. THUBURN. 1994. Mean climate and transience in the tropics of the UGAMP GCM: Sensitivity to convective parameterization. *Q.J.R. Meteorol. Soc.* **120**: 881–922.

TIEDTKE, M. 1989. A comprehensive mass flux scheme for cumulus parameterization in large-scale models. *Mon. Wea. Rev.* **117**: 1779–1800.

WASHINGTON, W.M. and G.A. MEEHL. 1993. Greenhous sensitivity experiments with penetrative cumulus convection and tropical cirrus cloud effects. *Clim. Dyn.* **8**: 211–223.

YANAI, M.; S.K. ESBENSEN and J.-H. CHU. 1973. Determination of bulk properties of tropical cloud clusters from large-scale heat and moisture budgets. *J. Atmos. Sci.* **30**: 611–627.

——; J.-H. CHU and T.E. STARK. 1976. Response of deep and shallow tropical cumuli to large-scale processes. *J. Atmos. Sci.* **33**: 976–991.

ZHANG, G.J. and N.A. MCFARLANE. 1991. Convective stabilization in midlatitudes. *Mon. Wea. Rev.* **119**: 1915–1928.

—— and H.R. CHO. 1991. Parameterization of the vertical transport of momentum by cumulus clouds. Part II: Application. *J. Atmos. Sci.* **48**: 2448–2457.

——. 1994. Effects of cumulus convection on the simulated monsoon circulation in a general circulation model. *Mon. Wea. Rev.* **122**: 2022–2038

—— and N.A. MCFARLANE. 1995. Role of convective-scale momentum transport in climate simulations. *J. Geophys. Res.* **100**: 1417–1426.

The UVic Earth System Climate Model: Model Description, Climatology, and Applications to Past, Present and Future Climates

Andrew J. Weaver[1], Michael Eby[1], Edward C. Wiebe[1], Cecilia M. Bitz[2], Phil B. Duffy[4],
Tracy L. Ewen[1], Augustus F. Fanning[1], Marika M. Holland[3], Amy MacFadyen[1], H. Damon Matthews[1],
Katrin J. Meissner[1], Oleg Saenko[1], Andreas Schmittner[1], Huaxiao Wang[4] and Masakazu Yoshimori[1]

[1]School of Earth and Ocean Sciences, University of Victoria, Victoria, British Columbia
[2]Polar Science Center, Applied Physics Laboratory, Seattle, Washington
[3]National Center for Atmospheric Research, Boulder, Colorado
[4]Lawrence Livermore National Laboratory, Livermore, California

ABSTRACT   *A new earth system climate model of intermediate complexity has been developed and its climatology compared to observations. The UVic Earth System Climate Model consists of a three-dimensional ocean general circulation model coupled to a thermodynamic/dynamic sea-ice model, an energy-moisture balance atmospheric model with dynamical feedbacks, and a thermomechanical land-ice model. In order to keep the model computationally efficient a reduced complexity atmosphere model is used. Atmospheric heat and freshwater transports are parametrized through Fickian diffusion, and precipitation is assumed to occur when the relative humidity is greater than 85%. Moisture transport can also be accomplished through advection if desired. Precipitation over land is assumed to return instantaneously to the ocean via one of 33 observed river drainage basins. Ice and snow albedo feedbacks are included in the coupled model by locally increasing the prescribed latitudinal profile of the planetary albedo. The atmospheric model includes a parametrization of water vapour/ planetary longwave feedbacks, although the radiative forcing associated with changes in atmospheric CO_2 is prescribed as a modification of the planetary longwave radiative flux. A specified lapse rate is used to reduce the surface temperature over land where there is topography. The model uses prescribed present-day winds in its climatology, although a dynamical wind feedback is included which exploits a latitudinally-varying empirical relationship between atmospheric surface temperature and density. The ocean component of the coupled model is based on the Geophysical Fluid Dynamics Laboratory (GFDL) Modular Ocean Model 2.2, with a global resolution of 3.6° (zonal) by 1.8° (meridional) and 19 vertical levels, and includes an option for brine-rejection parametrization. The sea-ice component incorporates an elastic-viscous-plastic rheology to represent sea-ice dynamics and various options for the representation of sea-ice thermodynamics and thickness distribution. The systematic comparison of the coupled model with observations reveals good agreement, especially when moisture transport is accomplished through advection.*

Global warming simulations conducted using the model to explore the role of moisture advection reveal a climate sensitivity of 3.0°C for a doubling of CO_2, in line with other more comprehensive coupled models. Moisture advection, together with the wind feedback, leads to a transient simulation in which the meridional overturning in the North Atlantic initially weakens, but is eventually re-established to its initial strength once the radiative forcing is held fixed, as found in many coupled atmosphere General Circulation Models (GCMs). This is in contrast to experiments in which moisture transport is accomplished through diffusion whereby the overturning is re-established to a strength that is greater than its initial condition.

When applied to the climate of the Last Glacial Maximum (LGM), the model obtains tropical cooling (30°N – 30°S), relative to the present, of about 2.1°C over the ocean and 3.6°C over the land. These are generally cooler than CLIMAP estimates, but not as cool as some other reconstructions. This moderate cooling is consistent with alkenone reconstructions and a low to medium climate sensitivity to perturbations in radiative forcing. An amplification of the cooling occurs in the North Atlantic due to the weakening of North Atlantic Deep Water formation. Concurrent with this weakening is a shallowing of, and a more northward penetration of, Antarctic Bottom Water.

Climate models are usually evaluated by spinning them up under perpetual present-day forcing and comparing the model results with present-day observations. Implicit in this approach is the assumption that the present-day observations are in equilibrium with the present-day radiative forcing. The comparison of a long transient integration (starting at 6 KBP), forced by changing radiative forcing (solar, CO_2, orbital), with an equilibrium integration reveals substantial differences. Relative to the climatology from the present-day equilibrium integration, the global mean surface air and sea surface temperatures (SSTs) are 0.74°C and 0.55°C colder, respectively. Deep ocean temperatures are substantially cooler and southern hemisphere sea-ice cover is 22% greater, *although the North Atlantic conveyor remains remarkably stable in all cases. The differences are due to the long timescale memory of the deep ocean to climatic conditions which prevailed throughout the late Holocene. It is also demonstrated that a global warming simulation that starts from an equilibrium present-day climate (cold start) underestimates the global temperature increase at 2100 by 13% when compared to a transient simulation, under historical solar, CO_2 and orbital forcing, that is also extended out to 2100. This is larger (13% compared to 9.8%) than the difference from an analogous transient experiment which does not include historical changes in solar forcing. These results suggest that those groups that do not account for solar forcing changes over the twentieth century may slightly underestimate (~3% in our model) the projected warming by the year 2100.*

RÉSUMÉ   *[Traduit par la rédaction] Un nouveau modèle climatique du système terrestre de complexité intermédiaire a été mis au point et sa climatologie a été comparée aux observations. Le modèle climatique UVic du système terrestre consiste en un modèle tridimensionnel de circulation océanique générale couplé à un modèle de glace marine thermodynamique/dynamique, à un modèle atmosphérique à bilans d'énergie et d'humidité avec rétroactions dynamiques et à un modèle thermomécanique des glaces des terres émergées. L'utilisation d'un modèle atmosphérique de complexité réduite permet de maintenir l'efficacité de calcul du modèle. Les transports de chaleur atmosphérique et d'eau douce sont paramétrés en utilisant la diffusion fickienne et l'on suppose que des précipitations se produisent lorsque l'humidité relative dépasse 85%. Le transport d'humidité peut également être réalisé par advection, si désiré. Les précipitations au-dessus de la terre ferme sont supposées retourner instantanément à l'océan par l'un des 33 bassins hydrographiques qui sont observés. Les rétroactions des albédos de la glace et de la neige sont inclus dans le modèle couplé en augmentant localement le profil latitudinal prescrit de l'albédo planétaire. Le modèle atmosphérique comporte un paramétrage des rétroactions de la vapeur d'eau/ondes longues planétaires, même si le forçage radiatif associé avec les changements dans le CO_2 atmosphérique est défini comme une modification du flux radiatif planétaire d'ondes longues. Un gradient thermique vertical fixe sert à réduire la température de surface au-dessus de la terre ferme dans les situations avec relief topographique. Le modèle utilise dans sa climatologie des vents prescrits à partir de la situation présente, quoiqu'il incorpore une rétroaction dynamique du vent qui exploite une relation empirique variant en fonction de la latitude entre la température atmosphérique de surface et la densité. La composante océanographique du modèle couplé est basée sur le modèle océanographique modulaire géophysique (GFDL) 2.2, avec une résolution globale de 3,6° (zonale) par 1,8° (méridionale) et 19 niveaux verticaux. Elle inclut une option pour le paramétrage du rejet d'eau salée. La composante glace marine contient une rhéologie élastique-plastique-visqueuse pour représenter la dynamique de la glace marine et diverses options pour la représentation de la thermodynamique de la glace marine et sa distribution d'épaisseur. La comparaison systématique du modèle couplé avec les observations révèle un bon accord, surtout lorsque le transport d'humidité est effectué par advection.*

Les simulations de réchauffement planétaire effectuées en utilisant le modèle pour explorer le rôle de l'advection d'humidité révèlent une sensibilité climatique de 3,0°C avec le doublement du CO_2, en conformité avec d'autres modèles couplés plus complets. L'advection d'humidité avec la rétroaction du vent mène à une simulation transitoire dans laquelle le brassage méridional dans l'Atlantique Nord s'affaiblit d'abord, mais qui se rétablit éventuellement à sa force initiale une fois que le forçage radiatif est maintenu constant, tout comme cela se produit avec plusieurs modèles couplés de la circulation atmosphérique générale (GCM). Ceci contraste avec les expériences dans lesquelles le transport d'humidité est accompli par diffusion et qui démontrent que le brassage est rétabli à une force supérieure à celle de sa valeur initiale.

Appliqué au climat du dernier maximum glaciaire (LGM), le modèle obtient un refroidissement tropical (30°N – 30°S), relativement au climat présent, d'environ 2,1°C au-dessus des océans et de 3,6°C au-dessus de la terre ferme. Ces valeurs sont généralement plus froides que les estimations CLIMAP, mais pas aussi froides que certaines autres reconstructions. Ce refroidissement modéré est compatible avec les reconstructions alkénone et une sensibilité climatique de faible à moyenne aux perturbations du forçage radiatif. Une amplification du refroidissement se produit dans l'Atlantique Nord à cause de l'affaiblissement de la formation d'eau profonde de l'Atlantique Nord. En même temps que se produit cet affaiblissement, il y a un amincissement de l'eau de fond de l'Antarctique ainsi qu'une pénétration plus orientée vers le nord de cette eau.

Les modèles climatiques sont habituellement évalués en les faisant tourner sous le forçage perpétuel actuel et en comparant les résultats des modèles avec les observations actuelles. Implicitement, on présume que les observations actuelles sont en équilibre avec le forçage radiatif actuel. La comparaison d'une intégration transitoire de longue durée (débutant il y a 6 000 ans) forcée en modifiant le forçage radiatif (solaire, CO_2 orbital) avec une intégration à l'équilibre révèle des différences substantielles. Par rapport à la climatologie dérivée de l'intégration de l'équilibre aux conditions actuelles, les températures globales moyennes de l'air à la surface et de la surface de l'eau sont respectivement plus froides de 0,74°C et 0,55°C. Les températures de la mer profonde sont substantiellement plus froides et le couvert de glace marine dans l'hémisphère sud est de 22% plus important, quoique le convoyeur de l'Atlantique Nord demeure remarquablement stable dans tous les cas. Les différences sont dues à la mémoire à long terme de la mer profonde des conditions climatiques qui ont prévalu à la fin de l'holocène. Il a également été démontré qu'une simulation du réchauffement planétaire qui démarre d'un climat à l'équilibre avec le climat actuel (démarrage à froid) sous-estime l'augmentation de la température mondiale en 2100 de 13% lorsqu'on la compare à une simulation transitoire soumise aux forçages historiques solaire, orbital et du CO_2 et qui a également été effectuée jusqu'en 2100. Ceci est plus grand (13% comparativement à 9,8%) que la différence d'une expérience transitoire analogue qui n'inclut pas les changements historiques dans le forçage solaire. Ces résultats suggèrent que les groupes qui ne tiennent pas compte des changements dans le forçage solaire au XXe siècle pourraient sous-estimer légèrement (~3% dans notre modèle) le réchauffement projeté en l'an 2100.

1 Introduction

Coupled atmosphere-ocean general circulation models (GCMs) are frequently used to understand both past, present and future climates and climate variability. The computational expense associated with these models, however, often precludes their use for undertaking extensive parameter sensitivity studies. While they must ultimately be used as the primary tool for undertaking climate projections on which policy will be based, it is important to conduct sensitivity studies in parallel using simpler models. Simple models, or models of intermediate complexity, allow one to explore the climate sensitivity associated with a particular process or component of the climate system over a wide range of parameters. In addition, they allow one to streamline the experiments that are performed using more complicated GCMs. These more idealized coupled models vary in complexity from simple one-dimensional energy balance/upwelling diffusion models (Wigley and Raper, 1987, 1992; Raper et al., 1996; Wigley, 1998), to zonally-averaged ocean/energy balance atmospheric models (Stocker et al., 1992); Stocker and Schmittner, 1997), to models with more sophisticated subcomponents (Fanning and Weaver, 1996; Petoukhov et al., 2000).

a Models of Intermediate Complexity

Simple and intermediate complexity climate models are designed with a particular class of scientific questions in mind. In the development of the model, only those processes and parametrizations are included which are deemed important in the quest to address the scientific questions of concern. For example, Wigley (1998) used an upwelling diffusion-energy balance climate model (see Kattenberg et al., 1996) to evaluate Kyoto Protocol implications for increases in global mean temperature and sea level. While such a simple climate model relies on climate sensitivity and ice-melt parameters obtained from coupled atmosphere-ocean GCMs, it nevertheless allows for a first-order analysis of the climatic consequences of various post-Kyoto emissions reductions. The results from this analysis suggested very minor effects on global mean temperature under a range of post-Kyoto emissions. Stocker and Schmittner (1997) used a three-basin zonally-averaged ocean circulation model coupled to a simple energy-balance atmospheric model (described in Stocker et al., 1992), to undertake a systematic parameter sensitivity study of the response of the North Atlantic thermohaline circulation to both the rate of increase and equilibrium concentration of atmospheric CO_2. They showed that, in general, the slower the rate of increase in atmospheric CO_2, the greater the equilibrium concentration required to permanently shut down the conveyor in the North Atlantic. While the actual critical thresholds that arose from this study would need verification by more complicated models, the study clearly illustrates the importance of the rate of CO_2 increase on the North Atlantic thermohaline circulation; a result difficult, if not impossible, to achieve with the computationally expensive present-generation coupled models.

The Climate and Biosphere Model (CLIMBER) group at the Potsdam Institute have taken the approach of building a climate model of intermediate complexity to examine climate change and variability with a sophisticated, albeit highly parametrized, atmospheric component. Their atmospheric model is based on the statistical-dynamical approach without resolving synoptic variability (Petoukhov et al., 2000; Ganopolski et al., 2001). Their three-basin, zonally-averaged ocean component is very similar to the ocean component of Stocker and Schmittner (1997) and they also incorporate a simple dynamic/thermodynamic ice model. The CLIMBER-2 model sacrifices resolution ($10°$ latitudinal by $51°$ zonal) and complexity for computational efficiency. This model has been used to investigate both the climate of the Last Glacial Maximum (LGM) (Ganopolski et al., 1998), as well as the cause for the collapse of the conveyer in global warming experiments (Rahmstorf and Ganopolski, 1999).

Fanning and Weaver (1996) developed an energy-moisture balance model coupled to an ocean GCM and a thermodynamic sea-ice model. This model has been used to undertake a number of sensitivity studies including the role of sub-gridscale ocean mixing (and flux adjustments, Fanning and Weaver (1997a)) in global warming experiments (Wiebe and Weaver, 1999) and steric sea level rise (Weaver and Wiebe, 1999), and also the ocean's role in the LGM and Ordovician climates (Weaver et al., 1998; Poussart et al., 1999).

All of the simple or intermediate complexity climate models discussed above are free of explicit flux adjustments. Nevertheless, these same models necessarily include implicit adjustments through the fine tuning, within observational ranges, of internal model parameters in an attempt to reproduce the present-day climate.

One of the purposes of this paper is to document the climatology of, and parametrizations used, in a new intermediate complexity Earth System Climate Model (ESCM) which has substantially evolved from the earlier version described by Fanning and Weaver (1996). This new model incorporates a better representation of atmospheric processes, including the addition of moisture advection, orography and wind feedbacks, as well as more sophisticated sea-ice and snow models, and a dynamical ice-sheet model.

Our philosophy in building this model is to develop a tool with which to understand processes and feedbacks operating within the climate system on decadal and longer timescales. We include a full three-dimensional ocean GCM as we believe horizontal ocean gyre transports and subduction processes are fundamental to the stability and variability of the thermohaline circulation; we believe the latter to be important for climate/paleoclimate change/variability on long timescales. Our atmospheric model is simple, although we believe it captures the necessary surface balances and feedbacks of heat, freshwater and momentum. Atmospheric 'weather' is not calculated explicitly in our model, although in some applications its statistics have been added as a forcing. This allows our model to reach an equilibrium climate free of variability, both internal to the atmosphere, and excited in the ocean and sea-ice subcomponent models. The sophisticated sea-ice model was built and tested within the context of the ESCM and has since been included in the Canadian Centre for Climate Modelling and Analysis (CCCma) coupled Atmosphere-Ocean General Circulation Model (AOGCM). The glaciological model was developed at the University of British Columbia and coupled into our model while the inorganic chemistry subcomponent was built following the guidelines of the international Ocean Carbon-Cycle Model Intercomparison Project. By building the ESCM in a systematic fashion, we ensure that energy and moisture are conserved exactly as all parameters are consistent among all subcomponent models.

Our approach is complementary to that of major coupled AOGCM centres, and our model fills an important gap within the hierarchy of climate models (Claussen et al., 2001; Shackley et al., 1998). In fact, our model has been used, and is still being used, as a tool to examine the sensitivity of a particular process or subcomponent model across a wide range of parameters, in order to streamline the process of improving certain components of the CCCma coupled AOGCM. The complexity of the CCCma AOGCM is such that relatively few 'production runs' can be conducted, leaving systematic parameter sensitivity analyses to be conducted with the University of Victoria (UVic) ESCM.

b *Initialization of Coupled Models*
In developing coupled AOGCMs, it is common to integrate them to equilibrium under present-day radiative forcing (in the case of coupled atmosphere-ocean models) or using present-day surface boundary conditions (in the case of ocean- or atmosphere-only models). The evaluation process then involves a detailed comparison of model results with present-day observations (e.g., Oreskes and Belitz, 1994; Gates et al., 1996; Flato et al., 2001). Since the ocean, and especially its deep and weakly ventilated regions, have a long adjustment time to changes in radiative forcing, the present-day oceanic observations are not in equilibrium with the present forcing. As such, models which start from a control integration which is in equilibrium with its radiative forcing suffer from a warm bias (Weaver et al., 2000). Even if the ocean or coupled model were perfect, it would therefore not be possible to reproduce these observations under perpetual present-day model forcing. Many coupled atmosphere-ocean climate GCMs that have been used to examine the transient response of the climate system to increasing greenhouse gases, start from a present-day equilibrium climate (see Kattenberg et al., 1996). This equilibrium climate is usually obtained by separately spinning up the oceanic and atmospheric components of the coupled model under perpetual present-day forcing (see Weaver and Hughes (1996) for a review). Upon coupling, flux adjustments are often employed to avoid climate drift. It has been suggested that this approach leads to an apparent 'cold start' problem (Fichefet and Tricot, 1992; Hasselman et al., 1993; Cubasch et al., 1994; Keen and Murphy, 1997), and it has been argued that since the prior build up of radiative forcing has been neglected, the initial rate of warming is slower than it should otherwise be (see Kattenberg et al., 1996).

This apparent problem has been dealt with recently by most coupled model groups by accounting for the twentieth century build-up of atmospheric CO_2 through long integrations starting from pre-industrial times. Nevertheless, these same models evaluate their present-day climate obtained from present-day radiative forcing, assume that the present-day climate is representative of 1850, and then impose changes in radiative forcing to the present-day climate consistent with changes since 1850 (Mitchell et al., 1995). Some more recent efforts (e.g., HadCM3 – Gordon et al., 2000) have actually used pre-industrial levels of greenhouse gases in their coupled model initialization, although the evaluation problem still remains as insufficient data exist to validate properly the equilibrium model climate with the pre-industrial observational record. A second goal of this study is to quantify the memory of the present-day ocean to prior conditions through an examination of its response to 6000 years of changing radiative forcing. This should be viewed as an extension of Weaver et al. (2000) in that we consider here not only the effects of changing CO_2 and orbital forcing, but also changes in solar luminosity.

c *The Climate of the Last Glacial Maximum*
The issue of tropical cooling at the LGM represents a challenge for both the paleoclimate data and modelling communities. Conflicting evidence has emerged from the reconstruction community with tropical coral (Guilderson et al., 1994; Beck et al., 1997), noble gas (Stute et al., 1995), ice core (Thompson et al., 1995), and sediment (Curry and Oppo, 1997) records inferring tropical temperatures substantially colder than today. Alkenone records (Sikes and Keigwin, 1994; Bard et al., 1997; Lyle et al., 1992), on the other hand, have inferred tropical temperatures colder than CLIMAP Project Members (1976, 1981) reconstructions (which were close to today's) but only slightly colder than present.

Simulations with coupled models also disagree on estimates of tropical temperatures during the LGM. Ganoploski et al. (1998), using the CLIMBER-2 model, and Bush and Philander (1998), using oceanic and atmospheric GCMs that were coupled together and integrated under LGM forcing for 15 years (starting from a present-day initial condition), found much cooler tropical LGM temperatures than today. Weaver et al. (1998), using a coupled Oceanic General Circulation Model (OGCM) energy-moisture balance model and Broccoli (2000), using an Atmospheric General Circulation Model (AGCM) coupled to a mixed layer oceanic model, found tropical temperatures only slightly cooler than today and consistent with alkenone reconstructions (see also Crowley, 2000). A final goal of this paper is to re-examine this issue with our new ESCM.

d *Outline*
The outline of the rest of this paper is as follows: in the next section we provide a detailed description of the UVic ESCM, starting with the atmospheric component (Section 2a) then moving to the cryospheric components (sea ice, land ice and snow – Section 2b) and the oceanic component (Section 2c). In Section 3 we provide a comparison between the model clima-

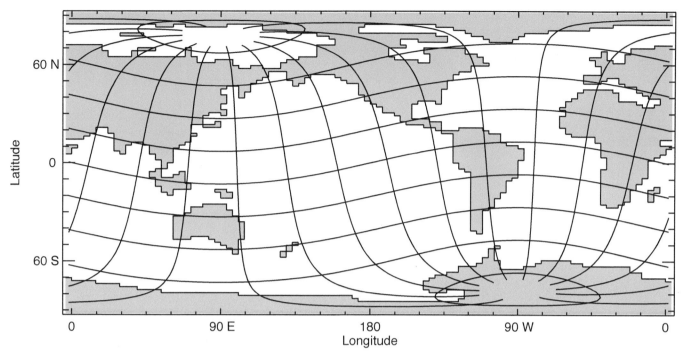

Fig. 1   Model grid showing lines of geographic latitude and longitude in the rotated frame of reference. Rotated lines of geographic latitude are every 20 degrees and lines of longitude are every 30 degrees.

tology and observations using a diffusive approximation for moisture transport. This approximation is relaxed in Section 4 when moisture advection is included and in Section 5 we perform global warming simulations with and without this moisture advection option. The climate of the LGM, and, in particular, the resulting Sea Surface Temperatures (SSTs) and Atlantic thermohaline circulation, are examined in Section 6. In Section 7 we perform experiments similar to those done by Weaver et al. (2000) in which we analyze the differences between the results from a transient experiment starting at 6 KBP and two experiments conducted with perpetual 1850 and 1998 radiative forcing. We also extend the transient integration out to 2100 to study the cold start problem. Our results are summarized in Section 8.

2 Model description

The coupled model consists of an energy-moisture balance atmospheric model, loosely based on Fanning and Weaver (1996), a dynamic-thermodynamic sea-ice model (Hibler, 1979; Hunke and Dukowicz, 1997; Bitz et al., 2001) and a primitive equation oceanic general circulation model (Pacanowski, 1995). The model is global in coverage with a spherical grid resolution of 3.6° (zonal) by 1.8° (meridional), similar to the resolution used in most coupled AOGCMs. A new feature of this model is that the poles in the ocean, atmosphere and sea-ice subcomponents have been rotated such that the North Pole is in Greenland while the South Pole remains in Antarctica (see Fig. 1). This simple Euler angle grid rotation (Appendix A) removes the problem of grid convergence near the North Pole, thereby eliminating the need for filtering or the placement of an artificial island at the pole in the ocean model.

The only modification from a "standard" spherically-gridded version of the model is that variables that are normally a function of grid latitude, such as the Coriolis parameter, are transformed so that they remain a function of geographic latitude.

a Atmospheric Model

The atmospheric model uses the Fanning and Weaver (1996) energy-moisture balance model as its starting point. The formulation of the vertically-integrated thermodynamic energy balance equations assumes a decreasing vertical distribution of energy and specific humidity (from surface values) with specified e-folding scale heights. The prognostic equations for momentum conservation are replaced by specified wind data, although dynamical wind feedbacks are also included (Section 2a1). The other major simplification to the atmosphere is the parametrization of atmospheric heat and moisture transport by diffusion, although moisture advection by the winds is also included as an option (Section 4).

The vertically-integrated atmospheric thermodynamic energy balance equation is given by:

$$\rho_a h_t c_{pa} \frac{\partial T_a}{\partial t} = Q_T + Q_{SW} C_A + Q_{LH}$$
$$+ Q_{LW} + Q_{SH} - Q_{PLW} \qquad (1)$$

where $\rho_a = 1.25$ kg m^{-3} is a constant surface air density, $h_t = 8.4$ km (Gill, 1982) is a constant representative scale height for temperatures, $c_{pa} = 1004$ J kg^{-1} K^{-1} is the specific heat of air at constant pressure, and T_a is the sea level air temperature. The terms on the right-hand side of Eq. (1) represent sources or sinks of heat for the atmosphere.

173

The heat transport term Q_T is parametrized by Fickian diffusion, and takes the form:

$$Q_T = \rho_a\, h_t\, c_{pa}\, \boldsymbol{\nabla} \cdot (\nu\, \boldsymbol{\nabla}\, T_a) \qquad (2)$$

where $\boldsymbol{\nabla}(\nu\, \boldsymbol{\nabla}\, T_a)$ is the diffusion operator (Appendix B) acting on T_a with diffusivity ν (Section 4; Fig. 23).

The incoming shortwave radiation at the top of the atmosphere is given by:

$$Q_{SW} = \frac{S_\odot}{4}\, I(1 - \alpha) \qquad (3)$$

where $S_\odot = 1368$ W m^{-2} is the solar constant and I is the annual distribution of insolation entering the top of the atmosphere (Berger, 1978). The planetary albedo α varies as a function of latitude and time of year to account for the effects of changes in solar zenith angle (Graves et al., 1993). $C_A = 0.3$ (da Silva et al., 1994) is an absorption coefficient which parametrizes the absorption of heat in the atmosphere by water vapour, dust, ozone and clouds etc. (Ramanathan et al., 1987). We refer the reader to Fanning and Weaver (1996) for an extensive parameter sensitivity study of the stand-alone Energy Moisture Balance Model (EMBM) to C_A.

The latent heat flux into the atmosphere takes the form:

$$Q_{LH} = \rho_o\, P \begin{cases} L_v & \text{if } P \text{ is rain} \\ L_s & \text{if } P \text{ is snow} \end{cases} \qquad (4)$$

where ρ_o is a reference density of water, $L_v = 2.50 \times 10^6$ J kg^{-1} is the latent heat of vapourization, $L_s = 2.84 \times 10^6$ J kg^{-1} is the latent heat of sublimation and P is the precipitation (in m s^{-1}).

The parametrization for outgoing planetary longwave radiation Q_{PLW} is taken from Thompson and Warren (1982) and depends on the surface relative humidity (r) and temperature (T_a). This term is modified to parametrize the radiative forcing associated with changes in atmospheric CO$_2$ concentration, and is given by:

$$\begin{aligned}
Q_{PLW} &= c_{00} + c_{01}\, r + c_{02}\, r^2 \\
&+ \left(c_{10} + c_{11}\, r + c_{12}\, r^2\right) T_a \\
&+ \left(c_{20} - c_{21}\, r + c_{22}\, r^2\right) T_a^2 \\
&+ \left(c_{30} + c_{31}\, r + c_{32}\, r^2\right) T_a^3 \\
&- \Delta F_{2x} \ln \frac{C(t)}{C_o}
\end{aligned} \qquad (5)$$

where $C(t)/C_o$ is the ratio of the prescribed atmospheric CO$_2$ concentration to a present day reference level ($C_o = 350$ ppm), and $\Delta F_{2x} = 5.77$ W m^{-2} corresponds to a specified radiative forcing of 4 W m^{-2} for a doubling of atmospheric CO$_2$ (Ramanathan et al., 1987). The constants in Eq. (5) are taken from Table 3 of Thompson and Warren (1982).

The longwave radiation emitted by the Earth's surface is absorbed by the atmosphere and re-emitted both upward and downward. The net upward longwave radiative flux into the atmosphere can be written as:

$$Q_{LW} = \varepsilon_s\, \sigma\, T_s^4 - \varepsilon_a\, \sigma\, T_a^4 \qquad (6)$$

where T_s is the surface temperature and $\varepsilon_s = 0.94$ and ε_a are the surface and atmospheric emissivities, and $\sigma = 5.67 \times 10^{-8}$ W m^{-2} K^{-4} is the Stefan-Boltzmann constant. The atmospheric emissivity is taken from Fig. 2a of Fanning and Weaver (1996), scaled by 0.98. A bulk parametrization is used for the calculation of the sensible heat flux,

$$Q_{SH} = \rho_a\, C_H\, c_{pa}\, U(T_s - T_a) \qquad (7)$$

where $C_H = 0.94\, C_E$ (Isemer et al., 1989) is the Stanton number and C_E is the Dalton number (see below), and U is the surface wind speed. We refer the reader to Fanning and Weaver (1996) for an extensive parameter sensitivity study of the stand-alone EMBM to $\varepsilon_s, \varepsilon_a, T_s$ and U.

The vertically-integrated moisture balance equation is given by,

$$\rho_a\, h_q \left\{ \frac{\partial q_a}{\partial t} - \boldsymbol{\nabla} \cdot (\kappa \boldsymbol{\nabla} q_a) \right\} = \rho_o(E - P) \qquad (8)$$

where $h_q = 1.8$ km (Peixoto and Oort, 1992) is a constant scale height for specific humidity, q_a is the surface specific humidity, κ is an eddy diffusivity (Section 4; Fig. 23), E is evaporation (in m s^{-1}) or sublimation and P is precipitation. Evaporation (or sublimation) is calculated from the following bulk formula:

$$E = \frac{\rho_a\, C_E\, U}{\rho_o} \left(q_s(T_s) - q_a\right) \qquad (9)$$

where C_E is the time-dependent Dalton number calculated according to Eq. (18) of Fanning and Weaver (1996) and $q_s(T_s)$ is the saturation specific humidity:

$$q_s(T) = c_1 \exp \begin{cases} \left(\dfrac{c_2\, T}{T + c_3}\right) & \text{if } T = T_i \\[2mm] \left(\dfrac{c_4\, T}{T + c_5}\right) & \text{otherwise} \end{cases} \qquad (10)$$

where T is the ocean (T_o) or ice (T_i) surface temperature, in the calculation of E (Eq. (9)), or the air temperature (T_a) in the calculation of precipitation P (Eq. (11)). The constants in Eq. (10) are defined as: $c_1 = 3.80$ g kg^{-1}, $c_2 = 21.87$, $c_3 = 265.5$ K, $c_4 = 17.67$ and $c_5 = 243.5$ K (Bolton, 1980). Precipitation is assumed to occur whenever the relative humidity r is greater than a specified threshold $r_{max} = 0.85$, and is given by:

$$P = \begin{cases} \dfrac{\rho_a\, h_q}{\rho_o\, \Delta t} \left(q_a - r_{max}\, q_s(T_a)\right) & \text{if } r > r_{max} \\[2mm] 0 & \text{otherwise} \end{cases} \qquad (11)$$

where Δt is the model time step and $q_s(T_a)$ is the saturation specific humidity at air temperature T_a.

No heat or moisture is stored on the land surface, so $E = 0$ and the heat radiated back to the atmosphere from the surface $(Q_{LW} + Q_{SH})$ must equal the shortwave radiation reaching the surface $(Q_{SW}(1 - C_A))$. Over land Eq. (1) becomes:

$$\rho_a \, h_a \, c_{p_a} \frac{\partial T_a}{\partial t} = Q_T + Q_{SW} + Q_{LH} - Q_{PLW}. \quad (12)$$

Precipitation that falls as rain or accumulated snow (Section 2b) that melts on land, is returned instantaneously to the ocean. The total runoff for a river basin is then divided between specified points (river outflows) and an even distribution along the basin's coastal points (see Fig. 2).

Orography is felt by the atmosphere in four ways (see Fig. 3). A globally-averaged lapse rate is used to reduce the model's apparent sea level temperature in calculating the following: the outgoing longwave radiation; the surface air temperature (SAT) dependent planetary co-albedo through the calculation of the areal fraction of terrestrial snow/ice; the saturation specific humidity to determine the amount of precipitation; whether the precipitation will fall as rain or snow. A beneficial consequence of adding orography is that realistic river basins can be used (compare Fig. 2 with Fig. 1 of Weaver et al. (1998)). Using the reduced temperature for calculating outgoing radiation makes the longwave radiation more dependent on topography than is seen in the observations, but it improves the simulated surface temperatures, which are otherwise generally too low over areas of high elevation.

The model includes a full annual cycle of solar insolation for both present and past orbital configurations (Berger, 1978). As such, it is capable of being used for both equilibrium paleoclimate applications (e.g., Weaver et al., 1998) or long transient paleoclimate integrations (e.g., Weaver et al., 2000). A monthly wind stress climatology, created from 40 years (1958–1998) of daily Kalnay et al. (1996) reanalysis data, is used to force the ocean and ice components of the coupled model. This wind stress climatology is also converted to wind speeds for use in the calculation of latent and sensible heat fluxes between the atmosphere and ice or ocean models.

1 PARAMETRIZATION OF WIND STRESS ANOMALIES

It has been suggested (Schiller et al., 1997; Fanning and Weaver, 1997b) that the dynamic response of the atmosphere to SST anomalies has a stabilizing effect on the Atlantic thermohaline circulation. In order to account for this effect in our non-dynamic energy-moisture balance model we follow the idea of Fanning and Weaver (1997b) and parametrize wind stress anomalies in terms of SAT anomalies. The starting point for this parametrization is a linear approximation between the density ρ and air temperature T_a near the surface:

$$\rho = a + bT_a. \quad (13)$$

As an example we show values of ρ and T_a at 45°N from the mean seasonal cycle of the 15-year European Centre for Medium-Range Weather Forecasts (ECMWF) reanalysis data Gibson et al. (1997) (abbreviated as ERA15) in Fig. 4. The correlation is similarly high at all other latitudes (not shown).

Using the equation of state for dry air it is now possible to express the sea-level pressure p in terms of air temperature T_a:

$$p = R \, \rho \, T_a = R(aT_a + bT_a^2), \quad (14)$$

where $R = 287$ J kg^{-1} K^{-1} is the gas constant for dry air. We use the geostrophic approximation, with Rayleigh damping near the equator, for the calculation of the near-surface wind velocities as follows:

$$-fv + \frac{1}{\rho_a \, r_e \cos\phi} \frac{\partial p}{\partial \lambda} + \frac{u}{\tau} = 0 \quad (15)$$

$$fu + \frac{1}{\rho_a \, r_e} \frac{\partial p}{\partial \phi} + \frac{v}{\tau} = 0 \quad (16)$$

where $\mathbf{V} = (u, v)$ is the wind velocity vector, and

$$\frac{1}{\tau} = \frac{1}{\tau_0} \exp\left[-\left(\phi/15^\circ\right)^2\right]. \quad (17)$$

Here $r_e = 6370$ km is the radius of the Earth, ϕ is the latitude, and λ is the longitude and $f = 2\Omega \sin(\phi)$ is the Coriolis parameter ($\Omega = 7.29 \times 10^{-5}$ s^{-1} is the Earth's rotation rate). The damping coefficient $\tau_0 = 3600$ s in Eq. (17) is assumed to be constant at all latitudes, although the entire Rayleigh damping term becomes negligible outside of the tropics. Solving Eqs (15) and (16) for the geostrophic velocities yields

$$u = \frac{1}{\left(f^2 + 1/\tau^2\right)\rho_a} \left(-\frac{1}{\tau r_e \cos\phi} \frac{\partial p}{\partial \lambda} - \frac{f}{r_e} \frac{\partial p}{\partial \phi}\right) \quad (18)$$

$$v = \frac{1}{\left(f^2 + 1/\tau^2\right)\rho_a} \left(\frac{f}{r_e \cos\phi} \frac{\partial p}{\partial \lambda} - \frac{1}{\tau r_e} \frac{\partial p}{\partial \phi}\right). \quad (19)$$

Surface wind velocities $\mathbf{V}_s$ are calculated through a rotation and contraction of the wind velocity vector as described in Fanning and Weaver (1997b), namely

$$\mathbf{V}_s = C_t \begin{pmatrix} \cos\psi & -\sin\psi \\ \sin\psi & \cos\psi \end{pmatrix} \mathbf{V}$$

where $C_t = 0.8$ is a contraction factor and $\psi = 20^\circ$ is a rotation angle (positive in the northern hemisphere and negative in the southern hemisphere). The wind stress vector $\boldsymbol{\tau}$ can then be derived from the bulk formula Gill (1982)

$$\boldsymbol{\tau} = c_D \, \rho_a \, |\mathbf{V}_s| \mathbf{V}_s \quad (20)$$

where $c_D = 1 \times 10^{-3}$ is a constant dimensionless drag coefficient.

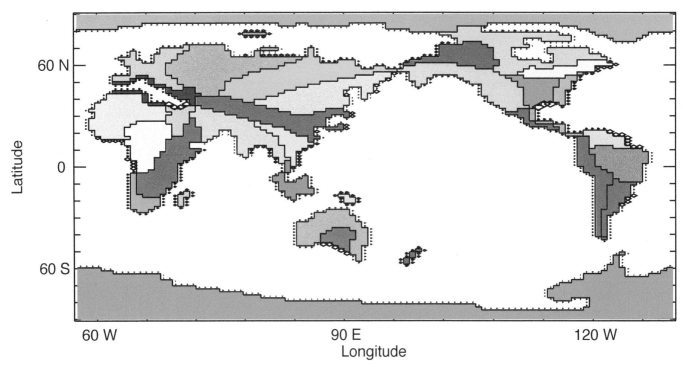

Fig. 2   River drainage basins: outlines indicate the 32 different river basins used and diamonds indicate discharge points. The size of the diamond indicates the proportion of total basin precipitation that is discharged at the point. Latitudes and longitudes are in the rotated coordinate system.

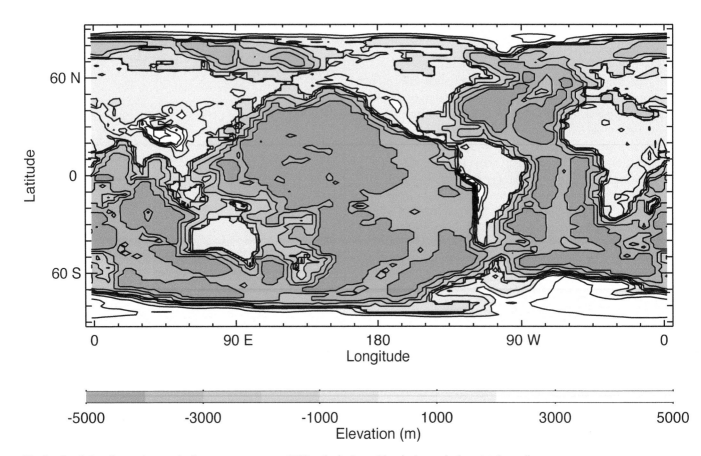

Fig. 3   Land elevations and ocean depths: contours are every 1000 m. Latitudes and longitudes are in the rotated coordinate system.

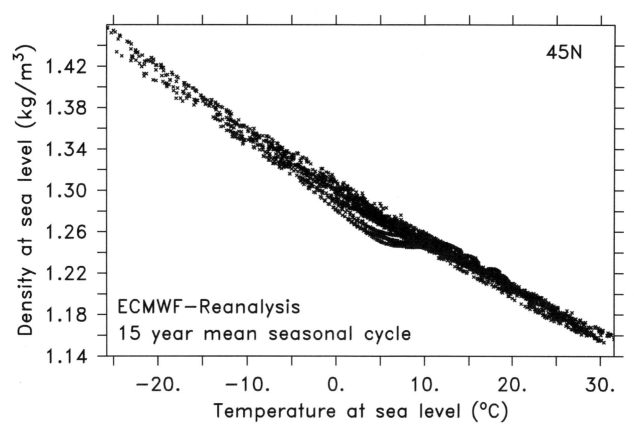

Fig. 4   Near-surface air density versus near-surface air temperature at 45°N from ERA15.

While one could use this wind stress approximation to provide mean winds, we only use it to calculate anomalous wind feedbacks that are then added to the mean observational wind field. That is, the model is integrated to equilibrium with specified present-day winds. Any perturbation from the present climate allows us to calculate SAT anomalies, and hence surface pressure anomalies (Eq. (14)). These surface pressure anomalies then allow us to calculate anomalous wind (Eqs (18) and (19)) and hence wind stress (Eq. (20)) anomalies which are added to the specified mean fields. This procedure leads to a first-order approximation of dynamical feedbacks associated with changing winds in a changing climate.

In the following we present an evaluation of this parametrization using results from an AGCM, in which wind velocities are explicitly computed. Initial tests showed that the use of global constants for the parameters a and b in Eq. (13) led to results which were not entirely satisfactory, mainly due to the fact that small errors in a and b lead to relatively large errors in the calculation of sea-level pressure in Eq. (14). We therefore allowed a and b to be functions of latitude, but not time. Their values at different latitudes, as calculated from the mean seasonal cycle from ERA15, are shown in Fig. 5. In order to estimate the uncertainty in a and b, we also recalculate them using National Centers for Environmental Prediction (NCEP) reanalysis data. While small differences exist (Fig. 5), the general form of the curves is remarkably similar for both reanalysis products.

For the evaluation we use sea-level pressure and near-surface temperature output from two runs (equilibrium control and $2 \times CO_2$ simulations) conducted using the CCCma second generation AGCM (McFarlane et al., 1992), coupled to a mixed layer ocean model. Geostrophic (frictional near the equator) velocities were computed from Eqs (18) and (19) using the GCM sea-level pressure data, and compared with the velocities arising from the parametrized sea-level pressure field obtained from Eq. (14) using the GCM near-surface temperature data as input. We did this for both the ERA15- and NCEP-derived values of a and b, although subsequent sections only use the ERA15 version of the parametrization. Figure 6 shows the zonally-averaged differences in geostrophic velocities between the $2 \times CO_2$ and control climates over each of the three major ocean basins, as well as over the total ocean. The parametrized geostrophic wind anomalies capture many of the large scale features of the anomalous geostrophic GCM wind fields, although the amplitude of the parametrized field is too large for u in the southern ocean. These differences highlight the limitations of the parametrization although the general shape of the curves is reasonably captured. Differences in the results from the NCEP- and ERA15-derived parametrizations are, in general, much smaller than their respective differences from the GCM wind field. Errors in the calculated surface wind stress would be larger due to the quadratic dependence of wind stress on the wind velocity in Eq. (20). Nevertheless, with these limitations in mind, we suggest that

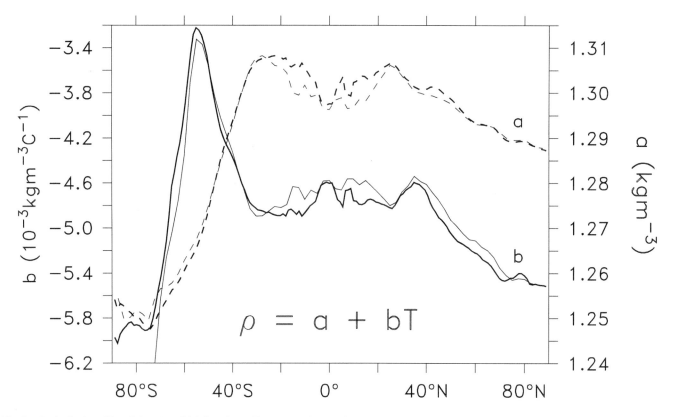

Fig. 5   Latitudinal profiles of slope b (solid, left scale) and intercept a (dashed, right scale) for the parametrization of surface air density as calculated by a linear regression from ERA15 (thick lines) and NCEP (thin lines).

the parametrization represents a first-order approximation of atmospheric wind feedbacks on the ocean circulation both through changes in sensible and latent heat exchange and changes in wind stress. The comparison of the red and green curves in Fig. 6 also allows us to argue that uncertainty in the physical basis of the parametrization is much larger than uncertainty in the individual constants a and b.

b Atmosphere–Ocean–Sea-Ice Coupling and the Sea-Ice and Snow Subcomponent Models

Our ESCM of intermediate complexity includes several options for the representation of sea-ice thermodynamics and thickness distribution. We start this section with a discussion of the standard sea-ice model which involves simple two-category (sea ice, open water) thermodynamics (Section 2b1i) and elastic-viscous plastic dynamics (Section 2b1ii). Other more sophisticated thermodynamic components have been developed such as a multi-layer thermodynamic model with heat capacity (Bitz and Lipscomb, 1999) and thickness distribution (Holland et al., 2001; Bitz et al., 2001). The more sophisticated thermodynamical treatment is discussed in Section 2b2 and the more sophisticated thickness distribution options are discussd in Section 2b3.

1 THE STANDARD SEA-ICE MODEL

i Thermodynamics

The standard representation of thermodynamics in the sea-ice model is the relatively simple zero layer formulation of Semtner (1976), with the lateral growth and melt parametrization of Hibler (1979). This assumes that the ice has no heat capacity and that the surface temperature is in instantaneous balance with the external forcing. The model predicts ice thickness (H_i), areal fraction (A_i) and ice surface temperature (T_i). The ice dynamics (Section 2b1ii) uses the elastic viscous plastic rheology developed by Hunke and Dukowicz (1997).

The change in ice thickness over a grid cell is given by,

$$\frac{\partial H_i}{\partial t} = \frac{(Q_b - Q_t)}{\rho_i L_f} - \frac{\rho_o}{\rho_i} E - L(H_i) \qquad (21)$$

where Q_b is the heat flux from the ocean, Q_t is the heat flux from the atmosphere (Q_t and Q_b are assumed to be positive downward), $\rho_i = 913$ and $\rho_o = 1035$ kg m^{-3} are representative values for the ice and water densities, respectively, $L_f = 3.34 \times 10^5$ J kg^{-1} is the latent heat of fusion of ice, E is sublimation given by Eq. (9) and $L(H_i)$ is the horizontal advection of H_i (Appendix B).

Since grid cells are allowed to be partially ice-covered, the total heat flux from the atmosphere is calculated for both the open water (Q_{to}) and ice covered (Q_{ti}) portions such that:

$$Q_t = (1 - A_i)Q_{to} + A_i Q_{ti} \qquad (22)$$

where A_i is the areal fraction of ice. Over the ocean:

$$Q_{to} = Q_{SW}(1 - C_A) - Q_{LW} - Q_{SH} - \rho_o L_v E \qquad (23)$$

and similarly over ice:

$$Q_{ti} = Q_{SW}(1 - C_A) - Q_{LW} - Q_{SH} - \rho_o L_s E \qquad (24)$$

where L_s is the latent heat of sublimation.

The heat flux from the ocean is calculated according to McPhee (1992)

$$Q_b = c_h u_\tau (T_f - T_o) \rho_o C_{po} \qquad (25)$$

where $c_h = 0.0058$ is an empirical constant, $u_\tau = 0.02$ m s^{-1} is the skin friction velocity, $C_{po} = 4044$ J kg^{-1} K^{-1} is the specific heat of sea water under ice at constant pressure. Note that Eq. (25) is equivalent to relaxing the temperature of the uppermost ocean grid box, with thickness $\Delta z = 50$ m, to the salinity-dependent freezing point $T_f(S_o)$, with a timescale of 5 days (here S_o is the salinity at the uppermost ocean level). The change in ice areal fraction depends on the growth rates over ice-covered and open water areas.

$$\frac{\partial A_i}{\partial t} = \begin{cases} \dfrac{(1 - A_i)}{H_0} \dfrac{(Q_b - Q_{to})}{\rho_i L_f} & \text{if } Q_{to} < Q_b \\ 0 & \text{otherwise} \end{cases}$$

$$+ \begin{cases} \dfrac{A_i}{2H_i}\left(\dfrac{(Q_b - Q_t)}{\rho_i L_f} - \dfrac{\rho_o}{\rho_i} E \right) & \text{if } Q_t > Q_b \\ 0 & \text{otherwise} \end{cases}$$

$$- L(A_i) \qquad (26)$$

where $H_0 = 0.01$ m is a demarcation thickness between thick and thin ice.

The first term on the right-hand side of Eq. (26) parametrizes the change in area under freezing conditions, where the open water portion $(1 - A_i)$ exponentially decays with time constant $H_0 \rho_i L_f/(Q_b - Q_{to})$. The second term parametrizes melting and assumes that all ice is uniformly distributed in thickness between 0 and $2H_i/A_i$. Assuming uniform melting over time Δt, this term opens up an area where thickness is less than $- \left(\frac{(Q_b - Q_t)}{\rho_i L_f} - E \right) \Delta t$.

The surface temperature is calculated by equating the conductive flux through the ice (and snow) with the energy flux at the surface

$$\frac{T_i - T_f}{\dfrac{H_i}{I_{cond}} + \dfrac{H_s}{S_{cond}}} = Q_{ti}(T_i) \qquad (27)$$

where $I_{cond} = 2.166$ W m^{-1} K^{-1} is a constant ice conductivity (and H_s and $S_{cond} = 0.31$ W m^{-1} K^{-1} are the thickness and conductivity of snow respectively). The maximum value of T_i is limited to the freezing temperature.

ii *Dynamics*

The ice dynamics are governed by the momentum balance:

$$m \frac{\partial \mathbf{u}}{\partial t} = \nabla \cdot \boldsymbol{\sigma} - mf\mathbf{k} \times \mathbf{u} + \boldsymbol{\tau}_a + \boldsymbol{\tau}_w - mg\nabla H \qquad (28)$$

where m is the ice mass per unit area, $\mathbf{u}$ is the ice velocity, $\boldsymbol{\sigma}$ is the internal stress tensor, f is the Coriolis parameter, $\mathbf{k}$ is the unit normal vector in the vertical, $\boldsymbol{\tau}_a$ and $\boldsymbol{\tau}_w$ are the atmospheric (wind) and oceanic stresses, H is the sea surface dynamic height, $g = 9.81$ m s^{-2} is the acceleration due to gravity. Ice acceleration and nonlinear advection terms are neglected. The internal stress tensor is solved from the elastic viscous plastic constitutive law described by Hunke and Dukowicz (1997):

$$\frac{1}{E} \frac{\partial \sigma_{ij}}{\partial t} + \frac{1}{2\eta} \sigma_{ij} + \frac{\eta - \zeta}{4\eta\zeta} \sigma_{kk} \delta_{ij} + \frac{P}{4\zeta} \delta_{ij} = \dot{\epsilon}_{ij} \qquad (29)$$

where E is Young's modulus (dependent on model resolution), η and ζ are shear and bulk viscosities (which depend on $\dot{\epsilon}_{ij}$ and P), P here is the pressure (which is a function of ice thickness and area), δ_{ij} is the Kronecker delta and $\dot{\epsilon}_{ij}$ is the strain rate tensor. This formulation adds an elastic component to the viscous plastic rheology described by Hibler (1979). It approximates the viscous-plastic solution on timescales associated with the wind forcing through the inclusion of artificial elastic waves that allow for an efficient numerical solution technique using an explicit timestep.

The oceanic stress is given by:

$$\boldsymbol{\tau}_w = \rho_o C_w |\mathbf{U}_w - \mathbf{u}|[(\mathbf{U}_w - \mathbf{u})\cos\theta + \mathbf{k} \times (\mathbf{U}_w - \mathbf{u})\sin\theta] \quad (30)$$

where $C_w = 0.0055$ is the water drag coefficient, $\mathbf{U}_w$ are the geostrophic ocean currents and $\theta = 25°$ is a turning angle. The tilt term $-mg\nabla H$ is calculated using geostrophy and approximating $\mathbf{U}_w$ by the second ocean level velocities.

2 SNOW

Precipitation falls as snow when the SAT (reduced with elevation at a prescribed lapse rate) falls below a critical value (usually −5°C). Snow allows for storage of moisture over land and, if complete melt does not occur, can be used as a primitive (non-dynamic) ice sheet model. Snow is not only important in creating a seasonal runoff but its presence also changes the value of the planetary albedo. The albedo is smoothly changed when the SAT falls below the critical value for snowfall (−5°C) and reaches it's maximum at −10°C. The albedo is also changed smoothly with the depth of snow, reaching a maximum when the average snow depth reaches 1 m. Over land, snow melt occurs once the SAT reaches the critical value for snowfall and melts at a rate of 0.5 cm d^{-1} °C^{-1}. The latent heat used to melt snow is taken from the atmosphere (see Eq. (4) with P interpreted as snow melt). Once snow accumulation is greater than 10 m, the model treats any further accumulation as rain. Optionally, for paleoclimate simulations snow thickness may be accumulated over land (up to some thickness limit).

Over sea ice snow is treated as part of the surface energy balance (Eq. (27)) with heat from the atmosphere preferentially melting snow and heat from the ocean preferentially melting ice. If all of the ice melts or there is no ice, snow is converted to water and placed in the ocean, with the necessary

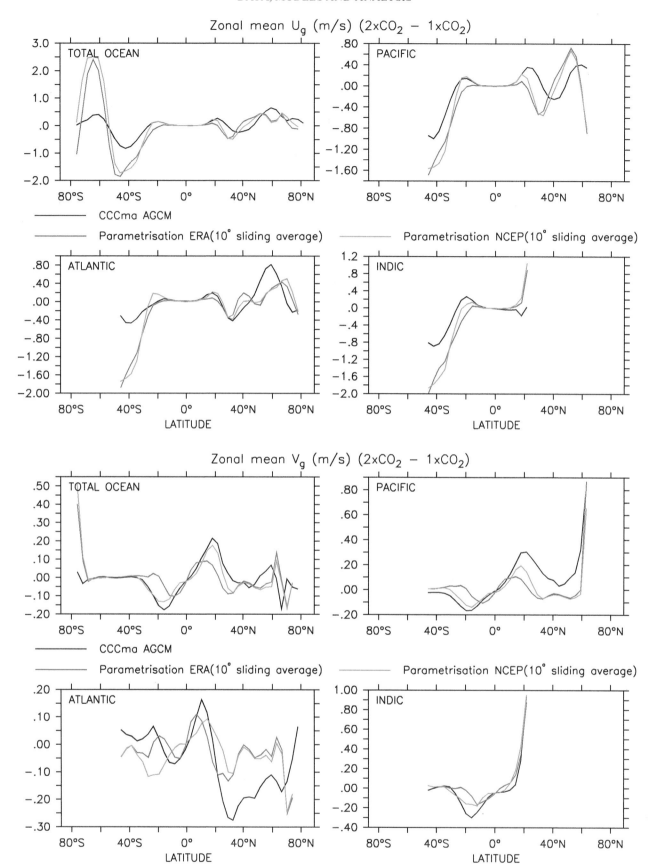

Fig. 6   Geostrophic (frictional near the equator) wind velocity anomalies for a doubling of CO_2. Zonal (upper four panels) and meridional (lower four panels) velocities were zonally averaged over the individual ocean basins and smoothed with a 10° sliding window to suppress small scale variability. Parameters a and b Eq. (13) are derived from ERA15 (red lines) and NCEP (green lines).

latent heat coming from the atmosphere. If the weight of accumulated snow is sufficient to push the ice-snow interface below the ocean surface, then the snow that would otherwise be submerged is converted to ice.

3 OTHER SEA-ICE MODEL OPTIONS

i *Ice thickness distribution*

A sub-gridscale Ice Thickness Distribution (ITD) (Thorndike et al., 1975) is incorporated into the model using the formulation of Hibler (1980) and Flato and Hibler (1995). This allows an arbitrary number of ice thickness categories within each model grid cell. The thermodynamic evolution of each category is independent of other categories with different fluxes computed at the ice-ocean and ice-atmosphere interfaces. The thickness distribution function is defined following Thorndike et al. (1975) as

$$\int_{h_1}^{h_2} g(h)dh = \frac{1}{\Gamma} \gamma(h_1, h_2) \tag{31}$$

where Γ is the total area of some fixed region, and $\gamma(h_1, h_2)$ is the area within Γ covered by ice of thickness h in the range $h_1 \leq h < h_2$. This distribution evolves according to

$$\frac{\partial g(h)}{\partial t} = -\nabla \cdot (\mathbf{u}g) - \frac{\partial}{\partial h}(f_g g) + \Psi + F_L, \tag{32}$$

where here f_g is the thermodynamic growth rate and F_L is the modification of the ITD due to lateral melting. ψ, the "redistribution" function, describes how ice is redistributed among the various thickness categories due, for example, to deformation events in which thin ice is ridged into thicker categories. A detailed description of the formulation of this function can be found in Flato and Hibler (1995). The compressive strength of the ice cover is computed as a function of the ITD, following Rothrock (1975). A description of the theory and numerical method used to solve for the compressive strength is discussed in Hibler (1980).

ii *Ice thickness distribution and multi-layer thermodynamics*

Another formulation of the ice-thickness distribution of Thorndike et al. (1975) has been developed with the UVic ESCM by Bitz et al. (2001) in which each category has a single thickness of ice which varies in response to ice growth/melt, advection, and deformation. This ice thickness distribution model has more in common with that shown in Rothrock (1986) where the interpolation process was eliminated and each category was allowed to vary freely in thickness for a one-year simulation. The number of categories is limited by constraining each category to lie between lower and upper thickness limits that are fixed. When a category outgrows its limits, the ice is transferred from one category to another, where it is merged with ice that may already exist by conserving volume, energy and area.

The thickness-distribution, $g(h)$, for M categories is written

$$g(h) = g_0 \, \delta(h) + \sum_{i=1}^{M} g_i \, \delta(h - H_i) \tag{33}$$

where M is the number of ice categories, $\delta(h)$ is the Dirac delta function, and g_i is the concentration within category i (category 0 is open water). The evolution of horizontal ice concentration and ice volume per unit area, V_i, for each category is governed by a pair of continuity equations

$$\frac{\partial g_i}{\partial t} = -\nabla \cdot (\mathbf{u}g_i) + \Psi_i + \mathcal{G}_i \tag{34}$$

and

$$\frac{\partial V_i}{\partial t} = -\nabla \cdot (\mathbf{u}V_i) + \Theta_i + \nu_i, \tag{35}$$

respectively, where $\mathbf{u}$ is the velocity of the ice, ψ_i and Θ_i are contributions by mechanical redistribution (Thorndike et al., 1975), and $\mathcal{G}i$ and ν_i are contributions by thermodynamic processes. Employing a vertical coordinate that is normalized by the sea-ice thickness, $z_i^* = z/h_i$, the energy of melting, $E_i(z_i^*)$ evolves according to

$$\frac{\partial E_i}{\partial t} = -\nabla \cdot (\mathbf{u}E_i) + \Pi_i + \varepsilon_i, \tag{36}$$

where $\Pi_i(z_i^*)$ is the contribution by mechanical redistribution and $\varepsilon_i(z_i^*)$ is the contribution by thermodynamic processes.

Vertical heat conduction and storage in the sea ice are governed by the heat equation. The vertical temperature profile is resolved in the ice in roughly 50-cm intervals, so the number of vertical layers N_i for category i depends on the thickness limits for that category. Following Bitz and Lipscomb (1999), the amount of energy needed to melt a unit volume of sea ice with salinity S at temperature T in Celsius, is equal to

$$q(S, T) = \rho_i c_i(-\mu S - T) + \rho_i L_f \left(1 + \frac{\mu S}{T}\right), \tag{37}$$

where $c_i = 2054$ J kg^{-1} K^{-1} and $L_f = 3.34 \times 10^5$ J kg^{-1} K^{-1} are the heat capacity and latent heat of fusion for fresh ice and $\mu = 0.054$°C psu^{-1} is the empirical constant from the linear approximation relating the melting temperature and the salinity of sea ice $T_m = -\mu S$. Thus the energy of melting for layer l in category i in Eq. (36) is

$$E_{i,l} = q(S_{i,l}, T_{i,l}) \frac{V_i}{N_i}. \tag{38}$$

In Eqs (37) and (38), the salinity profile is specified to vary from 0 psu at the top to 3.2 psu at the bottom according to the observations of Schwarzacher (1959), but the ice is assumed to be fresh for purposes of computing the freshwater exchange between sea ice and the atmosphere or ocean. For more details see Bitz et al. (2001).

4 CONTINENTAL ICE DYNAMICS MODEL AND COUPLING STRATEGY

Our ESCM includes the Continental Ice Dynamics Model (CIDM) that was developed at the University of British Columbia (Marshall, 1996; Marshall and Clarke, 1997a, 1997b). The reader is referred to Yoshimori et al. (2001) for a detailed evaluation of its performance and an application to the last glacial termination. Generally, glaciers can move due to the superposition of three mechanisms: 1) internal deformation of ice (creep); 2) basal sliding; and 3) subglacial sediment (bed) deformation (Alley, 1989; Paterson, 1994). Flow in most regions of an ice sheet is through internal deformation, whereas ice streams and surge lobes flow by the last two mechanisms. The CIDM models the first mechanism based on the vertically-integrated mass balance equation and three-dimensional momentum equations under the shallow ice approximation, in which the flow is approximated as vertical shear deformation with no horizontal shear deformation (Hutter, 1983, 1993). As an ice rheology in the constitutive equations, the traditional Glen's flow law is employed in which ice is treated as nonlinear (power law) viscous material (Glen, 1955, 1958; Paterson, 1994).

To investigate first-order continental ice dynamics and the coupled influence of land ice on the climate, we adopt a simple approach. Ice is approximated as isothermal and the vertically-integrated mass balance equation is written as

$$\frac{\partial H_i}{\partial t} + \mathbf{V}_h \cdot \left(\overline{\mathbf{v}}_h \, H_i \right) = M \qquad (39)$$

where H_i is ice thickness, $\overline{\mathbf{v}}_h$ is the vertically-averaged horizontal velocity, M is the net accumulation, t is time, and $\mathbf{V}_h$ is a horizontal gradient operator. The vertically-integrated horizontal volume flux is written as

$$\overline{\mathbf{v}}_h \, H_i = -\frac{2A(T)}{n+2} \left(\rho_i \, g \right)^n H_i^{n+2} \left| \mathbf{V}_h \, h_s \right|^{n-1} \mathbf{V}_h \, h_s \quad (40)$$

where ρ_i is ice density, g is the gravitational acceleration, h_s is the surface elevation. The flow law exponent n of Glen's flow law is set to 3, which it typical for ice sheet studies (Paterson, 1994). Therefore, the volume flux varies with ice thickness to the power 5 and surface slope to the power 3. Also, the flow law coefficient is given by

$$A(T) = m \, A_0 \, e^{(-Q/RT)} \qquad (41)$$

where T is the ice temperature, R is the ideal gas constant, A_0 is 1.14×10^{-5} Pa^{-3} yr^{-1} ($T < 263.15$ K) and 5.47×10^{10} Pa^{-3} yr^{-1} ($T \geq 263.15$ K), and Q, the activation energy, is 60 kJ mol^{-1}

($T < 263.15$ K) and 139 kJ mol^{-1} ($T \geq 263.15$ K). The enhancement factor m accounting for the softening of ice due to impurities, meltwater, and crystal fabric is set to 3 (van de Wal, 1999). We use a uniform (effective) ice temperature of $-15°$C representative of the lower ice column, where most internal deformation occurs. Although this value is too cold and makes the ice too stiff in some regions, it is a reasonable average of warm- and cold-based regions within ice rheology uncertainties.

In general, the bedrock responds to the varying ice load through an elastic deformation of the lithosphere and a viscous flow in the asthenosphere. The resulting uplift and depression of the bedrock is approximated by local displacement of mantle material, and neglect of the lithosphere flexure. A local response type isostatic adjustment is applied, in which the response is relaxed to a local isostatic equilibrium:

$$\frac{\partial h_b}{\partial t} = -\frac{1}{\tau} \left(h_b + r_\rho \, H_i - h_b^0 \right) \qquad (42)$$

where h_b is the bedrock elevation, $r_\rho = 0.277$ is the ratio of ice density to bedrock density, h_b^0 is the undisturbed (ice-free) equilibrium bedrock elevation, and τ is the relaxation time set to 3,000 years here (Peltier and Marshall, 1995; Tarasov and Peltier, 1997).

We do not permit basal sliding or ice stream development and, as a first step, we use the same resolution in the CIDM as the climate model ($3.6° \times 1.8°$), although we are working towards refining the resolution of the ice-sheet model and sub-gridscale topographic effects on ice-sheet evolution. Our current resolution is rather coarse for the proper representation of ice dynamics and mass balance, particularly in marginal-ice regions. However, this resolution permits easy and consistent coupling with the climate model, while providing a first look at the coupled problem.

In nature, ice accumulation occurs through snowfall, deposition of water vapour from the atmosphere (negligible), refreezing of melt water, and freezing of rain; while ice depletion occurs by sublimation, surface or basal melting, and calving in the ocean. For simplicity, we assume the dominant factors to be snowfall, surface melting and calving. Snowfall and surface melting are provided by the atmospheric model and we assume instantaneous calving and subsequent melting over the ocean, i.e., no ice is allowed over the ocean except sea ice.

Because of the slow response time of ice sheets relative to other climate system components, an asynchronous coupling strategy is adopted. When coupled, we integrate the climate model for 10 years, average the net surface accumulation over this period, and pass it to the CIDM as a boundary condition. The CIDM is then integrated for 2,000 years with a 10-year time step and the average ice thickness and surface elevation are returned to the climate model. This cycle is repeated until both models reach steady state.

As our ice sheet model is currently used only for equilibrium paleoclimate simulations, all precipitation occurring over land in the atmospheric model is passed instantaneously into the oceanic component (Fig. 2) regardless of whether it is rain or snow. This differs from the case where the climate model is run without the interactive CIDM, where snow remains on land until it melts. As mentioned above, to allow for ice-sheet growth the net snow accumulation is passed to the CIDM, although no melt water is passed to the oceanic component since the original source of the melt water (i.e., snow) is already passed to the ocean model when it falls on land. This is done to avoid making the ocean unstable through massive freshwater discharge owing to the asynchronous coupling which could allow the ice to be transported and piled up in an incompatibly warm location. The atmospheric model has no limitation on the amount of surface melting as long as ice exists. Therefore, it is possible that the ice accumulated in the 2,000-year CIDM integration segment could melt away during one atmosphere/ocean/sea-ice model integration segment (10 years). This strategy also allows arbitrary initial conditions for ice thickness. For example, this prevents the oceanic component becoming unstable due to flooding even when the LGM ice sheets are imposed as an initial condition in an incompatibly warm climate, which is inevitable for sensitivity experiments under perpetual forcing.

In summary, the net surface accumulation is calculated in the atmospheric model and passed to the CIDM. Ice thickness is then calculated in the CIDM and passed to the atmospheric model. There is no direct exchange of freshwater between the CIDM and the oceanic component. As a result, the total amount of moisture is strictly conserved within the climate model. Therefore, the model does not yet include the effects of the meltwater discharge on the ocean circulation, nor the effects of changes in continental ice volume on the ocean mean salinity. This latter issue is dealt with through modifying the initial global mean ocean salinity, according to proxy reconstructions, prior to integrating the model to equilibrium.

With these limitations in mind, it is very satisfying to find that simulated and observed present-day ice sheets agree well (Fig. 7) in both hemispheres (Yoshimori et al., 2001), although we tend to overestimate slightly their thickness. The maximum ice thickness near the Greenland summit is 3867 m in the model and about 3050 m in the observed field, whereas near Vostok in Antarctica, the simulated field is 4753 m thick versus 4150 m in observation. Differences between observed and simulated fields tend to be larger near margins due to the coarse resolution of our CIDM and hence its inability to capture sharp surface slopes there. These maximum values can be easily reduced by changing the ice dynamics flow parameter, although we have shown figures here that are consistent with those in Yoshimori et al. (2001).

c Ocean Model

The ocean component of the coupled model is the Geophysical Fluid Dynamics Laboratory (GFDL) Modular Ocean Model (MOM) version 2.2 (Pacanowski, 1995). MOM is based on the Navier Stokes equations subject to the Boussinesq and hydrostatic approximations. The momentum and tracer equations in spherical geometry are:

$$\frac{\partial u}{\partial t} = \frac{-1}{\rho_o \, r_e \cos \phi} \frac{\partial p}{\partial \lambda} + \frac{\partial}{\partial z}\left(A_v \frac{\partial u}{\partial z} \right) + \mathcal{F}(A_h, u, v)$$
$$+ \frac{uv \tan \phi}{r_e} + fv - L(u) \tag{43}$$

$$\frac{\partial v}{\partial t} = \frac{-1}{\rho_o \, r_e} \frac{\partial p}{\partial \phi} + \frac{\partial}{\partial z}\left(A_v \frac{\partial v}{\partial z} \right) + \mathcal{F}(A_h, v, -u)$$
$$- \frac{u^2 \tan \phi}{r_e} - fu - L(v) \tag{44}$$

$$\frac{\partial T}{\partial t} + L(T) = \frac{\partial}{\partial z}\left(k_v \frac{\partial T}{\partial z} \right) + \mathbf{\nabla}(k_h \mathbf{\nabla} T) \tag{45}$$

$$\frac{\partial S}{\partial t} + L(S) = \frac{\partial}{\partial z}\left(k_v \frac{\partial S}{\partial z} \right) + \mathbf{\nabla}(k_h \mathbf{\nabla} S) \tag{46}$$

$$\frac{\partial w}{\partial z} + \frac{1}{r_e \cos \phi}\left(\frac{\partial u}{\partial \lambda} + \frac{\partial(v \cos \phi)}{\partial \phi} \right) = 0 \tag{47}$$

$$\frac{\partial p}{\partial z} = -g\rho \tag{48}$$

where u, v and w are the velocity components in the zonal, meridional and vertical directions respectively, f is the Coriolis parameter, r_e is the radius of the earth, p is the pressure, ρ_o is a representative density for sea water, t is time, g is the acceleration due to gravity, and $A_v = 1 \times 10^{-3}$ m^2 s^{-1} is the vertical eddy viscosity. $\mathcal{F}$ is the horizontal Laplacian operator (Appendix B) representing the horizontal mixing of momentum given an eddy viscosity $A_h = 2.5 \times 10^5$ m^2 s^{-1} and L is the advection operator (Appendix B). The ocean density ρ of sea water is a nonlinear function of potential temperature (T), salinity (S) and pressure (UNESCO, 1981). A horizontal diffusivity of $k_h = 2 \times 10^3$ m^2 s^{-1} and a modified form of the Bryan and Lewis (1979) vertical distribution of vertical diffusivity, ranging from $k_v = 6 \times 10^{-5}$ m^2 s^{-1} in the thermocline to $k_v = 1.6 \times 10^{-4}$ m^2 s^{-1} in the deep ocean, are used.

While the climatology discussed in Section 3 is based on integrations using the standard horizontal/vertical diffusive approach, detailed sensitivity studies have also been done using different parametrizations of sub-gridscale mixing. For example, the Gent and McWilliams (1990) parametrization for mixing associated with mesoscale-eddies, in which the diffusion tensor is rotated along isopycnals and isopycnal thickness diffusion is introduced to account for the removal of potential energy from the stratification due to baroclinic instability, has been shown to improve certain aspects of our model climatology (Robitaille and Weaver, 1995; Wiebe and Weaver, 1999; Duffy et al., 1999, 2001). The model also incorporates a param-

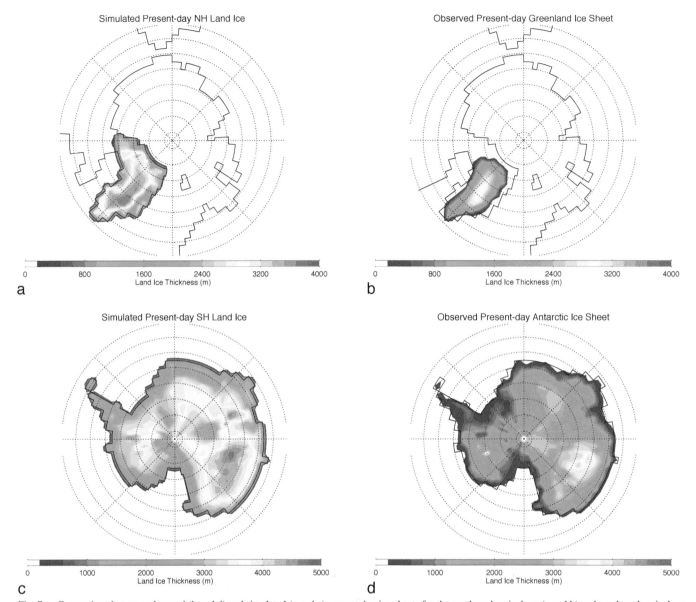

Fig. 7   Comparison between observed (b and d) and simulated (a and c) present-day ice sheets for the northern hemisphere (a and b) and southern hemisphere (c and d).

etrization of brine rejection during sea-ice formation (Duffy and Calderia, 1997) that substantially improves our representation of intermediate water properties (Duffy et al., 1999, 2001).

At the surface, the model is driven by both wind stress and surface buoyancy forcing. Since it is unphysical for ocean temperatures to fall below freezing, the ocean model always calculates the maximum amount of heat available in the first layer (Q_b; Eq. (25)) which may be given to either the atmosphere or sea ice. The net heat Q_H and implied salinity Q_S fluxes into the ocean are then,

$$Q_H = (1 - A_i)\, Q_{to} + A_i\, Q_b \qquad (49)$$

$$Q_S = \rho_o\, S^* \left(E - P - R + \frac{Q_b - Q_t}{\rho_o\, L_f} \right) \qquad (50)$$

where $S^* = 34.9$ psu is a representative salinity for the ocean and R is the runoff from land. Here Q_{to}, Q_t and Q_b are given by Eqs (23), (22) and (25), respectively, and all heat fluxes in Eqs (49) and (50) are defined as positive downward. As such, Q_b is always negative as the top level ocean temperature is always above the freezing point of water T_f (calculated according to Fofonoff and Millard (1983)).

The horizontal grid resolution is the same as that used in the atmosphere and sea-ice models and the vertical grid has 19 unequally spaced levels that vary smoothly in size, following a parabolic profile, from 50 m at the surface to 518 m at the deepest level (with bottom at 5396 m). The ocean bathymetry is taken from the Suarez and Takacs (1986) dataset. We used a raised cosine filter in both directions (latitude and longitude) to produce a weighted average for all of the bathymetry in a given grid box. In this approach, the data at the centre have the highest weight and the data at the edges have zero weight. In between, the weight is given a raised cosine functional form so that it slowly and smoothly varies from the highest weight unit it approaches zero. The sum of the weights is normalized to 1.0. Five islands were included: Madagascar, Australia, New Guinea, New Zealand and Spitsbergen. An unrealistic hole in the Lomonosov Ridge that was present in the Suarez and Takacs (1986) dataset (near the pole) was filled such that the maximum depth along the ridge was 2000 m. Due to the coarse resolution of our model, and since we require at least two tracer points across a strait for a baroclinic transport to exist, Fram, Hudson, Madagascar and the Gibraltar Straits are all wider than in reality. Bering Strait is not open in the version of the model discussed here. Flow over deep sills, such as over the Greenland-Iceland-Faroes rise are fundamental to processes involved in North Atlantic Deep Water (NADW) formation. As we neither resolve the small-scale troughs through these sills, nor currently parametrize their effects, we had to deepen Denmark Strait to 700 m and remove Iceland in order to improve the northward transport of heat in the North Atlantic.

d Coupling

Coupling between the atmosphere/sea-ice models and the ocean model is done every two ocean time steps, by which time the atmosphere and sea ice have completed four time steps. In the ocean we use a longer time step to integrate the tracer equations than the baroclinic velocity and barotropic vorticity equations. No time step splitting with depth (Bryan, 1984) is used.

Since each component uses a leapfrog time stepping scheme with an intermittent Forward Euler mixing time step, a special technique was used to calculate the fluxes exchanged between components to ensure that heat and freshwater were conserved exactly. If the number of time steps between mixing is limited to an even number, and the number of time steps that each subcomponent model completes between coupling is also an even number, then the average flux can be calculated by averaging the fluxes from every other time step (see Fig. 8). That is, only the fluxes that determined the final state of a particular subcomponent model at the time step of coupling are averaged. Coupling in this manner is extremely useful in finding any potential conservation errors within and between models. As an example, we show the total globally-averaged difference in heat and equivalent freshwater from the initial condition in a 100-year run (Fig. 9). Unlike salt and equivalent freshwater, which form a closed system in the model, heat is both gained from and lost to space. As such, the total heat calculation also subtracts any net heat to/from space since the start of the run. A least squares fit of the data shows a drift of about 1 TJ per 100 years or 31 W (6×10^{-14} W m^{-2}) for heat and about 0.5 Gg per 100 years or 15 mg s^{-1} (3×10^{-14} mg^{-1} s^{-1} m^{-2}) for freshwater, which is close to machine precision and most likely due to truncation error. In other words, the amount of drift globally in our coupled climate model is the equivalent of not accounting for a low wattage light bulb and a rapidly dripping faucet acting over the surface of the entire Earth. Ensuring that at least two time steps were completed for any model before coupling also increased the stability of the coupled model permitting the use of longer time steps.

3 Present-day climate simulation

As noted earlier, the standard version of our ESCM uses the standard representation of sea-ice (Section 2b1) thermo-dynamics (Section 2b1i) and dynamics (Section 2b1ii), as well as a diffusive representation for moisture transport. In this section we describe the control climatology of the ESCM under present-day radiative forcing. This typically involves 2000 years of integration by which time annually-averaged surface fluxes are close to zero. In the next section, we show the dramatic improvements that are realized when moisture transport is treated through advection instead of diffusion.

a Atmosphere

The equilibrium (after 2000 years) present-day atmospheric SAT and its difference from the annually-averaged NCEP Kalnay et al. (1996) reanalysis climatology (Fig. 10), shows good agreement with observations, especially on the zonal mean (Fig. 10). The global and annual mean model SAT is 13.9°C which differs from the NCEP climatology by only 0.12°C, and a similarly reasonable agreement with NCEP reanalysis data is seen in the seasonal cycle (e.g., Table 1). Of course there are local regions in the model where differences are larger. For example, over land, Antarctica and most high elevation regions are cooler than NCEP while the north-east coast of North America and Asia are warmer. Some of this discrepancy may be attributed to our use of a constant atmospheric lapse rate (6.5°C km^{-1}), but more likely the differences arise from our lack of vegetation and land surface feedbacks on SAT, as well as the lack of atmospheric dynamics (stationary waves, vertical motion etc. — see Section 4). That is, in our model, the planetary albedo is independent of land surface vegetation type and there is no latent or sensible heat exchange between the land and atmosphere.

Over the ocean, the North Atlantic is cooler than in the observations which, as we will discuss, is due to the slightly more extensive sea-ice extent than is observed in this region. The latter arises from the fact that NADW formation tends to occur slightly too far south, as is common in most ocean GCMs without an explicit representation of bottom boundary layers or a parametrization for flow over sills. In the Southern Ocean, the model is slightly too warm, although this feature can be improved with the inclusion of the Gent and McWilliams (1990) parametrization for mixing associated with mesoscale eddies (Duffy et al., 1999, 2001).

As noted by Fanning and Weaver (1996), representing moisture advection as a diffusive process is a useful simplification in attempting to develop an idealized atmospheric model suitable for long-timescale paleoclimate sensitivity studies, as the gross features of the precipitation and evaporation fields can be captured. Nevertheless, this approach has serious shortcomings as highlighted by the fact that tropical moisture transport, through surface trade winds and their subsequent

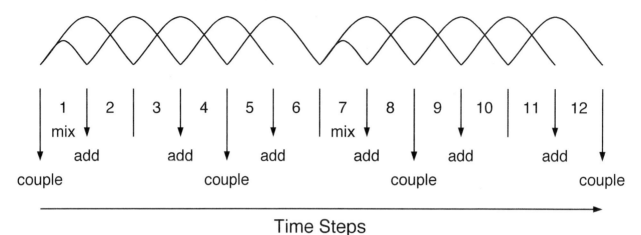

Fig. 8   Exact flux coupling for a leapfrog time step scheme. Coupling times (every 4 time steps in this example) are indicated by "couple"; "add" indicates times when the fluxes are added to produce the average flux (every other time step); "mix" indicates a mixing time step (every 6 time steps in this example).

convergence on the equator, is actually an up-gradient process. As discussed in Section 4 (see Fig. 23), we attempted to minimize these differences by allowing the moisture diffusion parameter to vary as a function of latitude.

When averaged globally, our total winter precipitation of 2.78 mm d^{-1}, and summer precipitation of 2.67 mm d^{-1}, are quite similar to both NCEP reanalysis data as well as the results from other coupled models (Table 2). Nevertheless, the two-dimensional field is far too diffuse (Fig. 11a), even in the zonal average (Fig. 11c), to be considered an accurate representation of observations (Fig. 11b). Our representation of evaporation, which does not rely on unresolved atmospheric processes, bears much better resemblance to the observed field (Fig. 12), although it slightly overestimates evaporation in the subtropical gyres. When combined (Fig. 13), the total evaporation minus precipitation field is more diffuse than in reality with the largest differences in the region of the Intertropical Convergence Zone (ITCZ) where moisture transport is up-gradient and so can't be represented by a down-gradient diffusive process.

b Ocean
Over the ocean, the atmospheric surface temperature field shown in Fig. 10a is tightly coupled to the simulated SST field (Fig. 14a), although it is smoother as anomalies tend to diffuse in the atmosphere. The simulated SST shows good agreement with the annual mean observations of Levitus and Boyer (1994) as illustrated in Fig. 14b, especially in the zonal mean (Fig. 14c), and the global average (18.3°C, about 0.05°C colder than that obtained from the Levitus climatology). Nevertheless, some large differences exist. In the southern Indian Ocean and near the mid-latitude coasts of the Pacific and Atlantic, simulated SSTs are too warm, while the Gulf of Alaska and Greenland-Iceland-Norwegian (GIN) Seas are too cold. The SST errors in the North Atlantic are largely attributable to the Gulf Stream separating too far north and not penetrating far enough into the GIN seas to keep them ice free. Similarly, in the Pacific, the Kuroshio separation is

too far north and not enough warm water reaches the Gulf of Alaska. These are common errors found in coarse resolution coupled models that do not employ flux adjustments.

The large positive difference between simulated and observed fields in the southern Indian Ocean is a consequence of the Antarctic Circumpolar Current being steered southward by the Kerguelen plateau and bringing relatively warm waters poleward. This tends to reduce the sea-ice extent simulated off the coast of Antarctica between the African and Australian continents. The warmer than observed areas off the west coasts of North and South America as well as Africa, coincide with areas of known marine stratocumulous cloud formation in the atmosphere. These are not treated in our atmospheric model, nor well, if at all in atmospheric GCMs, and so we overestimate the amount of incident solar radiation at the ocean surface, leading to warmer than observed SSTs. The lack of a cloud scheme may also account for discrepancies in other areas.

The simulated sea surface salinity (SSS) field (Fig. 15a) tends to be more uniform than observations (Figs 15b and 15c), although the global average is similar (34.75 psu simulated — a difference of only 0.04 psu from the Levitus et al. (1994) climatology). The most significant differences arise in the middle of subtropical gyres, where simulated salinities are too fresh; in the Arctic, where simulated salinities are too large; in the GIN Sea region, where salinities are too fresh; in the Antarctic Circumpolar Current, where salinities are too large. As noted in the next section, the surface salinity errors found in the subtropical gyre regions and the Arctic are largely eliminated when moisture transport is accomplished through advection rather than through diffusion. This follows since high pressure centres tend to be associated with descending air and surface divergence, advecting moisture away from the region. The diffusive representation of moisture transport, on the other hand, always transports moisture down gradient and hence even into high pressure centres such as those associated with the semipermanent subtropical and Arctic highs. The salinity discrepancy in the GIN Sea region is, as noted earlier,

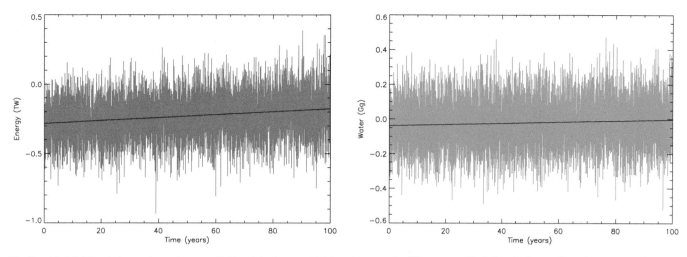

Fig. 9   Model drift relative to the total energy (left) and fresh water (right) at the start of a 100-year run. Black lines indicate a linear least squares fit to the totals.

once more associated with our inability to transport surface waters effectively across the Iceland-Faroes ridge, leading to NADW formation occurring too far to the south. We hope to alleviate this problem through the development of a new bottom boundary layer parametrization for flow over sills.

Finally, the saltier than observed waters in the Antarctic Circumpolar Current region arise from our use of a horizontal/ vertical ocean diffusion scheme. Although not in this standard version of our model, Robitaille and Weaver (1995), Duffy et al. (1999, 2001), and Wiebe and Weaver (1999) showed that when the diffusion tensor is rotated along isopycnals, and the Gent and McWilliams (1990) parametrization for mixing associated with mesoscale eddies is included, these discrepancies are greatly reduced. These parametrizations reduce spurious convection in the Southern Ocean, and so fresh surface waters are mixed less with more saline deeper waters. In addition, spurious meridional cross isopycnal diffusion of more saline subtropical waters into higher latitudes is eliminated. The net effect of including the Gent and McWilliams (1990) parametrization together with isopycnal mixing is to keep the surface ocean more fresh. Finally, as discussed in Duffy et al. (1999, 2001), a more realistic treatment of brine rejection, in which salinity is mixed to depth (which depends on a prescribed density contrast) during the process of sea-ice formation, improves the representation of the southern hemisphere water masses.

The global zonally-averaged potential temperature and salinity fields (Figs 16 and 17, respectively) show features typical of coarse resolution ocean models which do not incorporate isopycnal mixing and the Gent and McWilliams (1990) parametrization for mixing associated with mesoscale eddies. The thermocline is deeper than observed, so that positive potential temperature anomalies are evident (Fig. 16b). Similarly, in the subtropical gyre regions, which were too fresh at the surface (Fig. 15), subduction causes further freshening at depth (Fig. 17b). While the properties of the deep waters are reasonably well captured, the formation of rela-

tively fresh, Antarctic Intermediate Water (AAIW) is not captured. As noted earlier and discussed in Duffy et al. (1999), AAIW formation is much better simulated when an improved parametrization for sub-grid scale brine rejection, that occurs during the formation of sea ice around Antarctica, is incorporated (see also Section 3d). As we will see in Section 4, further improvements to the salinity distribution are realized when moisture transport is accomplished through advection rather than diffusion.

The meridional overturning in the Atlantic (Fig. 18a) indicates that about 18 Sv of NADW formation occurs in the model, although, as noted earlier, there is a tendency for it to form too far south. About 12 Sv of this NADW is transported across the equator into the South Atlantic. AABW (Fig. 18b), formed in the Ross and Wedell Seas, extends to about 30°N, filling the deep North Atlantic up to a depth of about 3000 m.

c *Sea Ice*

Even in its standard form, the sea-ice model used in the UVic ESCM is quite sophisticated when compared to those in use in most coupled atmosphere-ocean GCMs. When compared to these same models (Table 3), the resulting seasonal climatology of the UVic ESCM ranks very high (for both sea ice and snow cover combined and considered separately). Even so, the fact that NADW forms too far south (Fig. 18a) leads to a sea-ice cover that extends too far south in the North Atlantic, especially in winter (Fig. 19). Winds in the Arctic cause the sea ice to pile up on the northern coast of Canada and north-west Greenland where a maximum thickness of slightly over 4 m is simulated. Averaged over the entire Arctic Ocean, sea ice is about 1 m thick. In the southern hemisphere, sea ice typically forms to a thickness of 20 cm, consistent with observations, although it is thicker near the coast. Very little sea ice forms off the coast of Antarctica south of Australia due to the Antarctic Circumpolar Current being steered too far south by the Kerguelen plateau bringing relatively warm waters poleward.

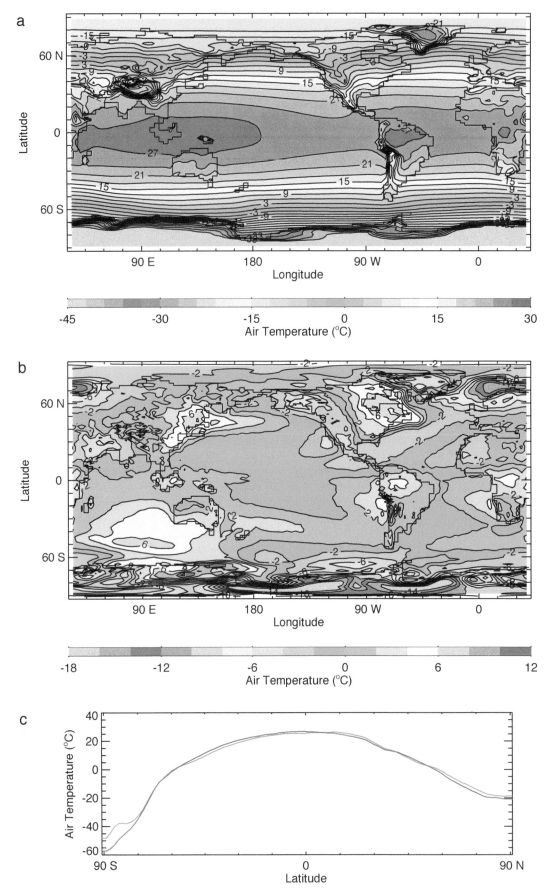

Fig. 10 Annually-averaged surface air temperature from a) model, b) model-NCEP and c) zonally-averaged model (red) and NCEP (green) climatologies.

TABLE 1. Coupled model simulations of globally-averaged SAT ranked by the average absolute error from observations (Gates et al., 1996).

Surface air temperature (°C)			
DJF	JJA	Error	Model
12.4	15.9	–	observations
12.3	15.7	0.15	UVic*
12.0	15.7	0.30	CCC
12.6	15.5	0.30	COLA*
12.1	15.3	0.45	CSIRO
13.0	15.6	0.45	GISS*
12.7	16.7	0.55	BMRC*
12.0	15.0	0.65	UKMO
11.0	15.2	1.1	MPI(LSG)
11.2	14.8	1.2	MPI(OPYC)
13.4	17.4	1.3	MRI
9.6	14.0	2.4	GFDL
15.5	19.6	3.4	NCAR*

*Indicates a model without flux adjustments.

TABLE 2. Coupled model simulations of globally-averaged precipitation ranked by the average absolute error from observations (Gates et al., 1996).

Precipitation (mm day^{-1})			
DJF	JJA	Error	Model
2.74	2.90	–	observations
2.72	2.86	0.030	CCC
2.79	2.92	0.035	BMRC*
2.73	2.82	0.045	CSIRO
2.64	2.73	0.14	MPI(OPYC)
2.78	2.67	0.14	UVic*
2.89	3.03	0.14	MRI
2.64	2.67	0.17	COLA*
3.02	3.09	0.24	UKMO
3.14	3.13	0.32	GISS*
2.39	2.50	0.38	GFDL
3.78	3.74	0.94	NCAR*

*Indicates a model without flux adjustments.

d Water Mass Formation

Here we extend the analysis of Duffy et al. (1999) by examining the advection and diffusion of three passive 'age' tracers in a version of the model that now incorporates the Duffy and Caldeira (1997) and Duffy et al. (1999, 2001) parametrization of brine rejection due to sea-ice growth, the Gent and McWilliams (1990) parametrization for mixing associated with mesoscale eddies, and five sea-ice categories (plus open water) (Bitz et al., 2001). The purpose of this experiment is to examine the major pathways of the global oceanic circulation within our model, and specifically to determine to what extent the model is able to capture the formation and recirculation of the NADW, AAIW and Antarctic Bottom Water (AABW) water masses. Another issue of interest is to see whether or not the model can simulate the so-called 'cold route' of the NADW return path through Drake Passage (Gordon, 1986, 1997) in the form of intermediate water, recently inferred from hydrographic data and inverse methods (Rintoul, 1991).

A novel feature of the model in this simulation is the parametrization of brine rejection due to sea-ice growth after Duffy et al. (1999), generalized for the case of multi-category sea ice. In the generalization we spread the brine under each sea-ice category according to its (thermodynamic) growth, after which both potential temperature and salinity are convectively mixed, under the category, if instability is detected. The resultant potential temperature and salinity profiles under each category within the grid cell (including open water) are averaged according to their areas to form mean grid cell profiles. Further details are given in Saenko et al. (2001).

The 'age' tracers (T_r) are governed by the same dynamics as potential temperature and salinity, except for an additional term (Q), i.e.,

$$\frac{\partial T_r}{\partial t} = -\text{advective}(T_r) + \text{diffusive}(T_r) +$$

$$\text{convective}(T_r) + Q. \tag{51}$$

The Q term takes the following form

$$Q = -\frac{T_r}{\tau} \quad \text{for regions of tracer release} \tag{52}$$

$$Q = \frac{1}{y_r} \quad \text{otherwise} \tag{53}$$

where the relaxation parameter τ is set to 30 days and y_r is a parameter set numerically equal to the number of seconds in a year, so that T_r corresponds to time in years since tracer release. The boundary conditions applied to Eq. (51) are zero flux at all boundaries, including the ocean-air interface, and initially T_r is set to zero over the whole ocean domain. In the ocean model, the vertical, isopycnal and isopycnal thickness diffusivities were set to $k_v = 0.7 \times 10^{-4}$, $k_i = 2.0 \times 10^3$, $k_t = 2 \times 10^3 \text{ m}^2 \text{ s}^{-1}$, respectively. In the Duffy and Caldeira (1997) and Duffy et al. (1999, 2001) approach to parametrize vertical mixing of brine associated with sea-ice formation, rejected salt is spread uniformly to a depth defined by a specified potential density difference between the level of salt spreading and the surface. In this experiment we choose this specified potential density difference as $\Delta\rho = 0.2 \text{ kg m}^{-3}$. A complete description of the generalization of this scheme to multi-category sea ice, as well as a sensitivity analysis of our model to the choice of $\Delta\rho$, can be found in Saenko et al. (2001).

The model was initially spun up for 2000 years until a statistical equilibrium was reached. At this point, three age tracers were released in the GIN Sea, North Pacific and Ross/Weddell Seas, respectively (Fig. 20), and the model integration was continued for a further 500 years. The results are illustrated through the 27°W section plots (through the Atlantic Ocean) of Fig. 21. Due to the nature of the boundary conditions on the age tracer, regions where the age tracer has not penetrated are indicated by the 500-year contour.

The model clearly simulates the sinking and spreading of NADW below the permanent thermohaline to a depth of

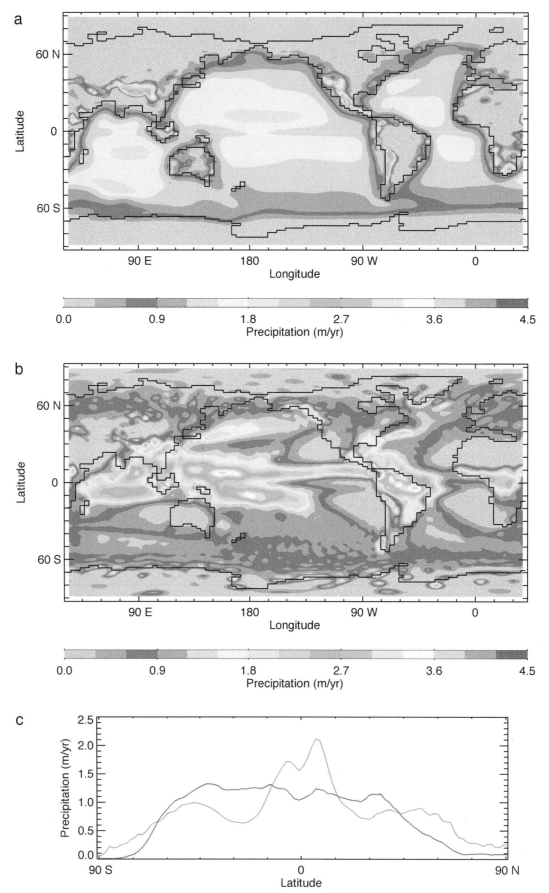

Fig. 11   Annually-averaged precipitation from a) model, b) NCEP and c) zonally-averaged model (red) and NCEP (green) climatologies.

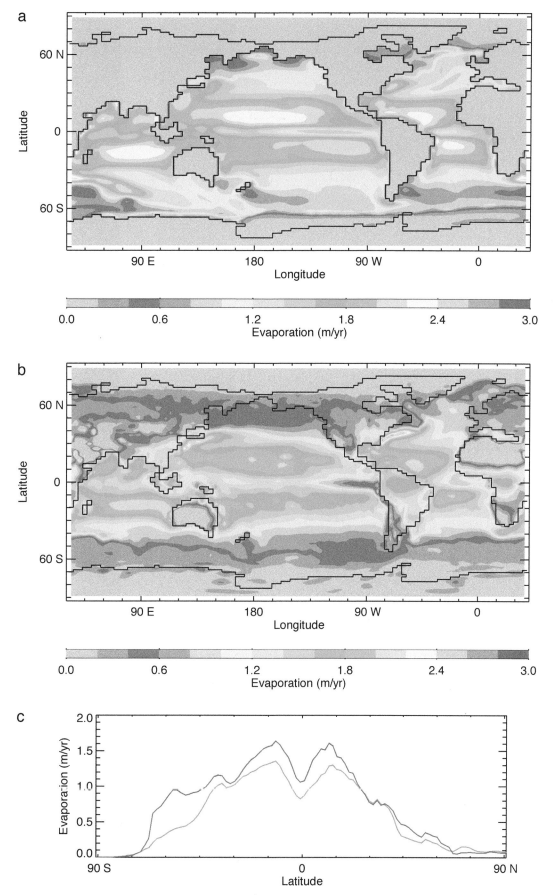

Fig. 12   Annually-averaged evaporation from a) model, b) NCEP and c) zonally-averaged model (red) and NCEP (green) climatologies.

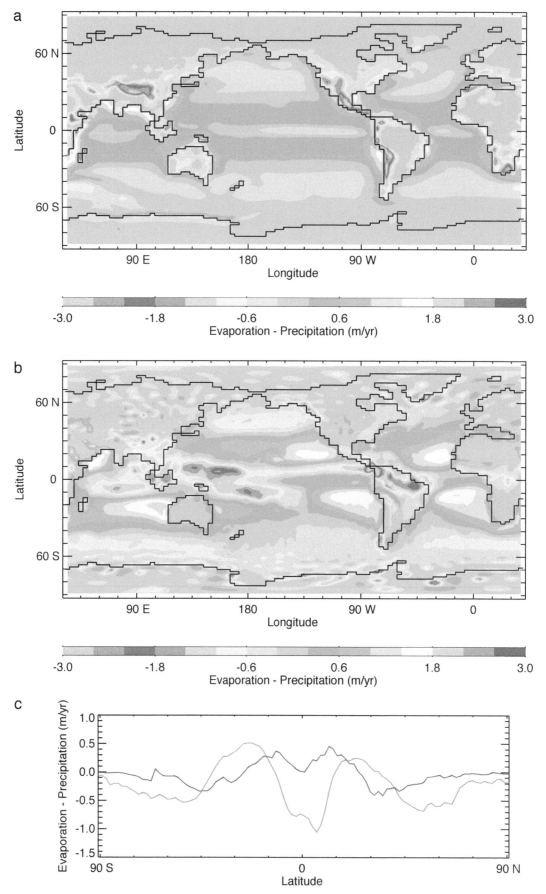

Fig. 13 Annually-averaged evaporation minus precipitation from a) model, b) NCEP and c) zonally-averaged model (red) and NCEP (green) climatologies.

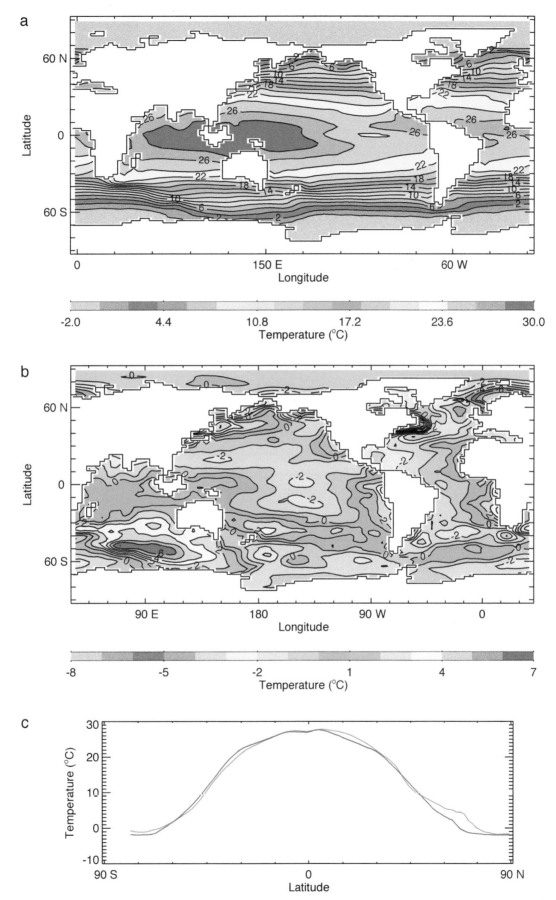

Fig. 14   Annually-averaged sea surface temperature from a) model, b) model-Levitus and c) zonally-averaged model (red) and Levitus (green) climatologies.

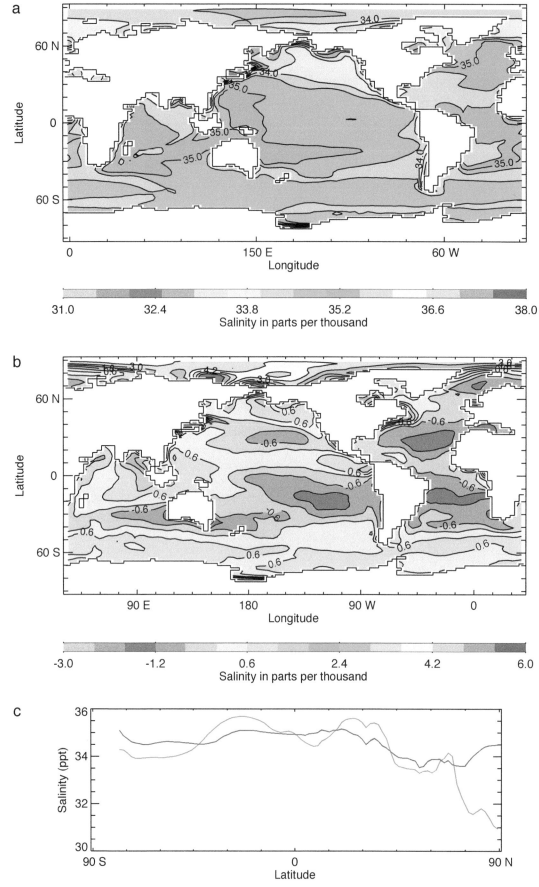

Fig. 15 Annually-averaged sea surface salinity from a) model, b) model-Levitus and c) zonally-averaged model (red) and Levitus (green) climatologies.

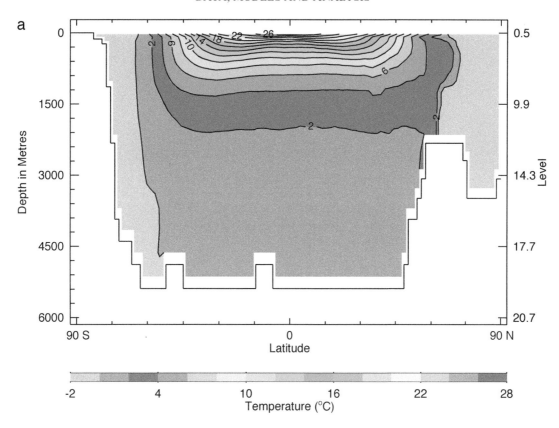

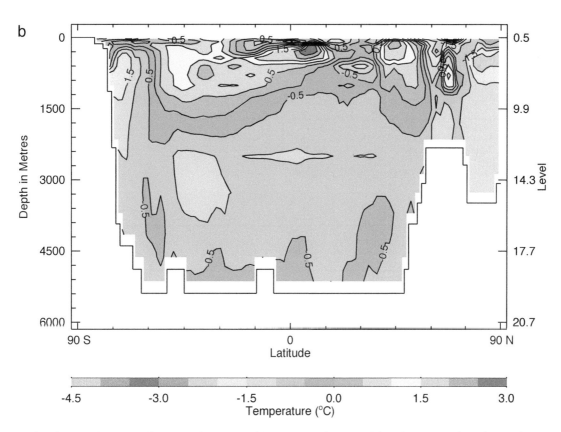

Fig. 16   Annual mean zonally-averaged ocean potential temperature from a) model and b) model-Levitus climatologies.

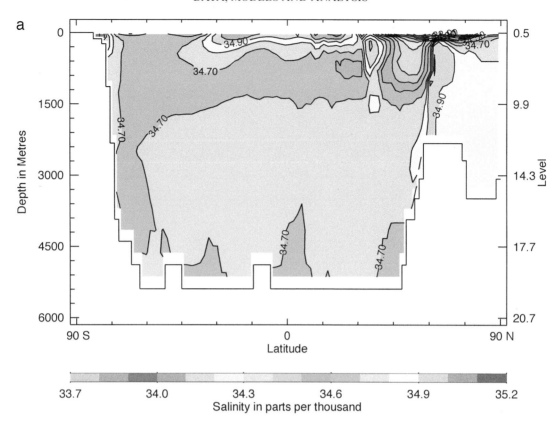

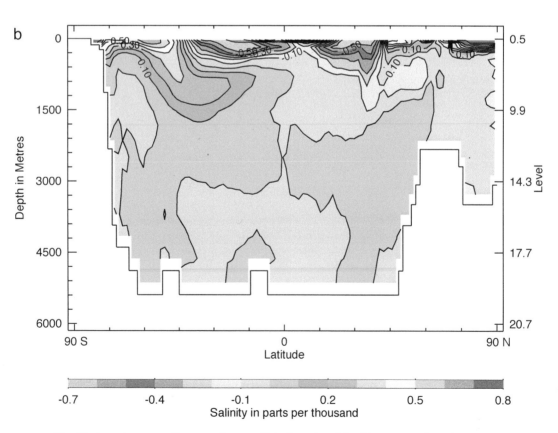

Fig. 17   Annual mean zonally-averaged ocean salinity from a) model and b) model-Levitus climatologies.

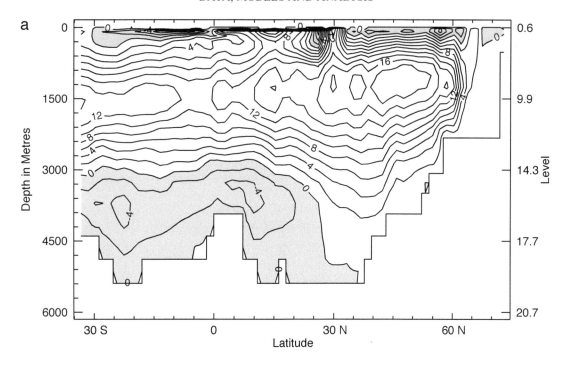

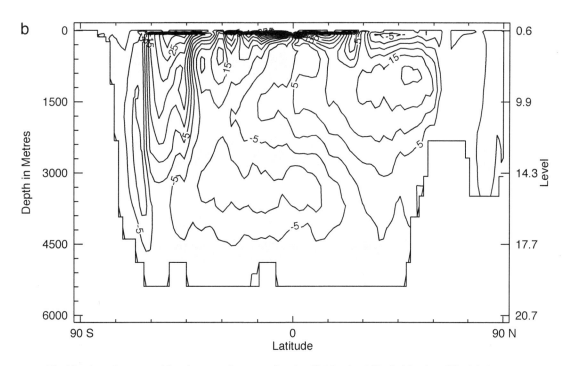

Fig. 18   Annual mean meridional overturning streamfunction (Sv) for the a) North Atlantic and b) global oceans.

about 3000 m where it overrides the AABW (Figs 21a and 21b). A detailed analysis indicates that the NADW flows into the Antarctic Circumpolar Current, primarily in a deep western boundary undercurrent, crosses the Indian Ocean with much of it entering the deep Pacific through two paths (one on either side of New Zealand) where it upwells. The remainder proceeds with the Antarctic Circumpolar Current to recirculate through the Drake Passage. The waters which upwell

in the North Pacific can be tracked by the North Pacific age tracer release experiment shown in Fig. 21c. Surface waters of the North Pacific leave the basin through two routes: the first and most important is via the Indonesian throughflow while the second is through Drake Passage. In the latter case, these North Pacific waters enter the South Atlantic in the form of AAIW which supports the results of the inverse modelling study of Rintoul (1991). Furthermore, as indicated in

TABLE 3.   Coupled model simulations of hemispheric ice and snow cover (10^6 km^2) ranked by the average absolute error from observations (Gates et al., 1996).

Northern Hemisphere				Southern Hemisphere			
ice cover		snow cover		ice cover			
DJF	JJA	DJF	JJA	JJA	DJF	Error	Model
14.8	10.7	44.7	7.8	16.4	6.2	–	observed
14.0	7.1	42.2	12.2	14.7	8.1	2.5	UVic*
11.0	8.3	41.2	2.5	12.4	6.4	3.2	GISS*
9.7	7.1	42.9	11.5	12.2	7.5	3.3	CCC
10.2	5.3	35.3	5.1	18.0	5.9	4.0	UKMO
19.1	11.5	34.3	2.9	11.6	3.1	4.7	MRI
13.6	0.5	41.4	2.1	5.3	0.5	6.2	NCAR*
9.3	1.6	53.4	12.1	4.0	3.7	7.1	COLA*
16.0	12.7	64.4	10.0	24.7	16.0	7.2	GFDL
16.6	14.5	37.5	11.6	21.1	18.9	7.3	CSIRO
18.9	16.7	62.0	28.4	0.5	0.5	11.6	BMRC*

*Indicates a model without flux adjustments.

Fig. 21c, most of this intermediate water then converts into thermocline water as it moves northward into the North Atlantic. The remainder, however, converts into deep and bottom water in the South Atlantic and then exits towards the Indian Ocean, in further agreement with observational-based results of Rintoul (1991). As such, it appears that the model supports both the warm and cold water routes of Gordon (1986) and Rintoul (1991).

e *Inorganic Ocean Carbon Cycle*

As a first step towards incorporating a comprehensive carbon cycle model into the UVic ESCM (see Section 8), we have included an inorganic ocean carbon model. This allows us to evaluate our model against those participating in the International Geosphere-Biosphere Programme (IGBP) Ocean Carbon-Cycle Model Intercomparison Project (OCMIP). We have implemented the inorganic carbon component by closely following the protocols set out by OCMIP (Orr et al., 1999). Dissolved inorganic carbon (*DIC*) is modelled as a passive tracer subject to the following conservation equation (c.f. Eq. (45)):

$$\frac{\partial DIC}{\partial t} + L(DIC) = \frac{\partial}{\partial z}\left(k_v \frac{\partial DIC}{\partial z}\right) + \mathbf{V}(k_h \mathbf{V} DIC) + J + J_v \quad (54)$$

where the source/sink term J represents the air-sea flux of CO_2 and the virtual source/sink term J_v represents the changes in *DIC* due to evaporation, precipitation and runoff. The air–sea gas-exchange flux is modelled using the following equation:

$$F = k_w(C_{atm} - C_{surf}) \quad (55)$$

where C_{atm} and C_{surf} are calculated using the partial pressures of CO_2 in both the atmosphere and ocean surface layer, respectively. The CO_2 transfer velocity k_w is modified from Wanninkhof (1992) and is given by:

$$k_w = \left(1 - \gamma_{ice}\right)0.337|\mathbf{V}_s|^2\left(\frac{S_c}{660}\right)^{0.5} \quad (56)$$

where γ_{ice} is the fractional sea-ice cover (varying between 0.0 and 1.0). We assume that there is no gas exchange through the ice. The wind speed $|\mathbf{V}_s|$ is as discussed in Section 2a1, and the Schmidt number (S_c) is a function of model SST (Wanninkhof, 1992).

To illustrate the quasi-equilibrium surface fluxes of CO_2 in our model, we use the same version as in section 3d but with only one ice category (plus open water) i.e., the model incorporates the Duffy and Caldeira (1997) parametrization of brine rejection due to sea-ice growth and the Gent and McWilliams (1990) parametrization for mixing associated with mesoscale eddies. The model is initialized with a pre-scribed preindustrial atmospheric pCO$_2$ of 280 ppm and integrated for 1440 years. The resulting quasi-equilibrium air-sea fluxes (Fig. 22) show a net-flux of CO_2 into the atmosphere occurring as a broad-banded feature about the equator and along the western coasts of South America and Africa, in areas of near-shore upwelling. This outgassing also occurs in regions of the Argentinian and south-eastern Australian coasts. Our use of isopycnal mixing reduces spurious deep convective plumes in the Southern Ocean leading to a fairly homogeneous region of CO_2 uptake between 50–65°S. High uptake also occurs in the North Atlantic and to some extent in the North Pacific and well as off the south-western coast of Africa.

f *Summary*

In this section we have tried to demonstrate that the UVic ESCM, in its standard representation, provides a reasonable simulation of the current climate. We have alluded to, and explicitly shown how certain shortcomings can be improved on through the inclusion of more sophisticated and realistic parametrization of sub-gridscale processes. While we believe the precipitation field is one of the major weaknesses of the diffusive moisture transport formulation of our atmospheric model, we still conclude that the ESCM serves as a useful tool with which to investigate the importance of climate processes in past, present and future climate simulations (see Sections 6 and 7). Of course, a caveat that we need to add concerning our comparison with coupled atmosphere-ocean GCMs (Tables

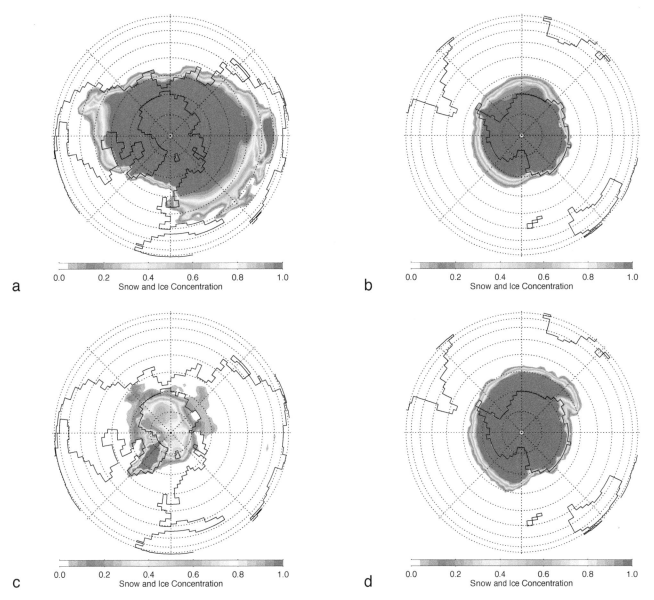

Fig. 19   Maximum sea-ice concentration. a) and c) show northern hemisphere polar projections in the boreal winter and summer, respectively. b) and d) show southern hemisphere polar projections in the austral summer and winter, respectively. Over the land, the contours indicate maximum concentration in snow coverage. The plots show the average concentration for the three-month period covering the summer (July–September) and winter (January–March). As such, they emphasize the maximum extent during any particular season.

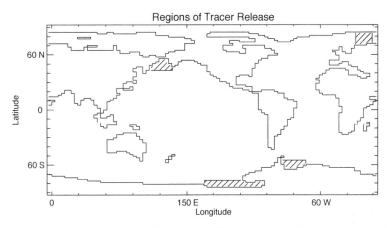

Fig. 20   Model domain: dashed lines in the GIN Seas, North Pacific, and Weddell and Ross Seas indicate where the passive "age" tracers were released.

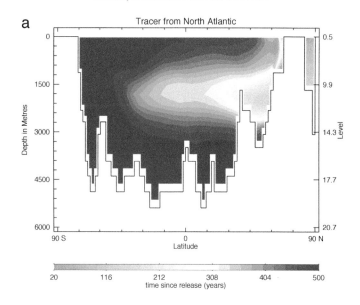

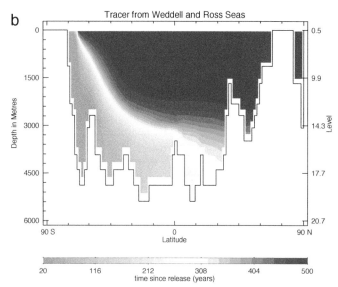

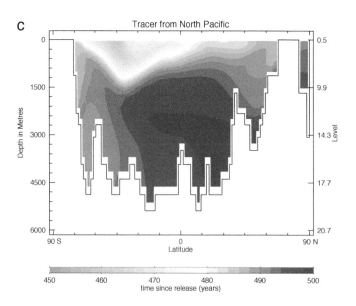

Fig. 21   A latitude-depth section at 27°W for the "age" tracer released from GIN Seas (a), Weddell and Ross Seas (b), and North Pacific (c).

Equilibrium air-sea CO₂ flux

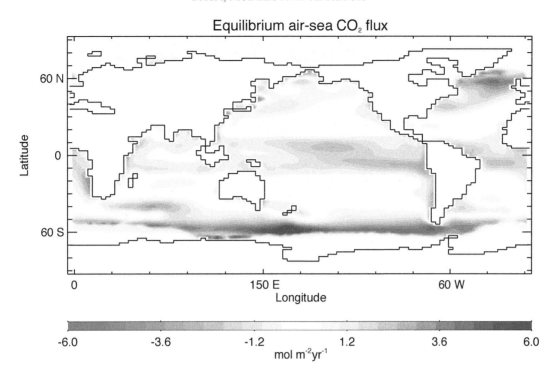

Fig. 22 Air-sea flux (Moles m⁻² yr⁻¹) of inorganic carbon after 1440 years of integration. Blue regions indicate areas of carbon uptake by the ocean whereas red areas indicate where the ocean is outgassing.

1–3) is that the indices we used represent relatively simple, albeit important, measures of model performance. In addition, most coupled modelling groups are currently using newer versions of their models than those evaluated in Gates et al. (1996). Finally, the simple atmospheric component of our model does not allow for internal variability, unlike more comprehensive coupled atmospherre-ocean GCMs.

In the next section, through the inclusion of advection in the treatment of moisture transport, we will show how the model leads to an extremely good simulation of the present-day precipitation field and surface salinity field, without compromising the climatology of the other fields discussed above. When combined with the wind feedback option, we will further determine (in Section 5) how the change of moisture transport from a diffusive to an advective formulation affects the response of the climate system in global warming experiments.

4 Moisture advection

As noted in Section 3, the precipitation field simulated by the coupled model when diffusive transport is used has serious limitations, especially near the ITCZ where moisture transport in nature is up-gradient. To improve our precipitation field, we exploit the observation that most of the atmospheric moisture transport happens within the planetary boundary layer, allowing us to use the surface wind field to accomplish advection. In this section we discuss the rather dramatic improvements that arise when we include moisture advection in our coupled model. In Section 5 we examine the effect that the inclusion of moisture advection has in global warming simulations where the wind feedback is also allowed to operate.

In order to include moisture advection we rewrite Eq. (8) as

$$\rho_a h_q \left\{ \frac{\partial q_a}{\partial t} - \nabla \cdot \left(\kappa \nabla q_a \right) + \beta \nabla \cdot \left(\mathbf{u} q_a \right) \right\} = \rho_o (E - P) \quad (57)$$

where the new parameter β relates the vertically-averaged advective moisture transport to the surface advective transport ($\beta = 0.4$) and $\mathbf{u}$ is the surface wind. In the absence of the representation of explicit cloud processes, another parametrization must be included in order to limit the local release of latent heat during strong precipitation events in regions of converging winds. Rather than locally changing the planetary albedo and reducing incoming solar radiation (associated with deep clouds for example) we adopted a modified thermal diffusive coefficient $v + v^*$ so that the atmospheric heat transport (Eq. (2)), becomes:

$$Q_T = \rho_a h_t c_{pa} \nabla \cdot ((v + v^*) \nabla T_a). \quad (58)$$

The effect of the v^* term in Eq. (58) is to balance the latent heat associated with condensation arising from convergence of moisture by advection. We parametrize v^* as

$$v^* = \begin{cases} 0 & P < P_{min} \\ v_0 \dfrac{P - P_{min}}{P_{max} - P_{min}} & P_{min} \le P \le P_{max} \\ v_0 & P > P_{max} \end{cases} \quad (59)$$

where $v_0 = 60 \times 10^6$ m² s⁻¹, $P_{min} = 2.0$ m yr⁻¹ and $P_{max} = 4.0$ m yr⁻¹ are empirical constants. This localized diffusive process is only active with the advective transport of moisture option and is only evoked in regions of strong moisture convergence. The use of moisture advection also necessitates the use of a larger thermal diffusive transport coefficient v in tropical regions, due to enhanced latent heat release, although the coefficient of moisture diffusion is substantially reduced (Fig. 23).

When moisture advection is included in the model, the representation of precipitation (Fig. 24) is substantially improved over the case where moisture transport is only accomplished through diffusion (Fig. 11). The most notable improvements occur near the ITCZ, where moisture transport is up-gradient, and near monsoon regions. The mid-latitude North Atlantic and Pacific are regions where the precipitation is still slightly underestimated, although the spatial pattern of precipitation over the ocean is more realistic there. The precipitation over land, within the coupled model, still remains troublesome in tropical regions, most likely due to the lack of land surface processes. Nevertheless, all precipitation over land in these regions is instantaneously returned to the ocean as runoff.

Evaporation minus precipitation (Fig. 25) is also substantially improved over the case in which moisture transport is only accomplished by diffusion (Fig. 13). The global zonally-averaged field reasonably reproduces the NCEP climatology at most latitudes. The mid-latitudes in the northern hemisphere have too much net evaporation, although at these latitudes the two-dimensional field bears a fine resemblance with the NCEP climatology.

The climatology of the coupled model is not changed dramatically from that presented in Section 3. For example, the SAT map (Fig. 26a) is very similar to the analogous diffusive version (Fig. 26b), as well as the NCEP reanalysis data (Fig. 10b). The differences between the SAT fields obtained under the advective and diffusive treatments of moisture transport (Fig. 26b) are largely confined to regions where large differences exist in latent heat release in the atmosphere. The advective model tends to cause the monsoon and ITCZ regions to be slightly warmer, as precipitation is increased there, and the ocean desert regions of the eastern flanks of the subtropical gyres to be slightly cooler, as descending air from above causes divergent surface winds.

The same features seen in the SAT field are reflected in the SST field (Fig. 27), leading to much better agreement with the Levitus and Boyer (1994) observations (Fig. 14b). Notable areas of discrepancy between observed and modelled SST fields in Section 3 were found in the southern Indian Ocean and along the eastern boundary, near the mid-latitude coasts of the Pacific and Atlantic, where simulated SSTs were too warm, as well as the Gulf of Alaska and the GIN Seas which were too cold. While the western boundary separation points of the Kuroshio and Gulf Stream have not changed, the SSTs over much of the North Atlantic have cooled. In fact, in most of the global ocean the subtropical gyre regions, and particularly their eastern edge, have cooled bringing the simulated fields into better agreement with observations.

The improved representation of both precipitation and evaporation minus precipitation is naturally reflected in a better

distribution of the SSS field (Fig. 28). The SSS field obtained without advection for moisture transport (Section 3) did not readily reveal local maxima in the subtropical gyre regions of the world oceans nor did it show much east-west asymmetry in the Pacific (Fig. 15). When moisture transport is accomplished through advection, salinity maxima develop in the subtropical gyre regions and the observed east-west tropical Pacific asymmetry develops, with a relatively fresh western equatorial Pacific and a relatively saline eastern subtropical Pacific.

The water mass structure of the global oceans remains similar to that found in Section 3, although the potential temperature and salinity properties of the different water masses have changed slightly. Generally, the thermocline region is slightly colder through subduction arising in the now colder subtropical regions, in the advective case (Fig. 29a) and the deeper ocean is slightly warmer. Very little change is seen in the zonally-averaged salinity field (Fig. 29b). Once more, AAIW is not well represented using the horizontal/vertical representation of mixing — see our earlier discussion and Duffy et al. (1999, 2001).

The generally colder surface ocean in the North Atlantic (Fig. 27), and the warmer, deep ocean seen in Fig. 29a counteracts the slightly increased overturning in the North Atlantic in the advective case (Figs 30a and 30c), to give only a slight decrease in the northward heat transport in the North Atlantic Ocean. While the large-scale features of the meridional overturning are only slightly changed upon moving to advective moisture transport, the increase in NADW formation comes at the expense of a slight decrease in AABW intrusion into the deep North Atlantic. The Deacon Cell in the Southern Ocean, which is a prominent feature of both this experiment and the diffusive case, is not a feature of our coupled model when we include isopycnal mixing and the Gent and McWilliams (1990) representation of mixing associated with mesoscale eddies, as expected from the earlier analyses of Danabasoglu et al. (1994) and Robitaille and Weaver (1995).

Taken together, the improvements realized when moisture transport is parametrized through advection by winds far outweigh the small increase in computational expense. As such, future versions of our coupled model will include moisture advection as the standard option. In the next section we show how the moisture advection option can be combined with the wind stress feedback parametrization to provide for a first-order feedback accounting for changing moisture advection in a changing climate.

5 Global warming and the role of moisture advection

In the previous section we demonstrated how the inclusion of moisture advection substantially improved the climatology of our ESCM. In this section we now determine whether or not the improved parametrization of moisture transport has any effect on the transient behaviour of the system. In particular, we conduct four global warming sensitivity experiments using both diffusive and advective treatments of moisture transport with and without the wind feedback parametrization of section (2a1).

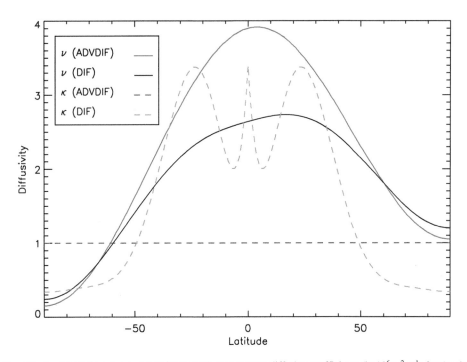

Fig. 23   Latitudinal profile of the heat ν (solid curves) and moisture κ (dashed curves) diffusion coefficients (in 10^6 m^2 s^{-1}) for the simulations using only diffusive moisture transport (DIF) and diffusive + advective moisture transport (ADVDIF).

The initial condition of the four experiments was the 1998 equilibrium climatology discussed in Sections 3 and 4 for diffusive and advective treatments of moisture transport, respectively. The atmospheric level of CO_2 was increased at a rate of 1% per year until 4× present day CO_2 was reached, at which point atmospheric concentrations were held fixed for 500 years of total integration.

Both the transient and quasi-equilibrium warming in all four experiments are nearly identical (Fig. 31a), suggesting very similar climate sensitivities of about 3.0°C (for a doubling of atmospheric CO_2), in all cases. After 500 years the warming relative to the initial condition in the four experiments is: 1) advective moisture transport — 6.00°C; 2) advective moisture transport with wind feedback — 6.01°C; 3) diffusive moisture transport — 6.23°C; 4) diffusive moisture transport with wind feedback — 6.26°C. The overall horizontal pattern of the warming is similar in all the experiments, with the exception of the region around the North Atlantic and over the subtropical gyre regions of the ocean (Fig. 32). All experiments show a tendency for more warming over land than over the ocean and an amplification of the warming at high latitudes. In all experiments the overturning in the North Atlantic initially weakens, although it re-establishes once the radiative forcing is held fixed (Fig. 31b).

In the two diffusive moisture transport experiments, the thermohaline circulation in the North Atlantic eventually re-established to a strength that was stronger than the 1998 initial condition, consistent with the results of Wiebe and Weaver (1999), obtained using an older version of our model. In the case of advective moisture transport, without the wind feedback, the thermohaline circulation declined further, in the

transient phase, but also re-equilibrated to a value slightly higher than in the 1998 inital condition. The advective moisture transport case with the wind feedback included closely tracked the analogous case without the feedback while the radiative forcing was changing, but re-established to a value that was almost identical to its initial condition (Fig. 31b). The differences seen in Fig. 32 in the northern North Atlantic are therefore simply explained by the different behaviour of the thermohaline circulation in the four experiments. In the diffusive transport case, where the overturning equilibrates at a stronger value, there is more melt back of sea ice leading to a subsequent positive albedo feedback on surface warming. In addition, there is an enhanced oceanic heat loss to the atmosphere due to the removal of the insulating sea-ice cover. In the case where the overturning equilibrates to a level similar to its initial condition, there is no additional positive feedback on the sea ice to enhance further the warming locally.

A fundamental question that remains concerns the physics responsible for the re-establishment of the thermohaline circulation to different strengths in the different experiments. As shown by Hughes and Weaver (1994), the strength of the overturning in the North Atlantic is proportional to the meridional gradient in the zonally-averaged depth-integrated steric height (DISH) from the tip of Africa to the latitude at which deep water formation occurs. Indeed, the overturning is also proportional to the meridional DISH gradient along the western boundary which is balanced by a current flowing down the meridional DISH gradient in a frictional boundary layer. As suggested in Hughes and Weaver (1994), perturbations to either the hydrological cycle (e.g., Fig. 33), heat balance or wind field which cause the DISH to change in either the

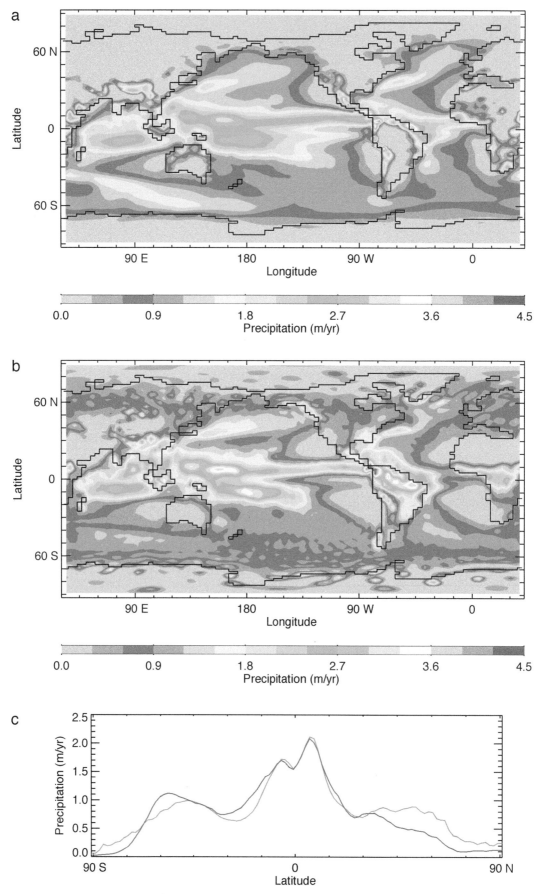

Fig. 24   Annual mean precipitation (m yr^{-1}) for a) model, b) NCEP, c) the zonally-averaged model (red) and NCEP (green) climatologies.

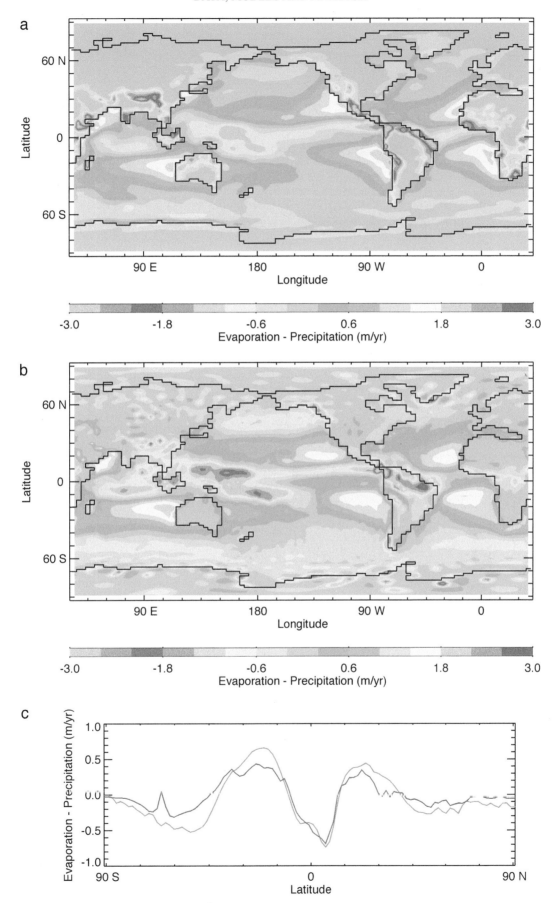

Fig. 25 Annual mean evaporation minus precipitation (m yr^{-1}) for a) model, b) NCEP, c) the zonally-averaged model (red) and NCEP (green) climatologies.

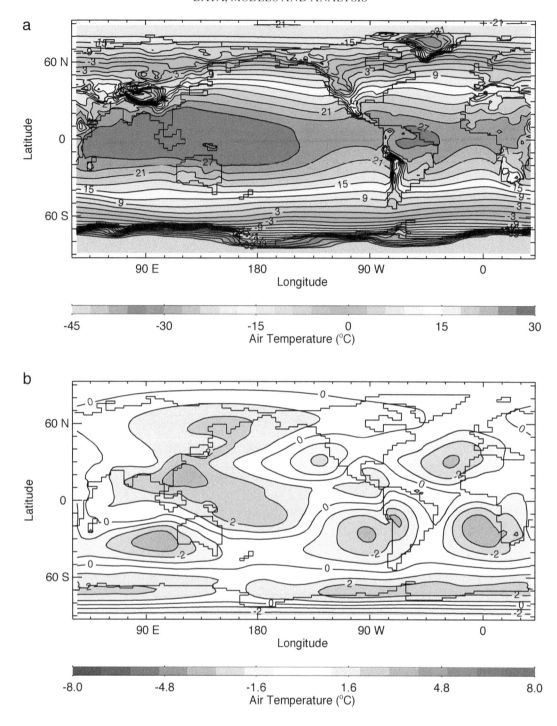

Fig. 26 a) Annual mean equilibrium surface air temperature (°C) using advection for moisture transport. b) The difference between a) and that obtained for purely diffusive moisture transport (Fig. 10a).

northern or southern hemisphere, subsequently cause the overturning circulation to change.

Wiebe and Weaver (1999), using the Fanning and Weaver (1996) version of this model in which moisture transport was parametrized through diffusion, conducted very similar experiments to those performed here. They too found that the North Atlantic overturning re-established to a strength greater than the initial condition and quantitatively analyzed why this was

so. In the transient case, they found that the initial increase in high latitude precipitation (as seen in Figs 33a and 34), and subsequent reduction of DISH gradient, was the cause of the initial thermohaline weakening (see Fig. 31b). As time went on, low latitude and South Atlantic heating increased the DISH and hence its meridional gradient so that the thermohaline circulation began to intensify to a new equilibrium that was stronger than its initial case (see also Manabe and Bryan (1985)).

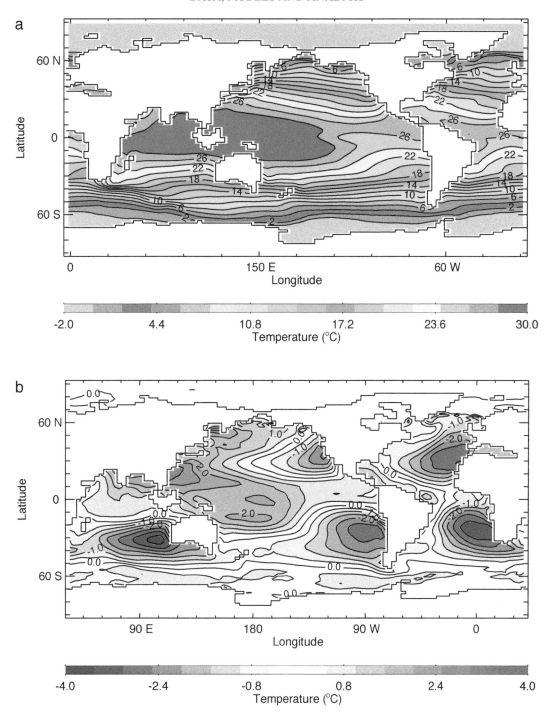

Fig. 27   a) Equilibrium sea surface temperature (°C) using advection for moisture transport. b) The difference between a) and that obtained for purely diffusive moisture transport (Fig. 14a).

Our two cases, with a diffusive representation of moisture transport, yield results that are very similar to those of Wiebe and Weaver (1999). When moisture advection is used without the wind feedback, the thermohaline circulation still equilibrates to a level which is slightly stronger than its initial condition. The fact that this is not the case when the wind feedback is used must arise from the improved representation of hydrological processes and their feedbacks in the atmos-

phere and ocean models. Figure 35a shows the meridional profile of the zonally-averaged DISH in the Atlantic Ocean for all the global warming runs and their respective initial conditions, relative to 1528 m. The difference between the zonally-averaged DISH at the latitude of the tip of Africa and the latitude of deep water formation is also shown as a function of the equilibrium overturning rate in Fig. 35b. The linear relationship between the strength of the overturning rate

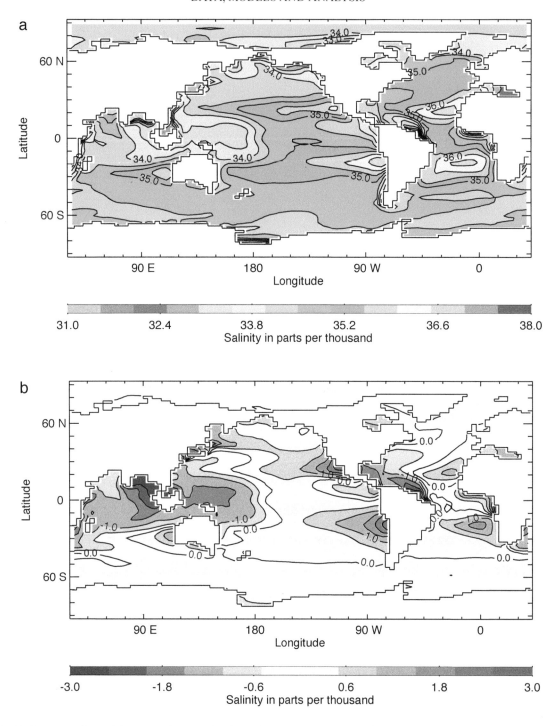

Fig. 28 a) Equilibrium sea surface salinity (psu) using advection for moisture transport. b) The difference between a) and that obtained for purely diffusive moisture transport (Fig. 15a).

and the meridional gradient in the DISH is verified in our present experiments. The inclusion of moisture advection and the wind feedback brings the results of our model more in line with those found in coupled atmosphere-ocean models where a re-establishment to the initial condition is most often seen (e.g., Kattenberg et al., 1996; Manabe and Stouffer, 1999). This is a very encouraging feature as it provides additional indirect validation of our wind feedback parametrization of section 2a1.

6 The climate of the last glacial maximum

As noted in the introduction, early CLIMAP Project Members (1976, 1981) attempts to reconstruct SSTs for the LGM around 21 KBP have suggested that relative to the present, global SSTs were on average 1.7°C cooler in August and 1.4°C cooler in February. These reconstructions further suggested that tropical SSTs were similar to those of the present climate whereas in the North Atlantic, SSTs were substantially

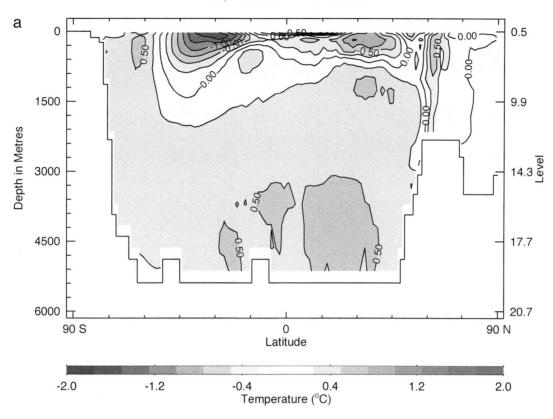

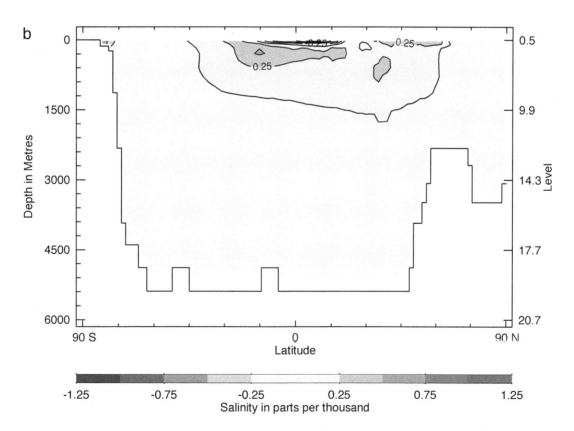

Fig. 29 The difference (advective minus diffusive) between the equilibrium global zonally-averaged a) potential temperature (°C) and b) salinity (psu) obtained under advective and diffusive representations of moisture transport.

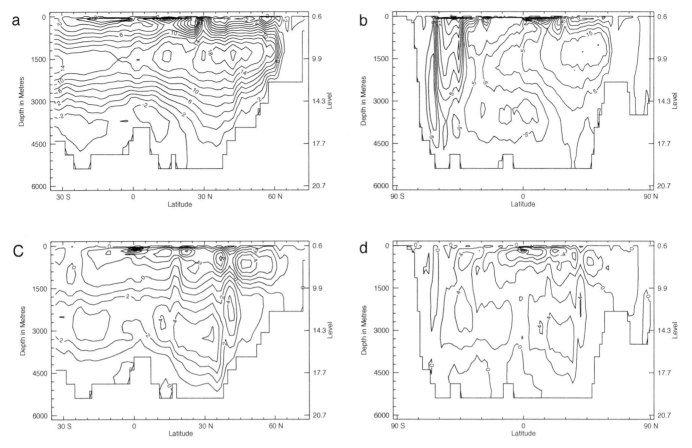

Fig. 30 Meridional overturning streamfunction (Sv) for a) the North Atlantic and b) global oceans at equilibrium for the case where advection is used for moisture transport. c) and d) show differences [a) minus Fig. 18a and b) minus Fig. 18b)] from the case when only diffusive moisture transport is used. The contour interval is 1 Sv in c), 2 Sv in a) and d), and 5 Sv in b).

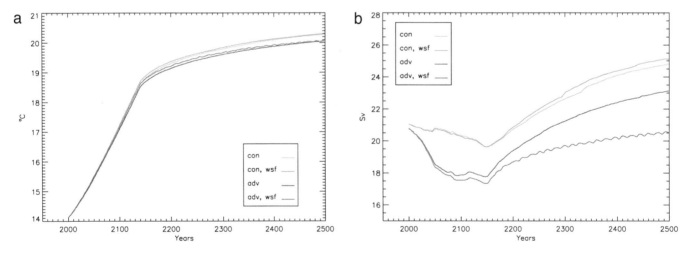

Fig. 31 Time series of a) global mean surface air temperature (°C) and b) maximum North Atlantic meridional overturning streamfunction (Sv) for the four experiments. Moisture transport is either diffusive ('con' in the legend; yellow, green) or advective ('adv' in the legend; blue, pink) and the wind feedback is either on ('wsf' in the legend; green, pink) or off (yellow, blue).

colder. While recent alkenone evidence (Sikes and Keigwin, 1994; Bard et al., 1997; Lyle et al., 1992) also supports tropical SSTs only slightly cooler at the LGM, additional evidence is contradictory. For example, coral records from

Barbados (Guilderson et al., 1994) and the south-west Pacific (Beck et al., 1997), ice core records from Peru (Thompson et al., 1995), noble gas measurements in Brazil (Stute et al., 1995) and ocean core records from the western equatorial

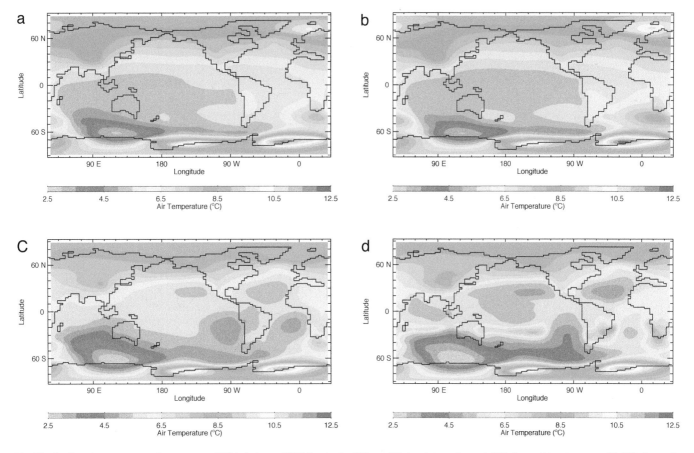

Fig. 32 Surface air temperature change at year 500 (relative to 1998) for the $4 \times CO_2$ equilibrium integrations: a) diffusive moisture transport; b) diffusive moisture transport with wind feedback; c) advective moisture transport; d) advective moisture transport with wind feedback.

Atlantic (Curry and Oppo, 1997) suggest LGM tropical temperatures were significantly below those at present.

Modelling studies of the LGM have provided conflicting results concerning the issue of tropical cooling. Most early simulations used atmospheric GCMs with either fixed SSTs or mixed layer ocean models at the lower boundary (Manabe and Broccoli, 1985; Hansen et al., 1984; Lautenschlager and Herterich, 1990; Kutzbach and Guetter, 1986; Crowley and Baum, 1997; Webb et al., 1997; Broccoli, 2000). Webb et al. (1997) obtained tropical SSTs that were about 5.5°C cooler than today (averaged over the region 16°N to 16°S), substantially larger than CLIMAP estimates, using an AGCM in which present-day oceanic heat transports were maintained. Broccoli (2000), on the other hand, obtained tropical SSTs that were 2.0°C cooler than present (averaged over the region 30°N to 30°S). Two previous studies using climate models of intermediate complexity (Ganopolski et al., 1998; Weaver et al., 1998) found their tropical temperatures to be cooler than CLIMAP, consistent with tropical alkenone reconstructions, but not as cool as some of the other reconstructions cited above. Bush and Philander (1998) on the other hand, using a coupled atmosphere-ocean GCM under LGM forcing, albeit for a short (15-year) integration by which time the deep ocean was still far from equilibrium, suggested 5°C tropical sea surface cooling.

Through a comprehensive analysis of the existing proxy record, Crowley (2000) questioned the validity of SST records obtained from ice-age coral proxies. In addition, he argued that small changes (on the order of 20%) in the assumed constant lapse rate, through which snow line proxies are projected to infer SSTs, could reconcile even CLIMAP data with tropical snow line proxies. Both he and Weaver et al. (1998) argue for more stable tropical SSTs which are colder than CLIMAP Project Members (1976, 1981), consistent with alkenone reconstructions (Sikes and Keigwin, 1994; Bard et al., 1997; Lyle et al., 1992), but not as cold as the borehole, coral and other reconstructions (Guilderson et al., 1994; Beck et al., 1997; Thompson et al., 1995; Stute et al., 1995; Curry and Oppo, 1997) mentioned above. He proposed a mid-range LGM climate sensitivity of 3.0°C for a doubling of CO_2, leading to a 2.5°C reduction in ice-age tropical SSTs relative to the present. Both of these are almost exactly as inferred in Weaver et al. (1998).

In this section we re-examine the issue of LGM tropical cooling using our ESCM. This should be viewed as an extension of the analysis of Weaver et al. (1998) in which an older and less sophisticated version of the model was used. In our first experiment we set the orbital parameters to 21 KBP values and lower the level of atmospheric CO_2 to 200 ppm. The topography on land was also raised through the use of the

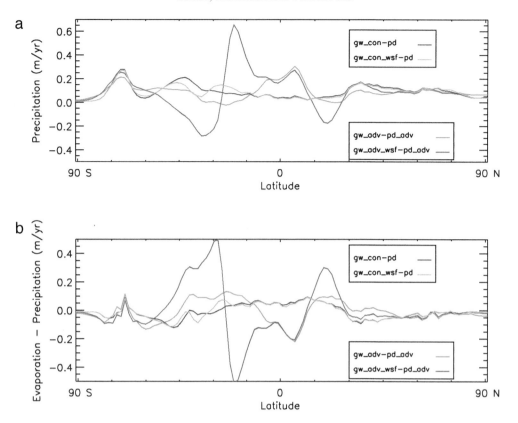

Fig. 33 Zonally-averaged a) precipitation and b) evaporation minus precipitation at year 500 (relative to 1998; 'pd' in legend) for the $4 \times CO_2$ equilibrium global warming ('gw' in legend) integrations. Moisture transport is either diffusive ('con' in legend; blue, orange) or advective ('adv' in legend; green, red) and the wind feedback is either on ('wsf' in legend; red, orange) or off (blue, green).

Peltier (1994) ICE4G reconstruction, and wherever continental ice sheets were present, the planetary co-albedo was locally reduced by 0.18. The coupled system was then integrated for 2000 years until an equilibrium was reached, using the present-day climate of Section 3 as the initial condition.

The global mean cooling for the LGM relative to the present is 3.6°C, of which 0.4°C arises solely due to the increased elevation associated with the prescribed ice-sheet thickness. The SAT difference (Fig. 36) shows a pronounced cooling around the region of the North Atlantic, where the prescribed ice sheets are present, that is amplified due to a weakening and shallowing of the meridional overturning, the southward expansion of sea ice and the subsequent sea-ice albedo feedback over the ocean. On average the northern hemisphere is 4.6°C colder than the present, whereas in the southern hemisphere, where there are fewer areas of land and ice sheets, this cooling is only 2.6°C. Our global cooling estimate lies within the range spanned by other AGCM and coupled model studies (e.g., Hansen et al., 1984; Lautenshclager and Herterich, 1990; Kutzbach and Guetter, 1986; Broccoli and Manabe, 1987; Broccoli, 2000; Hyde et al., 1989; Ganopolski et al., 1998; Bush and Philander, 1998).

As in Weaver et al. (1998), the equilibrium LGM North Atlantic overturning (Fig. 37a) is reduced in intensity and shallower in depth compared to the present-day climatology (Fig. 18a). For example, the maximum overturning at the LGM is ~10 Sv, about half of the 19–20 Sv found in the present-day equilibrium climate (Fig. 37b), and the base of the NADW layer is, on average, a few hundred metres shallower south of 25°N. AABW, originating in the Southern Ocean, penetrates further northward to about 25°N in the LGM case (Fig. 37a), filling the void left behind by the shallower NADW layer. The actual location of NADW formation did not change noticeably and North Pacific Intermediate Water (NPIW) did not intensify, consistent with Weaver et al. (1998).

The sea-ice distribution in the LGM experiment (solid line in Figs 36d and 36e) shows a rather dramatic seasonal variation in ice cover over the North Atlantic. In the winter, sea ice covers the Labrador Sea and extends across the North Atlantic to Great Britain (see Crowley and North (1991) pp. 49). In the summer, much of this sea ice melts away leaving most of the North Atlantic ice free, in accord with Hebbein et al. (1994). The maximum sea-ice extent at any time of the year (Fig. 38) shows that although we include a much more sophisticated dynamical/thermodynamical sea-ice model (Section 2b1), results are quite similar over the ocean to those of Weaver et al. (1998). In the southern hemisphere, our maximum sea-ice extent agrees well with the Crosta et al. (1998a, 1998b) proxy data, although we appear to have too little sea ice between Antarctica and Australia. As noted in section 3c, very little sea ice forms in this region in our model due to the Antarctic Circumpolar Current being steered too far south by the Kerguelen plateau and bringing relatively warm waters poleward.

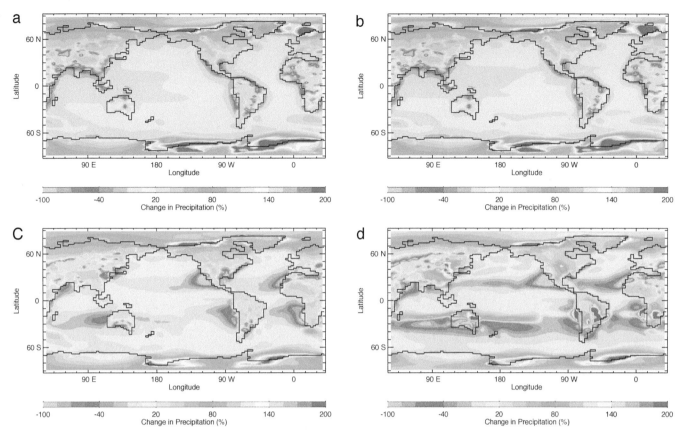

Fig. 34   Percentage change in precipitation at year 500 (relative to 1998) for the $4 \times CO_2$ equilibrium integrations: a) diffusive moisture transport; b) diffusive moisture transport with wind feedback; c) advective moisture transport; d) advective moisture transport with wind feedback.

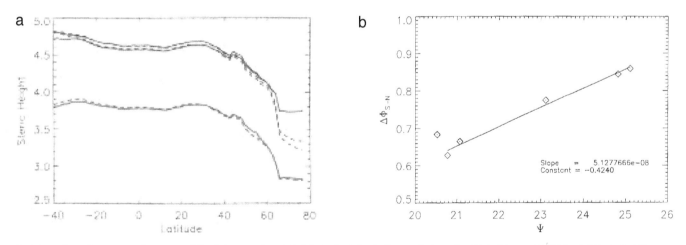

Fig. 35   a) Annual mean, zonally-averaged depth-integrated steric height in the Atlantic Ocean relative to level 10 of the model (1528 m). The dashed lines represent the integrations with diffusive moisture transport and the solid lines with advective moisture transport. Red curves are for the equilibrium present-day integrations whereas blue and purple lines are at year 500 (relative to 1998) for the global warming and global warming with wind feedback integrations, respectively. Units are in 10^{-6} kg m^{-1}. b) Difference between the zonally-averaged depth-integrated steric height in a) at the latitude of the tip of Africa (40°S) and the latitude of deep water formation (60°N), as a function of equilibrium overturning rate in Sv for all experiments. The equilibrium overturning rate for each experiment is taken from Fig. 31b, with the present day (red curves above) being represented by the 1998 initial conditions. The straight line is a least squares fit of all points.

The LGM SST field (Fig. 39a) and its difference from the present-day climate (Fig. 39b), show cooling everywhere in the ocean with a global and annual mean amount of 2.1°C. Averaged between 20°S and 20°N, tropical cooling is also

2.1°C, which is colder than that suggested by CLIMAP Project Members (1976, 1981), yet, consistent with alkenone reconstructions (Sikes and Keigwin, 1994; Bard et al., 1997; Lyle et al., 1992; Rostek et al., 1993), but not as cold as the

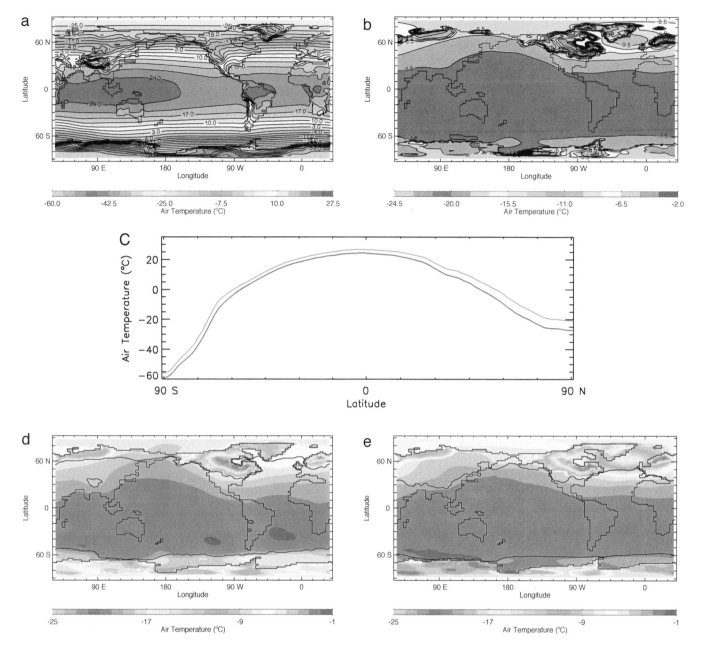

Fig. 36   Equilibrium annual mean surface air temperature for the a) LGM, b) LGM minus present-day (Fig. 10a) climatologies. c) zonally-averaged annual mean LGM (red) and present-day (green) surface air temperature profiles. d) summer (July to September) and e) winter (January to March) mean average surface air temperature differences for the LGM relative to the present-day climatology. The solid black lines in d) and e) indicate the seasonal average 10% concentration of sea ice over the ocean and snow coverage over land.

borehole, coral and other data mentioned earlier. The tropical cooling varies within and between basins in the 20°N–20°S band: from 2.9°–1.7°C in the Atlantic Ocean; 1.9°–2.3°C in the Indian Ocean; and 1.8°–2.6°C in the Pacific Ocean. The tropical Pacific is particularly interesting as there is an east-west asymmetry to the cooling with western tropical Pacific SSTs only about 1.8°–1.9°C cooler and eastern tropical Pacific SSTs 2.2°–2.3°C cooler than today, suggesting that LGM Pacific SSTs are similar to permanent La Niña conditions. The LGM SST cooling is amplified in the North Atlantic, associated with a weakening and shallowing of the conveyor,

with a maximum cooling of 8.0°C. Generally, the global SST fields are qualitatively similar to those found in Weaver et al. (1998) and are in excellent agreement with those inferred from alkenone reconstructions (Sikes and Keigwin, 1994; Bard et al., 1997; Lyle et al., 1992; Rostek et al., 1993; Ikehara et al., 1997; Schneider et al., 1995; Chapman et al., 1996).

The reduction and shallowing of the overturning in the North Atlantic, together with the surface cooling, leads to a redistribution of water mass properties in the global ocean. For example, the more northward penetration of AABW, which is cold relative to the displaced NADW, leads to a local maximum in

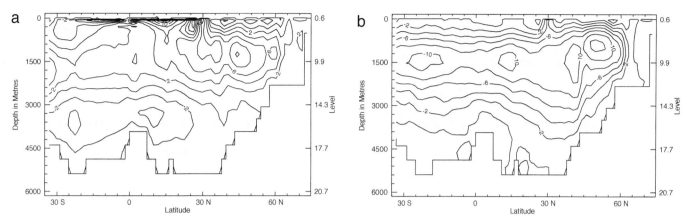

Fig. 37 a) North Atlantic meridional overturning streamfunction for the LGM equilibrium climate. b) Difference between a) and that obtained for the present-day climate (Fig. 18a). The contour interval is 2 Sv.

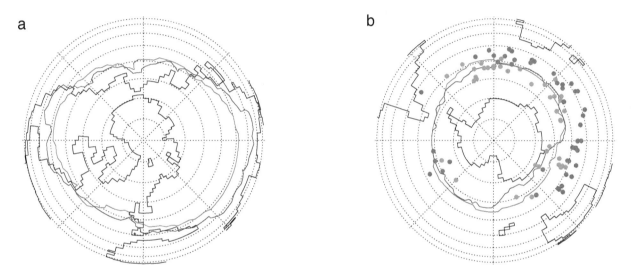

Fig. 38 Maximum equilibrium annually-averaged extent of sea ice (over the ocean) and snow (over the land) for the LGM climatology (blue) in the a) northern and b) southern hemispheres. The pink line represents the same field for Weaver et al. (1998). In the southern hemisphere, the dots indicate where the Crosta et al. (1998a, 1998b) paleo reconstruction of maximum sea-ice extent suggests the presence (green) and absence (red) of sea ice.

cooling in the deep North Atlantic around 30°N (Fig. 40b). Maximum cooling, of course, occurs in the thermocline regions of the global oceans (Fig. 40a), although most of the deep ocean is still 0.7°–0.9°C cooler than present day-climatology.

In the region of the NADW water mass (Fig. 40b), there appears to be a minimum in the zonally-averaged cooling relative to water masses below and above. Figure 41 shows the rather fascinating result that the minimum is the result of a cooling anomaly in the eastern Atlantic, up to 4.4°C at 600 m, being balanced by warm anomalies (up to 0.9°C) in the western Atlantic, most notably in the Labrador Sea and near equatorial regions. The analysis of the LGM fields suggests that the warm anomaly that develops in the Labrador Sea is due to the sea-ice cover insulating the ocean from the atmosphere. As such, convection is reduced relative to the present and relatively warm subsurface waters are not mixed with relatively cold surface waters, leaving a warm anomaly behind. In the equatorial region, there is a substantial drop in the cross-equa-

torial transport (see near-vertical contours in Fig. 37a in the upper 2000 m near the equator) of heat and salt. Relatively warm subsurface equatorial waters are therefore not efficiently transported out of the region towards the North Atlantic, leaving a warm anomaly behind. This same process leads to the local maximum in cooling in the eastern Atlantic, as the reduced thermohaline circulation and hence Gulf Stream and North Atlantic drift, does not bring in as much heat to the region.

In order to compare model-derived and CLIMAP SST estimates, we subtract the CLIMAP present day and LGM reconstructions for February (winter) and August (summer), and do the same for the model fields (here winter and summer are defined as January through March and July through September, respectively). The difference between these proxy and model difference fields (Fig. 42) reveals that in both the summer and the winter, we find significantly more cooling in the subtropical North and South Pacific and South Atlantic. These subtropical gyre regions are in fact places where CLIMAP data

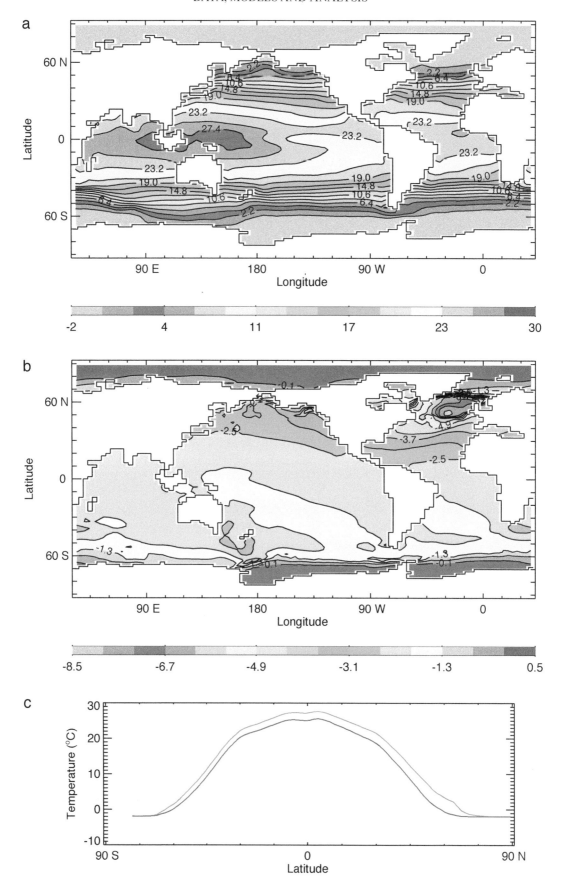

Fig. 39   Equilibrium annually-averaged sea surface temperature for: a) the LGM; b) the difference between the LGM and the present (LGM – PD); c) zonally-averaged LGM (red) and PD (green) equilibrium sea surface temperature climatologies.

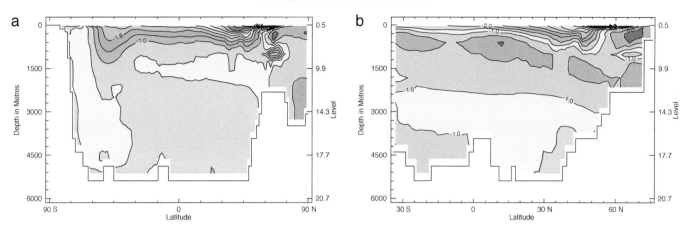

Fig. 40   LGM minus present-day annual zonally-averaged ocean potential temperature difference (°C) a) global and b) North Atlantic.

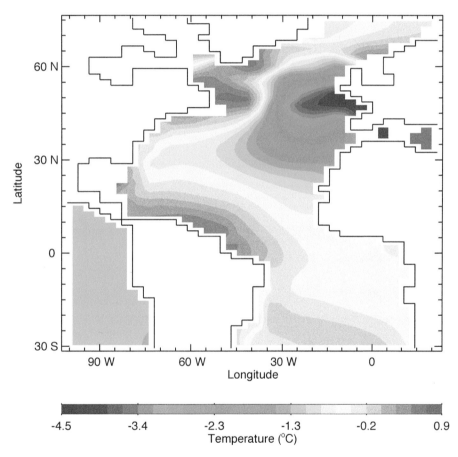

Fig. 41   Difference between LGM and present-day potential temperature at a depth of 600 m.

suggest a modest LGM warming relative to the present. Over much of the North Atlantic our model-derived LGM SST is less than that inferred by CLIMAP, although a portion of this difference is certainly related to the slightly cooler SSTs in our present-day climate (Fig. 14b). This follows since the possible LGM cooling is bounded below by the freezing point of water, and over much of the high latitude North Atlantic sea ice is present at least in the winter. In the Southern Ocean we also find a general tendency for the model to yield less

cooling than CLIMAP, although there are large regions where the model fields are in fact cooler than CLIMAP. Averaged over the band 20°S to 20°N, our model-derived SSTs are 0.9°C colder than CLIMAP in the summer and 1.5°C cooler in the winter.

In order to see whether or not the ICE4G reconstruction was a stable initial condition within our coupled model, an additional integration was performed in which the specified ice sheets were allowed to melt or grow. Snow was allowed

217

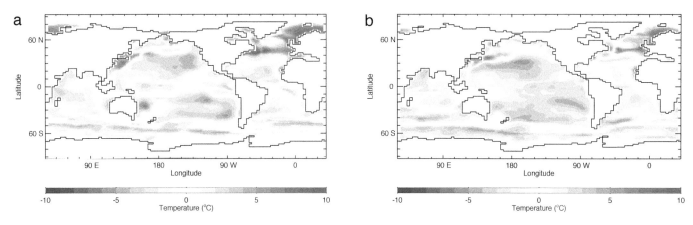

Fig. 42   Anomaly maps showing the difference between the model and CLIMAP Project Members (1976, 1981) sea surface temperature changes between the LGM and the present: a) summer; b) winter.

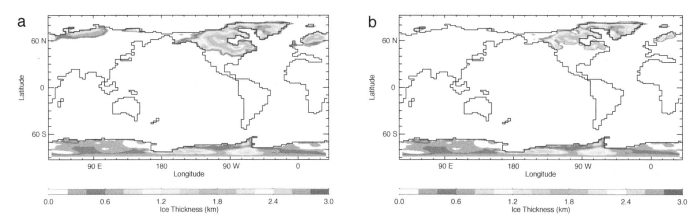

Fig. 43   LGM annually-averaged land ice thickness greater than the present-day bed topography: a) initial condition (reconstruction); and b) at equilibrium after release from the initial condition.

to accumulate to a maximum of 10 m on top of the existing ice sheet and further increase in snowfall raised the height of the ice sheet by the additional amount. Similarly, once all the snow on the ice sheet melted, the ice sheet was also allowed to melt at the same melting rate (Section 2b). As we did not include ice-sheet dynamics in this experiment (but see Section 2bb and Yoshimori et al. (2001)), the ice sheet was not allowed to grow any more than the ICE4G reconstruction and snow was not allowed to accumulate to more than 10 m. Any excess snow fall on the ice sheet was directly passed to the ocean through the appropriate river drainage basin. Freshwater fluxes associated with ice-sheet growth and melt were not passed to the ocean model as we were focused solely on thermodynamical feedbacks, although snow melt/accumulation was accounted for in the freshwater budget (as discussed in Section 2b).

When the ice sheet is allowed to adjust, as discussed above, it reduces in extent in many regions. Most notably, the Laurentide ice sheet in central North America melts back slightly, whereas the northern Eurasian ice sheet almost completely disappears. With respect to the latter, there has been some question as to whether or not the ICE4G reconstruction

of Peltier (1994) is in fact realistic there (e.g., Sher, 1995). The LGM global mean SAT cooling, relative to the present, drops by 0.6°C to 3.0°C from the case where the ice sheet is held fixed for two reasons. First, in regions where the ice sheet melts away, the surface temperature drops through elevation effects (accounting for 0.25°C of the decrease). Second, in regions where ice sheets existed year round, there was a continual albedo feedback which would not be present in the summer if the ice sheet were to melt away (accounting for 0.35°C of the decrease). Nevertheless, it is reassuring that the major portions of both the Laurentide and Fennoscandian ice sheets, as well as the Greenland and Antarctic ice sheets, remain stable. The timescale for ice-sheet evolution is, of course, many thousands of years so we should not expect the ICE4G reconstruction to be in equilibrium with the LGM radiative forcing, but rather to be in a continual state of adjustment as it moves towards the last glacial termination. As such, we would not expect our equilibrium climatology to agree precisely with the Peltier (1994) reconstruction.

For completeness we now test what role, if any, the inclusion of the wind feedback parametrization has on the LGM climate. To accomplish this task we calculate SAT anomalies, defined as departures from the present-day SAT field (Fig. 10a), in an experiment that starts at the end of the LGM experiment with prescribed ice sheets. The coupled model is integrated to a new equilibrium with the wind feedback continually being calculated (Section 2a1). The equilibrium wind stress anomaly field (Fig. 44a) reveals enhanced westerlies in the mid-latitude belt of the southern hemisphere, arising from a greater meridional surface pressure gradient, and an anticyclonic anomaly centred over the North Atlantic. The anticylonic anomaly develops in response to local cooling associated with a reduced overturning as in (Manabe and Stouffer, 1988; Fanning and Weaver, 1997b). Globally, the resulting wind anomalies have a negligible effect on LGM cooling (Fig. 44b) contributing only an additional 0.03°C. A similar negligible effect was found by Weaver et al. (1998).

While the present model is much more sophisticated than the one used by Weaver et al. (1998), there are still many feedbacks which have not been accounted for in our simulations. For example, the radiative forcing associated with atmospheric aerosols and clouds are both ignored. In the case of aerosols, it is not clear what the spatial pattern or sign and magnitude of this radiative forcing would be in terms of both aerosol type and distribution. Estimates from Peru (Thompson et al., 1995) suggest that LGM atmospheric dust levels may have been 200 times the present whereas in Antarctica they may have been 4–6/10–30 times the present for marine/continental aerosols, respectively (Petit et al., 1981). In the case of clouds, one would expect a reduced cloud cover in the drier LGM atmosphere (Hansen et al., 1984), although the incorporation of this effect in our model would be difficult and perhaps inconsistent with its present level of sophistication. Vegetation feedbacks and subsequent albedo changes through the expansion of dryland vegetation in Australia and Africa and the conversion of conifer to tundra in Europe and Siberia have also been suggested to enhance the cooling there (Crowley and Baum, 1997). As noted in Section 8, we are presently attempting to incorporate land surface and dynamic vegetation subcomponent models into our ESCM of intermediate complexity, but it is not yet possible to account for their effects. Finally, we have not accounted for the estimated 121 ± 5 m drop in sea level at the LGM (Fairbanks, 1989) through our use of present-day land-sea geometry in all experiments. This was done intentionally, however, so that we might perform a clean comparison with analogous present-day integrations. The LGM global mean salinity increase of about 1 psu arising from the reduced sea level, has been shown to have little effect on the climate of the LGM (Weaver et al., 1998).

Nevertheless, with these limitations in mind, our LGM experiments serve as a useful validation tool for the ESCM. Several paleo proxy studies (Boyle and Keigwin, 1987; Lea and Boyle, 1990) have inferred a shallowing and others (Curry and Lohmann, 1983; Oppo and Fairbanks, 1987; Charles and Fairbanks, 1992; Charles et al., 1996) a weakening of NADW at the LGM, both of which are borne out in the model results. In addition, a recent study (Rutberg et al., 2000) has inferred a reduced flux of NADW into the Southern Ocean, a feature also seen in our model solutions.

Our LGM climate cooling of 3.6°C relative to the present is within the range spanned by previous modelling studies. If we also account for the fact that global sea level was about 120 m lower than today (Fairbanks, 1989), and note that our model incorporates a specified lapse rate of 6.5°C km^{-1}, then our LGM SAT cooling over land is 4.4°C, so our global SAT cooling is 3.8°C. Averaged between 30°S and 30°N, tropical cooling is about 2.1°C over the oceans, and again, if we account for the sea level drop, 3.6°C over the land. These are generally cooler than CLIMAP estimates, but not as cool as some other reconstructions. Globally our model SSTs are in good agreement with alkenone reconstructions and are generally cooler than those of CLIMAP in the tropical ocean, while not as cool in the North Atlantic. Our model results also suggest that CLIMAP estimates in the subtropical gyres may be in error, especially since they infer a net LGM warming there relative to the present. As noted by Crowley (2000) and Weaver et al. (1998) it is possible for us to reconcile a more stable tropics (colder than CLIMAP, consistent with alkenone reconstructions but not as cold as the borehole, coral and other reconstructions mentioned earlier) with the results of our, and other, modelling studies if the climate sensitivity in nature is low-to-moderate and CLIMAP temperatures represent a slight underestimation of LGM cooling.

7 Climate model evaluation and the long timescale memory of the ocean

The most common method of evaluating an ocean or coupled climate model is to spin it up under perpetual present-day forcing and subsequently to compare the model climatology with present-day observations. Implicit in this approach is the assumption that the present-day observations are in equilibrium with the present-day forcing. The validity of this assumption is questionable as it is well known that the ocean, with its large thermal inertia, takes hundreds to thousands of years to equilibrate to its prescribed surface forcing. As such, changes in radiative forcing over recent Earth history are likely to have a signature in at least the deep ocean.

Weaver et al. (2000) provided some support for this notion through the analysis of integrations conducted using an earlier version of our coupled model. They compared the results from a transient integration forced by time-varying atmospheric CO_2 and orbital forcing, starting at 6 KBP, with equilibrium simulations obtained under perpetual 1850 (preindustrial) and 1998 (present-day) atmospheric CO_2 concentrations and orbital parameters. Substantial differences were found between the equilibrium climatologies and the transient simulation, even at 1850 (in weakly ventilated regions), prior to any significant changes in atmospheric CO_2. When compared to the present-day equilibrium climatology, differences were large. The global mean SATs and SSTs were ~0.5°C and ~0.4°C colder, respectively, deep ocean temperatures were substantially cooler, southern hemisphere sea-ice cover was 38% larger, and the North Atlantic conveyor 16% weaker in the transient case. These differences were due to the long timescale memory of the deep ocean to climatic conditions which prevailed throughout the late Holocene, as well as to its large thermal inertia.

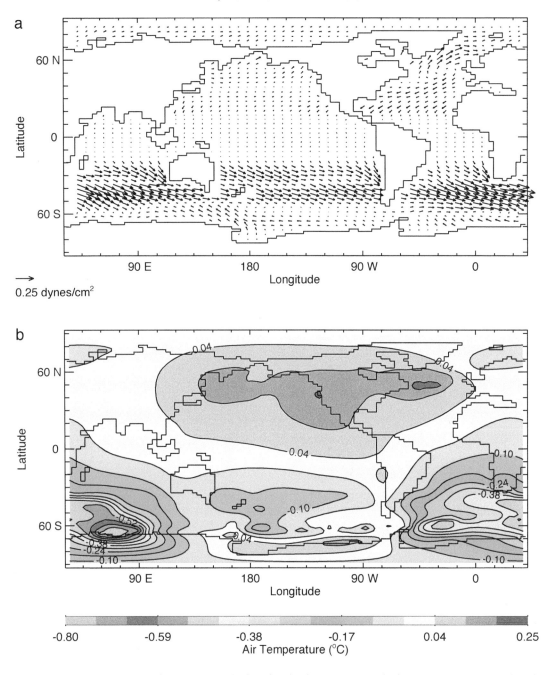

Fig. 44   a) Equilibrium LGM wind stress anomaly from the parametrization of section 2a1 and b) the equilbrium annually-averaged surface air temperature difference between a model with the parametrization and one without. Every second vector is shown in a).

The coupled modelling community has in some sense realized this potential problem, although it has been framed within a different context. It has been shown that climate models which do not account for the twentieth century build-up of atmospheric CO_2 through long integrations from, say, 1850, suffer from a cold start problem (Fichefet and Tricot, 1992; Hasselmann et al., 1993; Cubasch et al., 1994; Keen and Murphy, 1997), whereby the initial rate of warming is slower than it should otherwise be (see Kattenberg et al., 1996). This follows since the initial conditions for such experiments are in radiative equilibrium with their forcing, whereas a transient

experiment starting from preindustrial times has a warming commitment (Keen and Murphy, 1997), as it has yet to reach radiative equilibrium. Using the same earlier version of our model, Weaver et al. (2000) also demonstrated that a cold start global warming simulation (starting from a 1998 equilibrium climatology) underestimates the global temperature increase at 2100 by ~10%.

In this section we extend the analysis of Weaver et al. (2000) in two ways. First, we allow for changes in solar forcing through the use of the Lean et al. (1995) reconstruction of solar irradiance; second we conduct the experiments with an improved model that includes more sophisticated sea-ice and atmospheric components. In particular, as sea ice plays an important role in amplifying the response of the climate system to radiative forcing perturbations, we are able to quantify whether or not the analysis of Weaver et al. (2000) is sensitive to the representation of sea-ice processes.

a Experimental Design

Seven experiments were conducted with the coupled model (Table 4). The first three experiments consisted of spinning up the model from rest to equilibrium under 6 KBP, 1850 and 1998 orbital parameters and atmospheric CO_2 levels, and a solar constant of 1368 W m^{-2}. Atmospheric CO_2 levels of 280 ppm (Barnola et al., 1987) were used in the preindustrial 6 KBP and 1850 cases, while a modern level of 365 ppm was used in the 1998 case. The radiative forcing associated with the orbital parameters at the three time slices (Fig. 45) shows insignificant change between 1850 and the present. As mentioned earlier, variations over longer timescales do not greatly affect the annual mean insolation but rather have more impact on the seasonal and latitudinal distribution of this insolation (Fig. 45). These seasonal and latitudinal changes can then be amplified within the climate system through, for example, changes in the intensity and location of wintertime oceanic convection, and changes in sea-ice extent (and subsequent albedo feedback). Increased northern hemisphere insolation and decreased southern hemisphere insolation in the northern hemisphere summer and winter, respectively, are apparent at 6 KBP relative to the present. Atmospheric CO_2 levels, on the other hand, have a more substantial influence on the annual mean radiative forcing. These three experiments are hereafter referred to as 6 KBP, preindustrial (PI) and present-day (PD), respectively (Table 4).

The PD and PI equilibrium climatologies were used as initial conditions for two global warming experiments representing the cold start (GWcs in Table 4) and what we term the 'lukewarm' start (GWlws in Table 4), respectively. In the case of the lukewarm start, the atmospheric CO_2 concentration $C(t)$ was increased according to

$$C(t) = C_{PI} e^{k(t - 1850)^a} \qquad (60)$$

where $C_{PI} = 280$ ppm, $k = 7.944 \times 10^{-8}$, $a = 3.0058$ and t is the year. This exponential fit was obtained from a two-parameter fit (k and a) using the observed 1850–1990 increase in atmospheric CO_2 and other greenhouse gases that was used by Mitchell et al. (1995) in the United Kingdom Meteorologial Office (UKMO) global warming simulations. The parameter a was subsequently retained although k was recalculated by demanding that the CO_2 levels at 1998 match observations (365 ppm). The GWlws experiment, as well as the GWcs experiment were both extended out to year 2500 with atmospheric CO_2 increasing at 1% per year, until it reached levels 4 times the present (about year 2138), at which time atmospheric CO_2 levels were held fixed. The wind feedback of Section 2a1 was used in all these experiments.

The results from these experiments are compared at year 2100 with those obtained from two transient experiments which began at 6 KBP. From 6 KBP to 1610, these two transient experiments were identical with atmospheric CO_2 fixed at 280 ppm and orbital forcing varying according to Berger (1978). At 1610 they diverged, with one retaining a fixed solar constant and the other allowing it to vary (Fig. 46). From 1850 to 1998 both experiments allowed atmospheric CO_2 to vary according to Eq. (60). The transient constant and varying solar forcing experiments at 1850 (PItr and PItr ΔS_0, respectively) as well as at 1998 (PDtr and PDtr ΔS_0, respectively) were subsequently compared to the equilibrium climatologies PI and PD in order to determine the extent to which the climate system retains a memory to prior radiative forcing. These two experiments were also integrated to 2100 with CO_2 increasing as in GWlws and GWcs and are denoted GWtr and GWtr ΔS_0 in Table 4, respectively. In the case of GWtr ΔS_0, the solar constant after year 1998 was held fixed at 1368 W m^{-2}, while in both cases, the wind feedback was once more in effect. The GWtr run was further extended to year 2500, as in the case of GWlws and GWcs, with atmospheric CO_2 increasing at 1% per year, until it reached levels 4 times the present (about year 2138), at which time atmospheric CO_2 levels were held fixed.

b Transient Versus Equilibrium Climatologies at 1850 and 1998

Rather than repeat the analysis of Weaver et al. (2000) in full detail, we briefly describe the differences that arise from the use of our new model as well as the varying solar constant. All the major conclusions of Weaver et al. (2000) hold for the present experiments, although the exact magnitude of changes between the transient and equilibrium climatologies are slightly different.

In the case of constant solar forcing, the SAT differences at 1850 and 1998 between the transient and equilibrium climatologies (Figs 47a and 47b) are similar to those found in Weaver et al. (2000) (their Figs 3b and 3c) even though the models used are quite different. At 1850, the transient climatology is only 0.01°C cooler than the equilibrium climatology (Table 4), increasing to 0.61°C cooler at 1998. These compare with analogous values of 0.04°C and 0.46°C, respectively, found in Weaver et al. (2000). In all cases the differences are largest at high latitudes, due to amplification by the ice albedo feedback. In addition, changes are much larger in 1998 than in 1850 due to the fact that the climate system has yet to equilibrate to the build-up of atmospheric CO_2 over the twentieth century.

When solar forcing is allowed to vary (Figs 47c and 47d), the differences at 1850 are much larger. In the equilibrium PI experiment, a solar constant of 1368 W m^{-2} was used, whereas for the period 1610–1850, the Lean et al. (1995) reconstruction gives a constant which at all times is between 1364 and 1367 W m^{-2} (Fig. 46). As such, the climate during the entire period is colder than the corresponding transient integration with a fixed solar constant, so that at 1850, the PItr

TABLE 4. The table shows a summary of diagnostics from the model experiments. The second column gives the experiment while subsequent columns portray results from the equilibrium simulations (rows 1 through 3; the row number is given in the first column) as well as differences between the transient and the equilibrium experiments at 1850 (row 4) and 1998 (row 6). Rows 5 and 7 are analogous to rows 4 and 6 except solar forcing is allowed to vary. Row 8 shows the difference at 2100 between the cold start experiment (GWcs) and the continuation of the transient experiment to year 2100 (GWtr), while row 9 shows the same thing but for the case with varying solar forcing. Rows 10 and 11 are the same as rows 8 and 9 but for the 'luke-warm' start experiment (GWlws). The cold start, lukewarm start, and transient experiment differences from their initial condition (PD, PDlws and PDtr, respectively) at 2100 are given in rows 12–14, respectively. Row 15 is the same as row 14 but for the case of varying solar forcing. Global and Atlantic mean potential temperature and salinity are given in columns 2 to 4. As salinity is conserved in the coupled model, changes in global mean salinity are small and hence not shown. Minor differences would be reflected in northern (column 9) and southern (column 10) hemisphere sea-ice areas or changes in atmospheric storage. The global mean sea surface and surface air temperature are listed in columns 6 and 7 while the maximum value of the North Atlantic overturning streamfunction is given in column 8. Note that in column 2, experiments with PI, PD, or GW as a prefix imply results are taken at year 1850, 1998 and 2100, respectively.

#	Experiment	Global Mean T (°C)	Atlantic Mean T (°C)	Atlantic Mean S (psu)	Global Mean SST (°C)	Global Mean SAT (°C)	Maximum Atlantic ψ (Sv)	NH ice area 10^6 km^2	SH ice area 10^6 km^2
1	6 KBP	3.41	5.30	34.85	17.38	12.73	20.84	11.21	13.92
2	PI (1850)	3.43	5.33	34.85	17.44	12.76	20.51	14.03	11.64
3	PD (1998)	4.00	5.96	34.87	18.38	14.08	20.96	10.55	11.09
4	PItr-PI	−0.02	0.011	0.0020	−0.02	−0.01	−0.09	−0.07	0.18
5	PItrΔS_0-PI	−0.09	−0.083	0.0025	−0.24	−0.30	0.04	0.22	0.69
6	PDtr-PD	−0.50	−0.53	−0.011	−0.45	−0.61	−1.27	0.49	2.15
7	PDtrΔS_0-PD	−0.57	−0.62	−0.013	−0.55	−0.74	−1.09	0.59	2.49
8	GWcs-GWtr	0.37	0.35	0.0028	0.22	0.30	0.70	−0.17	−1.76
9	GWcs-GWtrΔS_0	0.41	0.41	0.0044	0.26	0.35	0.81	−0.03	−0.21
10	GWlws-GWtr	0.012	0.001	−0.0010	0.0032	0.0018	−0.01	−0.01	−0.03
11	GWlws-GWtrΔS_0	0.059	0.056	0.00059	0.038	0.053	0.10	−0.03	−0.21
12	GWcs-PD	0.41	0.40	0.01	2.24	3.05	−0.81	−2.30	−3.39
13	GWlws-PDlws	0.54	0.58	0.02	2.47	3.35	−0.30	−2.63	−3.79
14	GWtr-PDtr	0.55	0.58	0.02	2.47	3.35	−0.24	−2.62	−3.77
15	GWtrΔS_0-PDtrΔS_0	0.57	0.61	0.02	2.54	3.44	−0.53	−2.69	−3.94

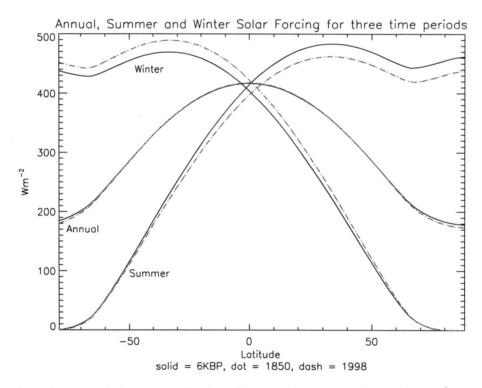

Fig. 45 Zonally-averaged annual, summer and winter mean profiles of top of the atmosphere incoming solar radiation (W m^{-2}) at 6 KBP (solid black), 1850 (red dotted), 1998 (blue dashed).

ΔS_0 experiment is 0.30°C cooler than PI and 0.29°C cooler than PItr (Table 4; Fig. 48a). The difference between PItr ΔS_0 and PItr reduces by 1998 as the solar constant increases from the end of the Little Ice Age through to the end of the twenti- eth century (Fig. 48c). Once more, differences are largest at high latitudes where sea ice is present. The general reduction in the differences between the two transient experiments (PItr ΔS_0 and PItr) from 1850 and 1998 is also reflected in the SST

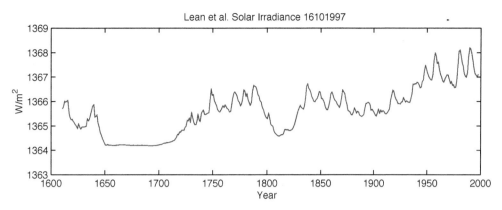

Fig. 46   Lean et al. (1995) solar irradiance (1610–1997) used in the transient integration with varying solar forcing.

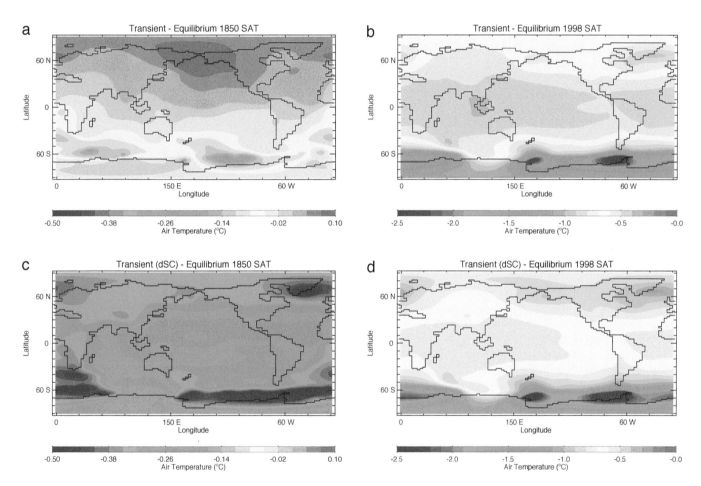

Fig. 47   Surface air temperature (°C) difference between the transient (without varying solar forcing) and equilibrium model simulations at 1850; b) as in a) but at 1998; c), d) as in a), b), respectively, but including varying solar forcing.

field (Figs 48b and 48d). Once more, the largest differences tend to be associated with regions near sea-ice boundaries where strong feedbacks exist.

The generally cooler transient climate tends to support more extensive sea ice in both the northern and southern hemispheres, as well as a slightly reduced overturning in the North Atlantic (Fig. 49; Table 4), consistent with the results of Weaver et al. (2000). In the case of the meridional overturning, however, it is interesting to note that the conveyor remains remarkably stable in all experiments through the 6000 years of integration. Differences between the 6 KBP, 1850 and 1998 equilibria are small, as are the differences between the corresponding transient and equilibrium cases (Fig. 49).

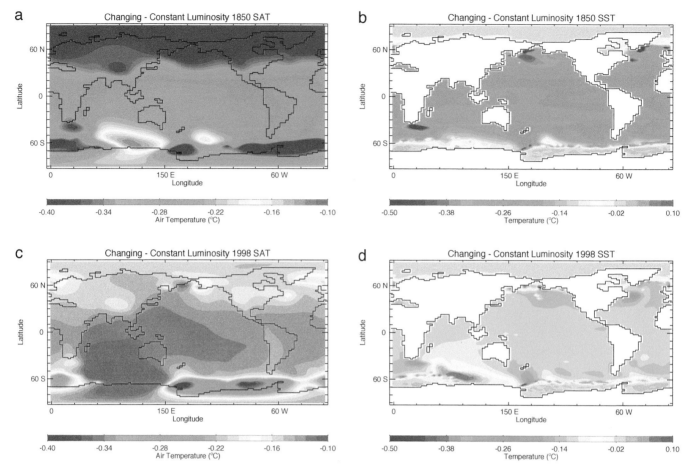

Fig. 48 Differences (°C) between the two transient integrations with and without solar forcing: a) surface air temperature at 1850; b) sea surface temperature at 1850; c) surface air temperature at 1998; d) sea surface temperature at 1998.

As was the case in Weaver et al. (2000), the most significant differences between the 1850 and 1998 equilibrium and transient experiments are associated with sea-ice processes and deep ocean convection. These differences are greatest at 1998, by which time the radiative forcing associated with atmospheric CO_2 has increased substantially, relative to preindustrial times. Deep ocean water mass properties (Fig. 50), which retain a memory of the radiative forcing throughout the late Holocene, emphasize the differences between the transient and equilibrium cases especially in the presence of a time varying solar constant.

In summary, our transient versus equilibrium results tend to support the conclusions of Weaver et al. (2000). The inclusion of more sophisticated representations of both sea ice and the atmosphere tended to reduce differences slightly, relative to those found by Weaver et al. (2000), when the solar constant was held fixed. Allowing the solar constant to vary over the last 400 years, however, tended to accentuate the differences between transient and equilibrium integrations, especially at 1998. Taken together, our analysis underlines a potential problem in the evaluation of present-day ocean only and coupled atmosphere-ocean general circulation models. In developing these models, it is common to integrate them to

equilibrium under present-day radiative forcing (in the case of coupled atmosphere-ocean models) or using present-day surface boundary conditions (in the case of ocean only models). The evaluation process then involves a detailed comparison of model results with present-day observations. Since the ocean, and especially its deep and weakly ventilated regions, have a long adjustment period to changes in radiative forcing, the present-day oceanic observations are not in equilibrium with the present forcing. Even if the ocean or coupled model were perfect, it would therefore not be possible to reproduce these observations under perpetual present-day model forcing.

c *Global Warming Simulations*

We now determine whether or not global warming simulations are sensitive to the approach used to initiate them. Four experiments were performed starting at either the PD (cold start—GWcs), PI (lukewarm start—GWlws), PDtr (warm start—GWtr) or PDtr ΔS_0 (warm start—GWtr ΔS_0) initial conditions. The latter two can be viewed as an extension of the transient integration starting at 6 KBP, although in the PDtr ΔS_0 case, the solar constant was held fixed at 1368 W m^{-2}. These four experiments were all integrated out to year 2100 with atmospheric CO_2 increasing at a rate of 1% per

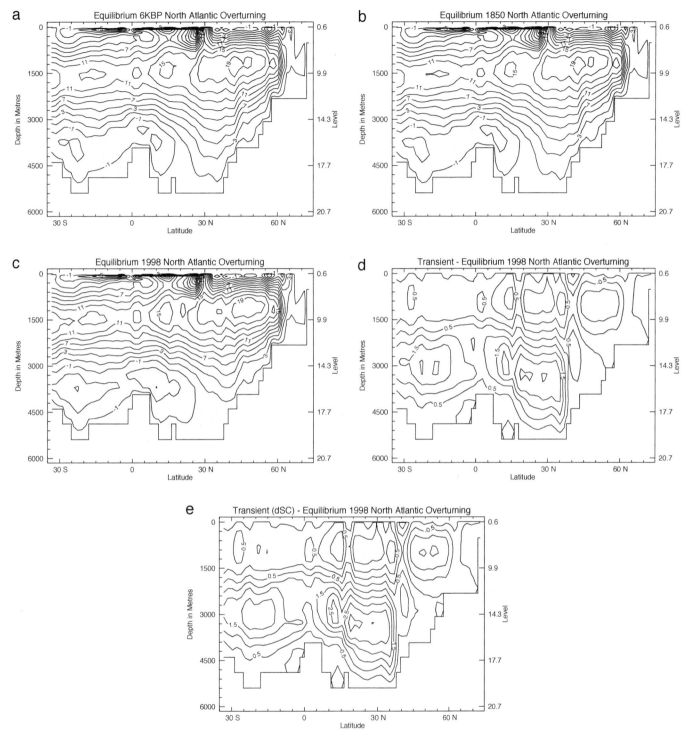

Fig. 49 Meridional overturning streamfunction (Sv) in the North Atlantic for: a) equilibrium 6 KBP climate; b) equilibrium 1850 climate; c) equilibrium 1998 climate; d) transient (no varying of solar forcing) minus equilibrium 1998 climate; e) transient (with solar forcing) minus equilibrium 1998 climate. The contour interval is 2 Sv in a)–c) and 0.5 Sv in d)–e).

year. Only the GWtr and GWcs were further extended to year 2500 (under a $4 \times CO_2$ stabilization scenario).

The projected results at year 2100 for the GWtr, GWtr ΔS_0 and GWlws experiments reveal globally-averaged SATs and SSTs (Fig. 51), sea-ice coverage (Fig. 52), steric sea level rise (Fig. 53) and strength of the North Atlantic overturning (Fig. 54), that are nearly indistinguishable from each other. The actual change relative to their respective initial conditions (Table 4), further reveals that the GWlws and GWtr experiments lead to almost identical projected changes at the year 2100. The GWtr ΔS_0 experiment, on the other hand, leads to a slightly greater projected warming and reduction in

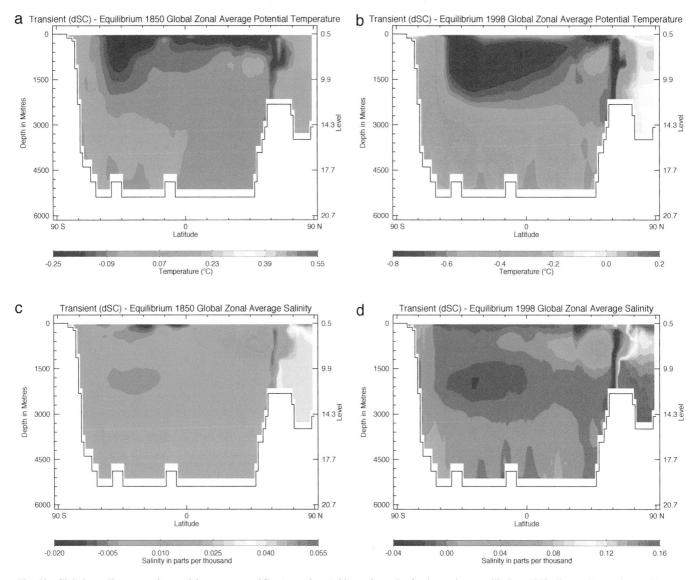

Fig. 50   Global zonally-averaged potential temperature (°C): a) transient (with varying solar forcing) minus equilibrium 1850 climate; b) transient (with varying solar forcing) minus equilibrium 1998 climate. c), d) as in a), b) but for global zonally-averaged salinity (psu).

sea-ice extent as the initial condition at 1998 was slightly cooler than in the GWlws and GWtr cases. During the further 102 years of integration, some of the warming commitment inherent in the GWtr ΔS_0 case is realized (recall that from 1610 to 1998 the solar constant was almost always below 1368 W m^{-2} and that after 1998 it was held fixed at 1368 W m^{-2}). These initial results suggest that projections of global warming conducted by international modelling groups are not affected, at least to first order, by the fact that they have not accounted for the radiative forcing prior to 1850. Nevertheless, these same results suggest that were changes in solar variability accounted for, future projections of global warming would be slightly higher than if they were not.

Since the GWtr, GWtr ΔS_0 and GWlws experiments lead to similar results, we only extended the GWtr experiment out to 2500 for comparison with the GWcs experiment. The differences between the cold start and long transient runs are ini-

tially much larger than those documented earlier, although by the year 2500, much of the warming potential inherent in the long transient run is realized (Figs 51, 52). The fact that the deep ocean is in equilibrium with the 1998 radiative forcing in GWcs greatly reduces the thermal expansion of the ocean by the year 2500 relative to the GWtr case (Fig. 53). This follows since the timescale for equilibration of the deep ocean to radiative forcing perturbations is on the millennial diffusive timescale. The net result is that the GWtr case produces a projected greater warming, loss of sea ice and sea level rise than the GWcs case (Table 4), due to the warmer initial condition of GWcs.

The meridional overturning streamfunction remains relatively stable in all cases (Figs 54, 55), with only a ~1 Sv weakening occurring during the phase of the integration where the radiative CO_2 forcing was changing. Once the forcing is held fixed, the North Atlantic overturning re-equilibrates to a

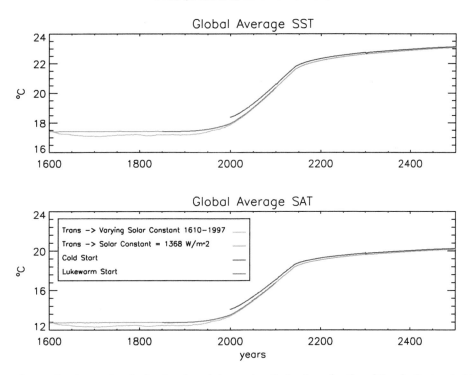

Fig. 51 Globally-averaged sea surface temperature (top) and surface air temperature (bottom) as a function of time for the transient integrations with varying (orange) and constant (green) solar forcing. The lukewarm start integration (pink) and cold start integration (blue) are also shown.

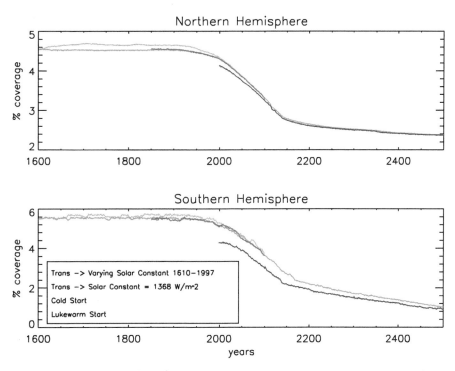

Fig. 52 Northern (top) and southern (bottom) hemispheric sea-ice coverage expressed as a fraction of the total hemispheric area for the transient integrations with varying (orange) and constant (green) solar forcing. The lukewarm start integration (pink) and cold start integration (blue) are also shown.

magnitude slightly larger than its initial condition (Fig. 54). As noted in Section 5 and Wiebe and Weaver (1999), this arises from an eventual increase in the zonally-averaged meridional gradient of the depth-integrated steric height in the North Atlantic. As further noted in Section 5, the inclusion of wind feedback together with the treatment of moisture transport by advection, causes the overturning to equilibrate to a value similar to its initial condition.

227

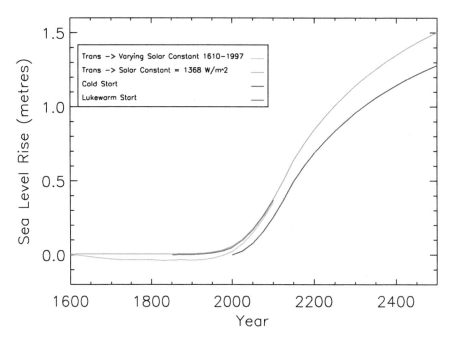

Fig. 53   Sea level rise (m) due to thermal expansion relative to the initial condition as a function of time for the transient integrations with varying (pink) and constant (blue) solar forcing. The lukewarm start integration (green) and cold start integration (red) are also shown.

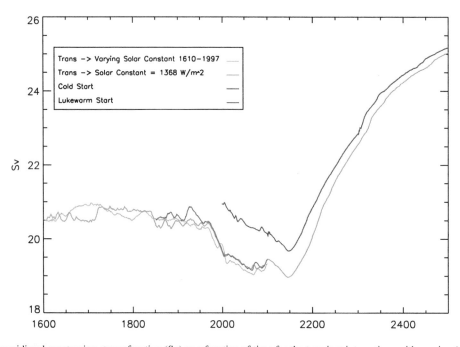

Fig. 54   North Atlantic meridional overturning streamfunction (Sv) as a function of time for the transient integrations with varying (pink) and constant (blue) solar forcing. The lukewarm start integration (pink) and cold start integration (blue) are also shown.

The projected strength of the overturning at the year 2100 is shown in Figs 55a–55d for the GWcs, GWlws, GWtr and GWtr ΔS_0 experiments, respectively. Evidently, the magnitude of the differences between the experiments is small (Figs 55e–55h) in all cases, suggesting a rather passive role of the ocean in terms of local feedbacks to the climate warming. Sea-ice feedbacks between the cold start and different tran-

sient runs, however, lead to substantial differences in the SAT change at year 2100 (Fig. 56).

The cold start case (GWcs) leads to global mean SATs and SSTs at 2100 that are 0.30°C and 0.23°C warmer, respectively, (Fig. 56c; Table 4) than for the corresponding transient case (GWtr). As a consequence, there is less sea ice in both the southern and northern hemispheres in GWcs relative to

228

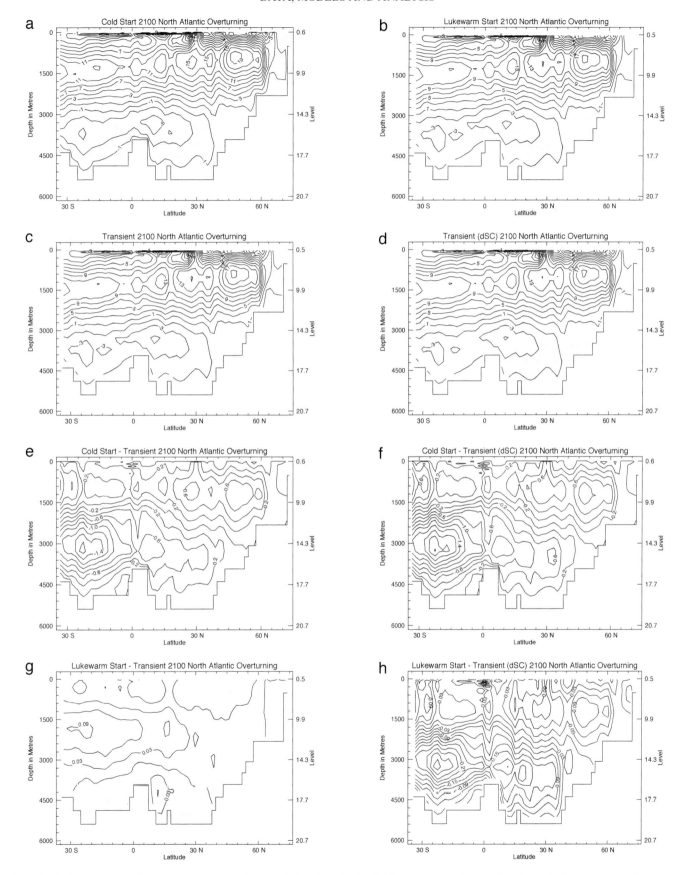

Fig. 55   North Atlantic meridional overturning streamfunction (Sv) at 2100 for the global warming experiments: a) cold start; b) lukewarm start; c) transient integration (constant solar forcing); d) transient integration (varying solar forcing); e) difference between a) and c); f) difference between a) and d); g) difference between b) and c); h) difference between b) and d). The contour interval is 2 Sv in a)–d), 0.2 Sv in e)–f), and 0.03 Sv in g)–h).

229

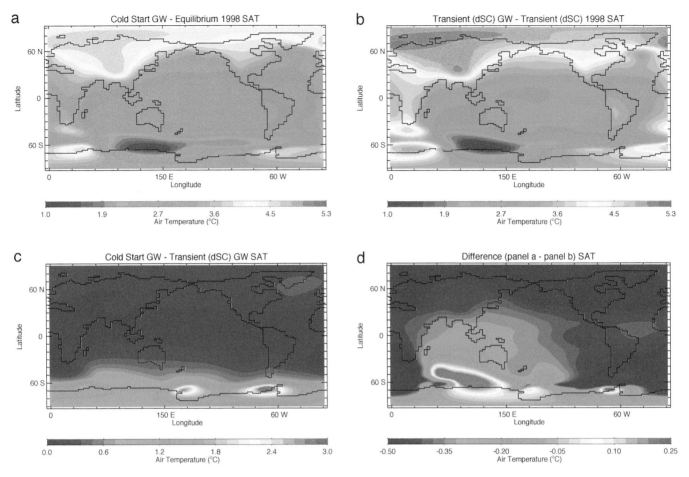

Fig. 56 Surface air temperature difference (°C) at 2100 relative to 1998 for: a) the cold start experiment; and b) the transient experiment with varying solar forcing; c) the difference between the year 2100 cold start and transient climates; d) panel a) minus panel b) which shows the difference in projected global warming relative to 1998.

GWtr (Table 4). In terms of a global mean SAT change at 2100 relative to their respective initial conditions, GWcs produces a ~10% smaller perturbation (3.05°C — Table 4; Fig. 56a) than GWtr (3.35°C — Table 4) and GWtr ΔS_0 (3.44°C — Table 4; Fig. 56b). Ironically, the cold start experiment starts from a warmer initial condition and leads to a warmer projected climate than does the warm start experiment. Nevertheless, the resulting projection of global mean SAT change in GWcs is less than the projected change in GWtr and GWtr ΔS_0, due to the warming commitment present in the latter experiment which is partially realized in the future. The differences in the projected global mean SAT anomalies between GWcs and GWtr ΔS_0 (or GWtr) are fairly uniform over the globe (Fig. 56d), although there are a few local maxima present. These occur exclusively near the Antarctic sea-ice and Agulhas Retroflection regions.

In summary then, the new version of the model leads to very similar results to those found by Weaver et al. (2000), although some subtle differences exist due to the improved representation of sea ice in the present model. Consistent with the results of others (Fichefet and Tricot, 1992; Hasselmann et al., 1993; Cubasch et al., 1994; Keen and Murphy, 1997;

Weaver et al., 2000), we found that the initial rate of warming is slower in 'cold start' global warming experiments than in analogous experiments that start with preindustrial levels of atmospheric CO_2. This follows since the initial conditions for such experiments are in radiative equilibrium with their forcing (e.g., PD), whereas a transient experiment starting from preindustrial times (e.g., PDtr) has a warming commitment (Keen and Murphy, 1997), as it has yet to reach radiative equilibrium. As demonstrated by our global warming experiments, SATs at 2100, obtained by starting from the PD (equilibrium 1998) case (GWcs), are generally warmer than in the case where we extended the PDtr (transient 1998) integration (GWtr). This follows since the initial condition is warmer as it is in radiative equilibrium with the forcing. Nevertheless, the global warming relative to 1998 is ~10% less at 2100 in the cold start case (3.05°C) than in the warm start case (3.35°C), as the latter has a warming commitment which is slowly being realized.

When the effects of varying solar forcing are incorporated into the model, the projected change at 2100 in global mean SAT is larger than in the case where solar forcing is held fixed. This follows since the initial 1998 state of the transient model is colder, in the case where solar forcing is allowed to vary, as it has not reached radiative equilibrium with the solar forcing that slowly increased from preindustrial times through to the late twentieth century. In the subsequent 102 years of integration, some of the extra warming commitment inherent in this experiment is therefore realized. This result suggests that those coupled modelling groups that do not account for solar forcing changes over the twentieth century may slightly underestimate (3% in our model) the projected warming by the year 2100.

8 Summary

The UVic ESCM has evolved substantially from its early form as the coupled energy-moisture balance model/OGCM of Fanning and Weaver (1996). It now contains sophisticated sea-ice and land-ice subcomponent models and an improved representation of the atmosphere. By building the ESCM in a systematic fashion, in which all parameters are consistent between all subcomponent models and careful attention is given to the averaging of fluxes between coupling timesteps, we ensure that energy and moisture are conserved exactly. In this paper we have documented all those subcomponent models, as well as the technique used to couple them, and evaluated the climatology of our model against observations.

The model is able to capture the pathways of intermediate, deep and bottom waters as revealed through simulations involving the release of passive tracers. Also, the model simulates both the cold and warm water routes (Gordon, 1986; Rintoul, 1991) by which the upper layer water returns to the Atlantic Ocean to compensate for NADW production and export. The global distribution of sea and land ice also compares favourably with observations. This model evaluation procedure also clearly demonstrates the importance of including the process of moisture transport through advection instead of diffusion. When this is done, the simulated evaporation, precipitation and evaporation minus precipitation fields bear reasonable agreement with NCEP reanalysis data.

The inclusion of moisture advection, together with our new parametrization for wind feedbacks, has a significant influence on the transient response of our model to increasing greenhouse gases. In previous experiments conducted with our model using the diffusive approach for moisture transport, we found that the North Atlantic thermohaline circulation re-established to a level in the future which was stronger than its initial present-day rate. With the aforementioned parametrization included, the model re-establishes to the same rate as in the present-day climate, consistent with the results from coupled AOGCMs.

When applied to the LGM we find a global mean cooling of $3.8°C$, with tropical cooling over the oceans of about $2.1°C$ and over the land of about $4.4°C$. Over the ocean this cooling is in line with recent alkenone reconstructions, although it is cooler than CLIMAP. In the North Atlantic we find less cooling than CLIMAP and a conveyor which is both weaker and shallower than in the present, again consistent with recent reconstructions.

The ESCM was also used to re-examine the issue of climate model initialization. Our integrations with varying solar constant demonstrated the importance of solar changes during the last few centuries in projecting future climate change. Due to an inherent additional warming commitment, experiments which account for solar forcing changes lead to a greater projected global mean temperature change by 2100.

The examples given in this paper are both interesting from a scientific perspective as well as illustrative of the type of problems that can be addressed with our model. Our philosophy in building this model has been to develop a tool with which to understand processes and feedbacks operating within the climate system on long timescales. We believe that this approach is complementary to that of major coupled AOGCM centres, and that our model fills an important gap within the hierarchy of climate models (Claussen et al., 2000; Shackley et al., 1998). In fact, our model has been used, and is still being used, as a tool with which to examine the sensitivity of a particular process or subcomponent model across a wide range of parameters, in order to streamline the process of improving certain components of the CCCma coupled AOGCM.

We have not yet included a parametrization for clouds and their feedbacks in our model although given both the lack of understanding of their basic physics, as well as the simplicity of our atmospheric model, it is difficult to see how any meaningful scheme could be incorporated. Clouds have an important role in climate both through their absorption of longwave radiation and through scattering incoming solar radiation back to space. Even the vertical distribution of clouds is important as low clouds, in general, act to cool the planet and high clouds, in general, act to warm it. Different climate models use different approaches to parametrize clouds, many of which are crude at best, leading to cloud feedbacks that vary between models. For example Cess et al. (1996) summarize the cloud feedback from a wide range of atmospheric models. Some models give a positive feedback in global warming experiments (whereby longwave effects dominate), and some give a negative feedback (whereby shortwave effects dominate). While 10 out of the 18 models considered had a net positive feedback, the fact that its sign is largely uncertain suggests that our approach is a sensible one. Other caveats of course also apply to our model. In particular, our simple wind feedback scheme is at best a very crude approximation and we do not currently have a land surface scheme in our model. In addition, the fact that we do not explicitly capture atmospheric 'weather' and its chaotic nature means that we are missing an internal stochastic forcing of the complex climate system.

Our model development phase is not yet complete. We are currently in the process of implementing a biotic ocean carbon cycle model into our ESCM following OCMIP guidelines. In addition, we are in the process of incorporating new land surface, terrestrial carbon cycle and dynamic vegetation models. The land-surface scheme used will be a simple, one-layer model, computing the energy, moisture and snow balance at the surface as recently developed by P. Cox and K. Meisner. The terrestrial carbon cycle and dynamic vegetation model that we are using is the Hadley Centre TRIFFID (Top-down Representation of Interactive Foliage and Flora Including Dynamics; Cox et al. (2000); Cox (2001)). The immediate goal of our work is the development of a comprehensive ESCM with which to explore the mechanisms of quaternary climate change/variability. Our ultimate goal is the development of a tool which, when coupled to socioeconomic models, forms the basis of an Integrated Assessment approach to exploring the inter-relationship between climate change and climate change policy.

Acknowledgements

Over the years the model development and its applied research have been supported by numerous grants. We are grateful for funding support provided by the following Natural Sciences and Engineering Research Council of Canada (NSERC) programs, Strategic, Operating, WOCE, CSHD, Steacie and Equipment; the Meteorological Service of Canada (MSC)/ Canadian Institute for Climate Studies (CICS) Canadian Climate Research Network (CCRN); the MSC/NSERC Operating/Strategic subvention program; the NOAA Scripps Lamont Consortium on the Ocean's Role in Climate; the International Arctic Research Center in Fairbanks, Alaska; an IBM Shared University Research Grant; and the Canadian Climate Change Action Fund. We are also grateful to our colleagues at the Canadian Centre for Climate Modelling and Analysis for many years of interesting discussions and to the University of Victoria for infrastructure support. This model is available to the community at http://climate.uvic.ca/climate-lab/ model.html. We are grateful to Dr. Xavier Crosta for providing us with his LGM sea-ice extent reconstruction data used in Fig. 38a.

Appendix A – Grid Rotation

Any spherical grid rotation can be specified by defining three solid body rotations. The angles which define the rotations are usually referred to as Euler angles (Goldstein, 1950). First, define the Z axis to be through the poles such that the X-Y plane defines the equator and the X axis runs through the prime meridian. The angle Φ is defined as a rotation about the original Z axis. Angle Θ is defined as a rotation about the new X axis (after the first rotation) and angle Ψ is defined as a rotation about the final Z axis (Fig. 57).

Consider a globe with a clear sphere surrounding it, with only grid lines of latitude and longitude. By moving the outer sphere, the grid poles can be moved to line up with different points on the globe. Once the new poles are located, two of the rotation angles can be defined as follows. The definition for Φ is 90° minus the geographic longitude of the new North Pole. This rotates the Y axis under the new pole. To move the Z axis down, Θ is defined to be 90° minus the geographic latitude of the new North Pole. This places the original Z axis though the new North Pole position.

To define the grid completely, a third rotation about the new Z axis, must be specified. The rotated grid longitude of any point on the geographic grid is still undefined. To specify this last rotation, choose a point on the geographic grid (the globe) to locate the rotated grid's prime meridian. Set angle Ψ to zero and calculate the longitude of this point on the rotated grid. This longitude is the final angle Ψ, the angle needed to rotate the point back to the prime meridian. The definition of Ψ is usually not very important since the new grid longitude is arbitrary, but it does make a difference in defining exactly where the new grid starts. This may be important if it is desirable to line up grids for nesting. It may appear that all of the angle definitions are of the opposite sign to what they should be, but this comes from thinking about rotating the axes rather than rotating the rigid body.

Having defined the angles of rotation, an orthogonal transformation matrix can be written. The first rotation through an angle Φ about the Z axis (counterclockwise looking down the Z0 axis) is given by:

$$D = \begin{pmatrix} \cos\Phi & \sin\Phi & 0 \\ -\sin\Phi & \cos\Phi & 0 \\ 0 & 0 & 1 \end{pmatrix}$$

The second rotation through an angle Θ about the new X axis (counterclockwise looking down the X1 axis) is given by:

$$C = \begin{pmatrix} 1 & 0 & 0 \\ 0 & \cos\Theta & \sin\Theta \\ 0 & -\sin\Theta & \cos\Theta \end{pmatrix}$$

and the final rotation through an angle Ψ about the new Z axis (counterclockwise looking down the Z2 axis) is given by

$$B = \begin{pmatrix} \cos\Psi & \sin\Psi & 0 \\ -\sin\Psi & \cos\Psi & 0 \\ 0 & 0 & 1 \end{pmatrix}$$

Note that a total rotation matrix can be written as the product of the three rotations BCD (ordering is important). The total rotation matrix A is

$$\begin{pmatrix} \cos\Psi\cos\Phi - \cos\Theta\sin\Phi\sin\Psi & \cos\Psi\sin\Phi + \cos\Theta\cos\Phi\sin\Psi & \sin\Psi\sin\Theta \\ -\sin\Psi\cos\Phi - \cos\Theta\sin\Phi\cos\Psi & -\sin\Psi\sin\Phi + \cos\Theta\cos\Phi\cos\Psi & \cos\Psi\sin\Theta \\ \sin\Theta\sin\Phi & -\sin\Theta\cos\Phi & \cos\Theta \end{pmatrix}$$

Transforming points from the unrotated system to the rotated system (marked by primes) follows from

$$\begin{pmatrix} x' \\ y' \\ z' \end{pmatrix} = A \begin{pmatrix} x \\ y \\ z \end{pmatrix}$$

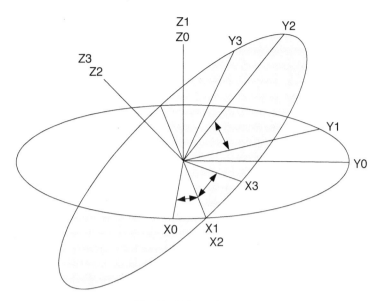

Fig. 57   Euler angles.

and transforming points from the rotated system (marked by primes) to the unrotated system follows from

$$\begin{pmatrix} x \\ y \\ z \end{pmatrix} = A^{-1} \begin{pmatrix} x' \\ y' \\ z' \end{pmatrix}$$

where the inverse transform A^{-1} is

$$\begin{pmatrix} \cos\Psi\cos\Phi - \cos\Theta\sin\Phi\sin\Psi & -\sin\Psi\cos\Phi - \cos\Theta\sin\Phi\cos\Psi & \sin\Theta\sin\Phi \\ \cos\Psi\sin\Phi + \cos\Theta\cos\Phi\sin\Psi & -\sin\Psi\sin\Phi + \cos\Theta\cos\Phi\cos\Psi & -\sin\Theta\cos\Phi \\ \sin\Theta\sin\Psi & \sin\Theta\cos\Psi & \cos\Theta \end{pmatrix}$$

Note that the inverse transform A^{-1} is equivalent to A with Ψ and Φ switched and all rotation angles made negative.

Appendix B – Operators in Spherical Coordinates

Tracer Diffusion

$$\mathbf{\nabla}\cdot(\xi\mathbf{\nabla}\chi) = \frac{1}{r_e^2\cos^2\phi}\frac{\partial}{\partial\lambda}\left(\xi\frac{\partial\chi}{\partial\lambda}\right)$$
$$+ \frac{1}{r_e^2\cos\phi}\frac{\partial}{\partial\phi}\left(\xi\cos\phi\frac{\partial\chi}{\partial\phi}\right)$$

Advection

$$L(\xi) = \frac{1}{r_e\cos\phi}\left(\frac{\partial(u\xi)}{\partial\lambda} + \frac{\partial(v\xi\cos\phi)}{\partial\phi}\right)$$
$$+ \frac{\partial(w\xi)}{\partial z}$$

Momentum Viscosity

$$\mathfrak{F}(A_h,\xi,\chi) = \mathbf{\nabla}\cdot(A_h\mathbf{\nabla}\xi) + A_h\left(\frac{1-\tan^2\phi}{r_e^2}\right)\xi$$
$$- A_h\frac{2\sin\phi}{r_e^2\cos^2\phi}\frac{\partial\chi}{\partial\lambda}$$

References

ALLEY, R.B. 1989. Water-pressure coupling of sliding and bed deformation: II. Velocity-depth profiles. *J. Glaciol.* **35**. 119–129.

BARD, E.; F. ROSTEK and C. SONZOGNI. 1997. Interhemispheric synchrony of the last deglaciation inferred from alkenone palaeothermometry. *Nature*, **385**: 707–710.

BARNOLA, J.M.; D. RAYNAUD, Y.S. KOROTKEVICH and C. LORIUS. 1987. Vostok ice core provides 160,000-year record of atmospheric CO_2. *Nature*, **329**: 408–414.

BECK, J.W.; J. RÉCY, F. TAYLOR, R.L. EDWARDS and G. CABIOCH. 1997. Abrupt changes in early holocene tropical sea surface temperature derived from coral records. *Nature*, **385**: 705–707.

BERGER, A.L. 1978. Long-term variations of daily insolation and Quaternary climate change. *J. Atmos. Sci.* **35**: 2362–2367.

BITZ, C.M. and W.H. LIPSCOMB. 1999. An energy-conserving thermodynamic model of sea ice. *J. Geophys. Res.* **104**: 15669–15677.

————; M.M. HOLLAND, A.J. WEAVER and M. EBY. 2001. Simulating the ice-thickness distribution in a coupled climate model. *J. Geophys. Res.* **106**: 2441–2464.

BOLTON, D. 1980. The computation of equivalent potential temperature. *Mon. Weather Rev.* **108**: 1046–1053.

BOYLE, E.A. and L.D. KEIGWIN. 1987. North Atlantic thermohaline circulation during the past 20,000 years linked to high-latitude surface temperature.

Nature, **330**: 35–40.

BROCCOLI, A.J. 2000. Tropical cooling at the last glacial maximum: An atmosphere – mixed layer ocean model simulation. *J. Clim.* **13**: 951–976.

——— and S. MANABE. 1987. The influence of continental ice, atmospheric CO_2, and land albedo on the climate of the last glacial maximum. *Clim. Dyn.* **1**: 87–99.

BRYAN, K. 1984. Accelerating the convergence to equilibrium of ocean-climate models. *J. Phys. Oceanogr.* **14**: 666–673.

——— and L. LEWIS. 1979. A water mass model of the world ocean. *J. Geophys. Res.* **84**: 311–337.

BUSH, A.B.G. and S.G.H. PHILANDER. 1998. The role of ocean-atmosphere interactions in tropical cooling during the last glacial maximum. *Science*, **279**: 1341–1344.

CESS, R.D.; M.H. ZANG, W.J. INGRAM, G.L. POTTER, V. ALEKSEEV, H.W. BARKER, E. COHEN-SOLAL, R.A. COLMAN, D.A. DAZLICH, A.D.D. GENIO, M.R. DIX, V. DYMNIKOV, M. ESCH, L.D. FOWLER, J.R. FRASER, V. GALIN, W.L. GATES, J.J. HACK, J.T. KIEHL, H. LETREUT, K.W. LOS, B.J. MCAVANEY, V.P. MELESHKO, J.J. MORCETTE, D.A. RANDALL, E. ROECKNER, J.F. ROYER, M.E. SCHLESINGER, P.V. SPORYSHEV, B. TIMBAL, E.M. VOLODIN, K.E. TAYLOR, W. WANG and R.T. WETHERALD. 1996. Cloud feedback in atmospheric general circulation models: An update. *J. Geophys. Res.* **101**: 12791–12794.

CHAPMAN, M.R.; N.J. SHACKLETON, M. ZHAO and G. EGLINTON. 1996. Faunal and alkenone reconstructions of North Atlantic surface hydrography and paleotemperature over the last 28 kyr. *Paleoceanogr.* **11**: 343–357.

CHARLES, C.D. and R.G. FAIRBANKS. 1992. Evidence from Southern Ocean sediments for the effect of North Atlantic deep-water flux on climate. *Nature*, **355**: 416–419.

———; J. LYNCH-STIEGLITZ, U.S. NINNEMANN and R.G. FAIRBANKS. 1996. Climate connections between the hemispheres revealed by deep sea sediment core/ice core correlations. *Earth Planet Sci. Lett.* **142**: 19–27.

CLAUSSEN, M.; L.A. MYSAK, A.J. WEAVER, M. CRUCIFIX, T. FICHEFET, M. LOUTRE, S.L. WEBER, J. ALCAMO, V.A. ALEXEEV, A. BERGER, R. CALOV, A. GANOPOLSKI, H. GOOSSE, G. LOHMAN, F. LUNKEIT, I. MOHKOV, V. PETOUKHOV, P. STONE and Z. WANG. 2001. Earth system models of intermediate complexity: Closing the gap in the spectrum of climate system models. *Clim. Dyn.* In press.

CLIMAP PROJECT MEMBERS. 1976. The surface of the ice-age earth. *Science,* **191**: 1131–1137.

———. 1981. Seasonal reconstructions of the earth's surface at the glacial maximum. Technical report, Geol. Soc. Am. Map Chart Ser. MC-36, pp. 1–18.

COX, P.M. 2001. Description of the "TRIFFID" Dynamic Global Vegetation Model. Technical Report Technical Note, HCTN24, Hadley Centre, 16 pp.

———; R.A. BETTS, C.D. JONES, S.A. SPALL and I.J. TOTTERDELL. 2000. Acceleration of global warming due to carbon cycle feedbacks in a coupled climate model. *Nature*, **408**: 184–187.

CROSTA, X.; J.J. PICHON and L.H. BURCKLE. 1998a. Application of modern analog technique to marine Antarctic diatoms: Reconstruction of maximum sea ice extent at the last glacial maximum. *Paleoceanogr.* **13**: 286–297.

———; ——— and ———. 1998b. Reappraisal of Antarctic seasonal sea ice at the last glacial maximum. *Geophys. Res. Lett.* **25**: 2703–2706.

CROWLEY, T.J. 2000. CLIMAP SSTs re-visited. *Clim. Dyn.* **16**: 241–255.

——— and G.R. NORTH. 1991. *Paleoclimatology.* Number 18 in Oxford Monographs on Geology and Geophysics. Oxford University Press. 339 pp.

——— and S. BAUM. 1997. Effect of vegetation on an ice-age climate model simulation. *J. Geophys. Res.* **102**: 16463–16480.

CUBASCH, U.; B.D. SANTER, A. HELLBACH, G. HEGERL, H. HOECK, E. MAIER-REIMER, U. MIKOLAJEWICZ, A. STÖSSEL and R. VOSS. 1994. Monte Carlo climate change forecasts with a global coupled ocean-atmosphere model. *Clim. Dyn.* **10**: 1–19.

CURRY, W.B. and G.P. LOHMANN. 1983. Reduced advection into Atlantic Ocean deep eastern basins during last glacial maximum. *Nature*, **308**: 317–342.

——— and D.W. OPPO. 1997. Synchronous high-frequency oscillations in tropical sea surface temperatures and North Atlantic deep water production during the last glacial cycles. *Paleoceanogr.* **12**: 1–14.

DA SILVA, A.M.; C. YOUNG and S. LEVITUS. 1994. Atlas of Surface Marine Data 1994: Vol. 3, Anomalies of Heat and Momentum Fluxes. Technical Report NOAA Atlas NESDIS 8, U. S. Dept. Commerce, NOAA. 413 pp.

DANABASOGLU, G.; J.C. MCWILLIAMS and P.R. GENT. 1994. The role of mesoscale tracer transports in the global ocean circulation. *Science*, **264**: 1123–1126.

DUFFY, P.B. and K. CALDEIRA. 1997. Sensitivity of simulated salinity in a three-dimensional ocean model to upper-ocean transport of salt from sea-ice formation. *J. Phys. Oceanogr.* **27**: 498–523.

———; M. EBY and A.J. WEAVER. 1999. Effects of sinking of salt rejected during formation of sea ice on results of a global ocean-atmosphere-sea ice climate model. *Geophys. Res. Lett.* **26**: 1739–1742.

———; ——— and ———. 2001. Climate model simulations of effects of increased atmospheric CO_2 and loss of sea ice on ocean tracer uptake. *J. Clim.* **14**: 520–532.

FAIRBANKS, R.G. 1989. A 17,000-year glacio-eustatic sea level record on the Younger Dryas event and deep-ocean circulation. *Nature*, **342**: 637–642.

FANNING, A.G. and A.J. WEAVER. 1996. An atmospheric energy-moisture model: Climatology, interpentadal climate change and coupling to an ocean general circulation model. *J. Geophys. Res.* **101**: 15111–15128.

——— and ———. 1997a. On the role of flux adjustments in an idealized coupled climate model. *Clim. Dyn.* **13**: 691–701.

——— and ———. 1997b. Temporal-geographical meltwater influences on the North Atlantic conveyor: Implications for the Younger Dryas. *Paleoceanogr.* **12**: 307–320.

FICHEFET, T. and C. TRICOT. 1992. Influence of the starting date of model integration on projections of greenhouse gas induced climate change. *Geophys. Res. Lett.* **19**: 1771–1774.

FLATO, G.M. and W.D. HIBLER. 1995. Ridging and strength in modelling the thickness distribution of Arctic sea ice. *J. Geophys. Res.* **100**: 18611–18626.

———; G.L. BOER, N.A. MCFARLANE, D. RAMSDEN, M.C. READER and A.J. WEAVER. 2001. The Canadian Climate Centre for Climate Modelling and Analysis global coupled model and its climate. *Clim. Dyn.* **16**: 451–467.

FOFONOFF, N.P. and R.C. MILLARD. 1983. Algorithms for computation of fundamental properties of seawater. Technical report, UNESCO Tech. Papers in Marine Sci. No. 44, Paris, 53 pp.

GANOPOLSKI, A.; S. RAHMSTORF, V. PETOUKHOV and M. CLAUSSEN. 1998. Simulation of modern and glacial climates with a coupled global model of intermediate complexity. *Nature*, **391**: 351–356.

———; V. PETOUKHOV, S. RAHMSTORF, V. BROVKIN, M. CLAUSSEN, A. ELISEEV and C. KUBATZKI. 2001. CLIMBER-2: A climate system model of intermediate complexity. Part II: Model sensitivity. *Clim. Dyn.* **17(10)**: 735–751.

GATES, W.L.; A. HENDERSON-SELLERS, G.J. BOER, C.K. FOLLAND, A. KITOH, B.J. MCAVANEY, F. SEMAZZI, N. SMITH, A.J. WEAVER and Q.C. ZENG. 1996. Climate models – evaluation. In: *Climate Change 1995 – The Science of Climate Change: Contribution of Working Group I to the Second Assessment Report of the Intergovernmental Panel on Climate Change*, J.T. Houghton, L.G. Meira Filho, B.A. Callander, N. Harris, A. Kattenburg, and K. Maskell (Eds), Cambridge University Press Cambridge, England. pp. 229–284.

GENT, P.R. and J.C. MCWILLIAMS. 1999. Isopycnal mixing in ocean circulation models. *J. Phys. Oceanogr.* **20**: 150–155.

GIBSON, J.K.; P. KÅLLBERG, S. UPPALA, A. NOMURA, A HERNANDEZ and E. SERRANO. 1997. ERA description. In: ECMWF Re-Analysis Project Report Series, Vol. 1. ECMWF, Reading, UK, 74 pp.

GILL, A.E. 1982. *Atmosphere-Ocean Dynamics*, Vol. 30 of Int. Geophys. Ser. Academic Press, New York, NY. 662 pp.

GLEN, J.W. 1955. The creep of polycrystalline ice. *Proc. Roy. Soc. Lon. A.* **228**: 519–538.

———. 1958. The flow law of ice. A discussion of the assumptions made in glacier theory, their experimental foundations and consequences. *IASH*, **47**: 171–183.

GOLDSTEIN, H. 1950. *Classical Mechanics.* Addison-Wesley, New York, 672 pp.

GORDON, A.L. 1986. Interocean exchange of thermocline water. *J. Geophys. Res.* **91**: 5037–5046.

———. 1997. Which is it: warm or cold route, or maybe both? *WOCE Newsletter*, **28**: 37–38.

GORDON, C.; C. COOPER, C.A. SENIOR, H. BANKS, J.M. GREGORY, T.C. JOHNS, J.F.B. MITCHELL and R.A. WOOD. 2000. The simulation of sst, sea ice extents and ocean heat transports in a version of the Hadley Centre coupled model without flux adjustments. *Clim. Dyn.* **16**: 147–168.

GRAVES, C.E.; W.H. LEE and G.R. NORTH. 1993. New parameterizations and sensitivities for simple climate models. *J. Geophys. Res.* **98**: 5025–5036.

GUILDERSON, T.P.; R.G. FAIRBANKS and J.L. RUBENSTONE. 1994. Tropical temperature variations since 20,000 years ago: Modulating interhemispheric climate change. *Science*, **263**: 663–665.

HANSEN, J.; A. LACIS, D. RIND, G. RUSSELL, P. STONE, I. FUNG, R. RUEDY and J. LERNER. 1984. Climate sensitivity: Analysis of feedback mechanisms. In: *Climate Processes and Climate Sensitivity*, J.E. Hansen and T. Takahashi (Eds) Vol. 29 of Geophysical Monograph, Am. Geophys. Union. pp. 130–163.

HASSELMANN, K.; R. SAUSEN, E. MAIER-REIMER and R. VOSS. 1993. On the cold start problem in transient simulations with coupled ocean-atmosphere models. *Clim. Dyn.* **9**: 53–61.

HEBBEIN, D.; T. DOKKEN, E. ANDERSON, M. HALD and A. ELVERHOL. 1994. Moisture supply for northern ice-sheet growth during the last glacial maximum. *Nature,* **370**: 357–359.

HIBLER, W.D. 1979. A dynamic thermodynamic sea ice model. *J. Phys. Oceanogr.* **9**: 815–846.

———. 1980. Modeling a variable thickness ice cover. *Mon. Weather Rev.* **108**: 1943–1973.

HOLLAND, M.M.; C.M. BITZ, M. EBY and A.J. WEAVER. 2001. The role of ice ocean interactions in the variability of the North Atlantic thermohaline circulation. *J. Clim.*, **14**: 656–675.

HUGHES, T.M.C. and A.J. WEAVER. 1994. Multiple equilibria of an asymmetric two-basin ocean model. *J. Phys. Oceanogr.* **24**: 619–637.

HUNKE, E.C. and J.K. DUKOWICZ. 1997. An elastic-viscous-plastic model for sea ice dynamics. *J. Phys. Oceanogr.* **27**: 1849–1867.

HUTTER, K. 1983. *Theoretical Glaciology: Material Science of Ice and the Mechanics of Glaciers and Ice Sheets.* Reidel Publishing Company, Dordrecht, Netherlands, 510 pp.

———. 1993. Thermo-mechanically coupled ice-sheet response – cold polythermal, temperate. *J. Glaciol.* **39**: 65–86.

HYDE, W.T.; T.J. CROWLEY, K.Y. KIM and G.R. NORTH. 1989. Comparison of GCM and energy balance model simulations of seasonal temperature changes over the past 18 000 years. *J. Clim.* **2**: 864–887.

IKEHARA, M.; K. KAWAMURA, N. OHKOUCHI, K. KIMOTO, M. MURAYAMA, T. NAKAMURA, T. OBA and A. TAIRA. 1997. Alkenone sea surface temperature in the Southern Ocean for the last two deglaciations. *Geophys. Res. Lett.* **24**: 679–682.

ISEMER, H.J.; J. WILLEBRAND and L. HASSE. 1989. Fine adjustment of large scale air-sea energy flux parameterizations by direct estimates of ocean heat transport. *J. Clim.* **2**: 1173–1184.

KALNAY, E.; M. KANAMITSU, R. KISTLER, W. COLLINS, D. DEAVEN, L. GANDIN, M. IREDELL, S. SAHA, G. WHITE, J. WOOLLEN, Y. ZHU, A. LEETMAA and R. REYNOLDS. 1996. The NCEP/NCAR 40 year reanalysis project. *Bull. Am. Meteorol. Soc.* **77**: 437–471.

KATTENBERG, A.; F. GIORGI, H. GRASSL, G.A. MEEHL. J.F.B. MITCHELL, R.J. STOUFFER, T. TOKIOKA, A.J. WEAVER and T.M.L. WIGLEY. 1996. Climate models-projections of future climate. In: *Climate Change 1995*, J.T. Houghton, L.G. Meira Filho, B.A. Callander, N. Harris, A. Kattenburg and K. Maskell (Eds). Cambridge University Press. pp. 285–357.

KEEN, A.B. and J.M. MURPHY. 1997. Influence of natural variability and the cold start problem on the simulated transient response to increasing CO_2. *Clim. Dyn.* **13**: 835–845.

KUTZBACH, J.E. and P.J. GUETTER. 1986. The influence of changing orbital parameters and surface boundary conditions on climate simulations for the past 18,000 years. *J. Atmos. Sci.* **43**: 1726–1759.

LAUTENSCHLAGER, M. and K. HERTERICH. 1990. Atmospheric response to ice age conditions: climatology near the earth's surface. *J. Geophys. Res.* **95**: 22547–22557.

LEA, D.W. and E.A. BOYLE. 1990. Foraminiferal reconstructions of barium distributions in water masses of the glacial ocean. *Paleoceanogr.* **5**: 719–742.

LEAN, J.L.; J. BEER and R. BRADLEY. 1995. Reconstruction of solar irradiance since 1610: Implications for climate change. *Geophys. Res. Lett.* **22**: 3195–3198.

LEVITUS, S. and T.P. BOYER. 1994. NOAA Atlas NESDIS 4, World Ocean Atlas 1994, Volume 4: Temperature. NOAA, U. S. Dept. Commerce. 117 pp.

———; R. BURGETT and T.P. BOYER. 1994. NOAA Atlas NESDIS 3, World Ocean Atlas 1994, Volume 3: Salinity. NOAA. U. S. Dept. Commerce. 99 pp.

LYLE, M.W.; F.G. PRAHL and M.A. SPARROW. 1992. Upwelling and productivity changes inferred from a temperature record in the central equatorial pacific. *Nature,* **355**: 812–815.

MANABE, S. and A.J. BROCCOLI. 1985. The influence of continental ice sheets on the climate of an ice age. *J. Geophys. Res.* **90**: 2167–2190.

——— and K. BRYAN. 1985. CO_2-induced change in a coupled ocean-atmosphere model and its paleoclimatic implications. *J. Geophys. Res.* **90**: 11689–11707.

——— and R.J. STOUFFER. 1988. Two stable equilibria of a coupled ocean-atmosphere model. *J. Clim.* **1**: 841–866.

——— and ———. 1999. Are two modes of thermohaline circulation stable? *Tellus,* **51A**: 400–411.

MARSHALL, S.J. 1996. Modelling Laurentide Ice Stream Thermomechanics. Ph.D. thesis, University of British Columbia, Vancouver, B.C., Canada, 256 pp.

——— and G.K.C. CLARKE. 1997a. A continuum mixture model of ice stream thermomechanics in the Laurentide ice sheet, 1. theory. *J. Geophys. Res.* **102**: 20599–20614.

——— and ———. 1997b. A continuum mixture model of ice stream thermomechanics in the Laurentide ice sheet, 2. application to the Hudson Strait ice stream. *J. Geophys. Res.* **102**: 20599–20614.

MCFARLANE, N.A.; G.J. BOER, J.P. BLANCHET and M. LAZARE. 1992. The Canadian Climate Centre second-generation circulation model and its equilibrium climate. *J. Clim.* **5**: 1013–1044.

MCPHEE, M.G. 1992. Turbulent heat flux in the upper ocean sea ice. *J. Geophys. Res.* **97**: 5365–5379.

MITCHELL, J.F.B.; T.C. JOHNS, J.M. GREGORY and S.F.B. TETT. 1995. Climate response to increasing levels of greenhouse gases and sulphate aerosols. *Nature,* **376**: 501–504.

OPPO, D.W. and G. FAIRBANKS. 1987. Variability in the deep and intermediate water circulation of the Atlantic Ocean during the past 25,000 years: Northern Hemisphere modulation of the Southern Ocean. *Earth Planet. Sci. Lett.* **86**: 1–15.

ORESKES, N.; K. SHRODER-FRECHETTE and K. BELITZ. 1994. Verification, validation, and confirmation of numerical models in the earth sciences. *Science*, **263**: 641–646.

ORR, J.C.; R. NAJJAR, C.L. SABINE and F. JOOS. 1999. Abiotic-HOWTO. Internal OCMIP Report, LSCE/CEA Saclay, Gif-sur-Yvette, France. 29 pp.

PACANOWSKI, R. 1995. MOM 2 Documentation User's Guide and Reference Manual, GFDL Ocean Group Technical Report. NOAA, GFDL. Princeton. 232 pp.

PATERSON, W.S.B. 1994. *The Physics of Glaciers.* third edition. Elsevier, New York. 480 pp.

PEIXOTO, J.P. and A.H. OORT. 1992. *Physics of Climate.* American Institute of Physics, New York. 520 pp.

PELTIER, W.R. 1994. Ice age paleotopography. *Science*, **265**: 195–201.

——— and S. MARSHALL. 1995. Coupled energy-balance/ice sheet model simulations of the glacial cycle: A possible connection between terminations and terrigenous dust. *J. Geophys. Res.* **100**: 14269–14289.

PETIT, J.R.; M. BRIAT and A. ROYER. 1981. Ice age aerosol content from east Antarctica ice core samples and past wind strength. *Nature*, **293**: 391–394.

PETOUKHOV, V.; A. GANOPOLSKI, V. BROVKIN, M. CLAUSSEN, A. ELISEEV, C. KUBATZKI and S. RAHMSTORF. 2000. CLIMBER-2: A climate system model of intermediate complexity. Part I: Model description and performance for present climate. *Clim. Dyn.* **16**: 1–17.

POUSSART, P.F.; A.J. WEAVER and C.R. BARNES. 1999. Late Ordovician glaciation under high atmospheric CO_2: A coupled model analysis. *Paleoceanogr.* **14**: 542–558.

RAHMSTORF, S. and A. GANOPOLSKI. 1999. Long-term warming scenarios computed with an efficient coupled climate model. *Clim. Change*, **43**: 353–367.

RAMANATHAN, V.; L. CALLIS, R. CESS, J. HANSEN, I. ISAKSEN, W. KUHN, A. LACIS, F. LUTHER, J. MAHLMAN, P. RECK and M. SCHLESINGER. 1987. Climate-chemical interactions and effects of changing atmospheric trace gases. *Rev. Geophys.* **25**: 1441–1482.

RAPER, S.C.B.; T.M.L. WIGLEY and R.A. WARRICK. 1996. Global sea-level rise: past and future. In: *Sea-Level Rise and Coastal Subsidence*, J.D. Milliman and B.U. Haq (Eds). Kluwer Academic Publishers. pp. 11–46.

RINTOUL, S.R., 1991. South Atlantic interbasin exchange. *J. Geophys. Res.* **96**: 2675–2692.

235

ROBITAILLE, D.Y. and A.J. WEAVER. 1995. Validation of sub-grid-scale mixing schemes using CFCs in a global ocean model. *Geophys. Res. Lett.* **22**: 2917–2920.

ROSTEK, F.; G. RUHLAND, F. BASSINOT, P.J. MUELLER, L.D. LABEYRIE, Y. LANCELOT and E. BARD. 1993. Reconstructing sea surface temperature and salinity using $\delta^1 8o$ and alkenone records. *Nature, 364*: 319–321.

ROTHROCK, D.A. 1975. The energetics of the plastic deformation of pack ice by ridging. *J. Geophys. Res.* **80**: 4514–4519.

———. 1986. Ice thickness distribution-measurement and theory. In: *The Geophysics of Sea Ice*, N. Untersteiner (Ed.) Vol 146, NATO ASI Series B, Physics, Plenum, New York, London. pp. 551–575.

RUTBERG, R.L.; S.R. HEMMING and S.L. GOLDSTEIN. 2000. Reduced North Atlantic deep water flux to the glacial Southern Ocean inferred from neodymium isotopes ratios. *Nature, 405*: 935–938.

SAENKO, O.; G.M. FLATO and A.J. WEAVER. 2001. Improved representation of sea-ice processes in climate models. ATMOSPHERE-OCEAN. In press.

SCHILLER, A.; U. MIKOLAJEWICZ and R. VOSS. 1997. The stability of the North Atlantic thermohaline circulation in a coupled ocean-atmosphere general model. *Clim. Dyn.* **13**: 325–347.

SCHNEIDER, R.R.; P.J. MULLER and G. RUHLAND. 1995. Late Quaternary surface circulation in the east equatorial South Atlantic: Evidence from alkenone sea surface temperatures. *Paleocenogr.* **10**: 197–219.

SCHWARZACHER, W. 1959. Pack ice studies in the Arctic Ocean. *J. Geophys. Res.* **64**: 2357–2367.

SEMTNER. A.J. 1976. A model for the thermodynamic growth of sea ice in numerical investigations of climate. *J. Phys. Oceanogr.* **6**: 379–389.

SHACKLEY, S.; P. YOUNG, S. PARKINSON and B. WYNNE. 1998. Uncertainty, complexity and concepts of good science in climate change modelling. *Clim. Change*, **38**: 159–205.

SHER, A. 1995. Is there any real evidence for a huge ice sheet in East Siberia? *Quat. Int.* **28**: 39–40.

SIKES, E.L. and L. KEIGWIN. 1994. Equatorial Atlantic sea surface temperature for the last 30 kyr: A comparison of U^{k}_{37} δ^{18} and foraminiferal assemblage temperature estimates. *Paleoceanogr.* **9**: 31–45.

STOCKER, T.F.; D.G. WRIGHT and L.A. MYSAK. 1992. A zonally averaged, coupled ocean-atmosphere model for paleoclimate studies. *J. Clim.* **5**: 773–797.

——— and A. SCHMITTNER. 1997. Influence of CO_2 emission rates on the stability of the thermohaline circulation. *Nature, 388*: 862–865.

STUTE, M.; M. FORSTER, H. FRISCHKORN, A. SEREJO, J.F. CLARK, P. SCHLOSSER, W.S. BROECKER and G. BONANI. 1995. Cooling of tropical Brazil (5°C) during the last glacial maximum. *Science, 269*: 379–383.

SUAREZ, M.J. and L.L. TAKACS. 1986. Global 5'×5' depth elevation. Technical report, National Geophysical Data Centre, NOAA, U.S. Dept. of Commerce, Code E/GC3, Boulder CO 80303.

TARASOV, L. and W.R. PELTIER. 1997. A high-resolution model of the 100 ka ice-age cycle. *Ann Galciol.* **25**: 58–65.

THOMPSON, L.G. E. MOSLEY-THOMPSON, M.E. DAVIS, P.N. LIN, K.A. HENDERSON, J. COLE-DAI, J.F. BOLZAN and K.B. LIU. 1995. Late glacial stage and holocene tropical ice core records from Huascarán, Peru. *Science, 269*: 46–50.

THOMPSON, S.L. and S.G. WARREN. 1982. Parameterization of outgoing infrared radiation derived from detailed radiative calculations. *J. Atmos. Sci.* **39**: 2667–2680.

THORNDIKE, A.S.; D.A. ROTHROCK, G.A. MAYKUT and R. COLONY. 1975. The thickness distribution of sea ice. *J. Geophys. Res.* **80**: 4501–4513.

UNESCO. 1981. Tenth report of the joint panel on oceanographic tables and standards. Technical report, UNESCO Tech. Papers in Marine Sci. No. 36, Paris, 24 pp.

VAN DE WAL, R.S.W. 1999. The importance of thermodynamics for modeling the volume of the Greenland ice sheet. *J. Geophys. Res.* **104**: 3887–3898.

WANNINKHOF, R. 1992. Relationship between wind speed and gas exchange over the ocean. *J. Geophys. Res.* **97**: 7373–7382.

WEAVER, A.J. and T.M.C. HUGHES. 1996. On the incompatibility of ocean and atmosphere models and the need for flux adjustments. *Clim. Dyn.* **12**: 141–170.

———; M. EBY, A.F. FANNING and E.C. WIEBE. 1998. Simulated influence of carbon dioxide, orbital forcing and ice sheets on the climate of the Last Glacial Maximum. *Nature, 394*: 847–853.

——— and E.C. WIEBE.1999. On the sensitivity of projected oceanic thermal expansions to the parameterisation of sub-grid scale ocean mixing. *Geophys. Res. Lett.* **26**: 3461–3464.

———; P.B. DUFFY, M. EDY and E.C. WIEBE. 2000. Evaluation of ocean and climate models using present-day observations and forcing. ATMOSPHERE-OCEAN, **38**: 271–301.

WEBB, R.S.; D.H. RIND and D. SIGMAN. 1997. Influence of ocean heat transport on the climate of the last glacial maximum. *Nature, 385*: 695–699.

WIEBE, E.C. and A.J. WEAVER. 1999. On the sensitivity of global warming experiments to the parameterisation of sub-grid-scale ocean mixing. *Clim. Dyn.* **15**: 875–893.

WIGLEY, T.M.L. 1998. The Kyoto Protocol: CO_2, CH_4, and climate implications. *Geophys. Res. Lett.* **25**: 2285–2288.

——— and S.C.B. RAPER. 1987. Thermal expansion of sea water associated with global warming. *Nature, 330*: 127–131.

——— and ———. 1992. Implications for climate and sea level of revised IPCC emissions scenarios. *Nature, 357*: 293–300.

YOSHIMORI, M.; A.J. WEAVER, S.J. MARSHALL and G.K.C. CLARKE. 2001. Glacial termination: Sensitivity to orbital and CO_2 forcing in a coupled climate system model. *Clim. Dyn.* **17**: 571–588.

Index

Page numbers in bold refer to tables. Page numbers in italics refer to figures.

abnormal climate indices 42, 44, 58; 1900–98 55–7, **55**, *56–7*, **56**; 1950–98 **55**, 57
absolute humidity, CFL system study 11, *16*, 18
accumulated precipitation 82, 116, 175
Acoustic Doppler Current Profiler (ADCP) 8
Adjusted Precipitation for Canada - Daily (APC1-Daily) dataset 79
Advanced Microwave Sounding Unit-B (AMSU-B) 107
Advanced Very High Resolution Radiometer (AVHRR) 8
advection 233; moisture 185, 201–3, 206–8, *211*, *213*; of thermodynamic quantities/chemical tracers, CanAM4 124–5; of
aerosols, CanAM4 125–6; optical properties 130, 131
age tracers 189, 197–8, *199*, *200*
air–sea flux of inorganic carbon 198, *201*
Alaskan Shelf, eddies in 13, 14
Amundsen 5, 13–14, 16, 18; scientific equipment used on 5, *6*
Amundsen Gulf (AG) 5; eddies in 13, 19; physical oceanography of 11–13; and southern Beaufort Sea, exchanges between 13; temperature anomalies over 17; upwelling in 13
annual precipitation 33, *33*; adjusted for each region *34*; for north of 55°N 34–5, *34*
Antarctic Bottom Water (AABW) 189, 197, 202, 212, 214
Antarctic Circumpolar Current 186, 187, 197, 212
Antarctic Intermediate Water (AAIW) 187, 189, 197, 202
anticyclonic eddies 14
Arakawa-Schubert scheme 146, 149
Arctic cod 7
Arctic Ocean, and global warming 4
Atmosphere-Ocean General Circulation Models (AOGCMs) 172
atmospheric boundary layer (ABL), CFL system study 11, *15*, 18, 19
atmospheric dust, in last glacial maximum 219
atmospheric forcing, effect on sea ice 4, 15, 16
Atmospheric General Circulation Model (AGCM) 172
Atmospheric General Circulation Model, third generation (AGCM3) 123; comparison with CanAM4 135; Taylor diagram for *135*
atmospheric heat transport 174, 201
atmospheric model, UVic Earth System Climate Model 173–8, 185–6
Atmospheric Model Intercomparison Project (AMIP) 134, 135
atmospheric temperature, CFL system study 11, *16*, 17
atmospheric thermodynamic energy balance equation 173
autocorrelation 41, 45, 59
Automated Voluntary Observation System (AVOS) 11
autoregression 45
autumn precipitation 33–4, *33*; for north of 55°N 35, *35*
AXYS technology Inc. 11

Banks Island 5, 16, 17; salinity and northward velocity near *8*
baroclinicity 17, 18

bathymetry 184–5
Beaufort Gyre (BG) 4, 5, 8, 9, 19
Beaufort Sea region (BSR): eddies in 14; ice motion anomalies in 15; relative vorticity in 9; sea level pressure over 15
biological sampling program, CFL system study 7
bootstrap resampling procedure 63
brine rejection 13, 14, 184, 187, 189, 198
bulk aerodynamic formulae 151
buoyancy sorting scheme 146

Campbell soil-water characteristic curve 94, 97, 101
Canada Basin 4, 13, 15, 19
Canadian Arctic Shelf Exchange Study (CASES) 4
Canadian Centre for Climate Modelling and Analysis (CCCma) 123, 172
Canadian Climate Centre general circulation model (CCC GCM) 145–7; downdraft ensemble 149–50; simulations, effect on 157–65; updraft ensemble 147–9
Canadian Earth System Model (CanESM) 123
Canadian Fourth Generation Atmospheric Global Climate Model (CanAM4) 123–4; advection of thermodynamic quantities and chemical tracers 124–5; aerosols 125–6; chemical reactions in **126**; cloud radiative effects from simulations with 138, *139*; comparison with AGCM3 135; experimental design 134–5; features **124**; land and ocean surface 134; layer clouds 126–9; mass fixer 140; mean distributions of clouds and precipitation 135–9; mean total cloud fraction for 137, *137*; moist convection 129–30; overlap of precipitation with clouds 141; radiation 130–2; surface fluxes 133; Taylor diagram for *135*; turbulence 132–3
Canadian Ice Service (CIS), Archive Documentation Series 9–10
Canadian Land Surface Scheme (CLASS) 93, 134; hydrology 94–5; new hydraulic parameterization scheme *98*; soil thermal regime in 95; *see also* peatland environments, hydraulic parameterization for
Canadian Meteorological Centre (CMC) 106
Cape Bathurst 19
carbon cycle model 198
carbon dioxide (CO_2) 172, 203, 211, 219, 221; and North Atlantic thermohaline circulation 171; quasi-equilibrium surface fluxes of 198; radiative forcing associated with 174, 224; transfer velocity 198
chemical tracer advection, CanAM4 124–5
Circumpolar Flaw Lead (CFL) system study 3–7; ArcticNet moorings 5, *7*; atmospheric coupling *15*, 18; atmospheric temperature and absolute humidity *16*, 18; climatology of 10–11, *12*, *13*; cyclone tracking 11, *14*; data and methods 7–11; drift stations 5, *6*, *7*; fast ice stations 5, *6*; MVP temperature and fluorescence section 8, *9*; open-water sample sites 5, *6*; physical system 11–18; Radiometrics MVP-3000A microwave profiling

radiometer measurements *13*; salinity and northward velocity near Banks Island *8*; sea-ice concentration by type 9, *11*; sea-ice minimum and ice mobility 8–9, *10*; seasonal mean air temperature, wind speed and sea level pressure anomalies 10, *12*

CLIMAP 208, 211, 213, 215, 217, 219

climate change indices *see* precipitation indices; temperature indices

climate extremes index (CEI) 42

climate warming, and precipitation 36

climatic trend analysis, data-related difficulties in 41

CLIMBER-2 model 171, 172

closure conditions 146, 150

cloud base closure condition 130

cloud base mass 130

cloud condensate 127, 132

cloud cover 17, 58; and diurnal temperature range 41, 48, 53; last glacial maximum 219

cloud droplet number concentration 126, 128

Cloud Feedback Model Intercomparison Project (CFMIP) 124; Observational Simulator Package (COSP) 136, *136*

cloud feedbacks 124, 231

cloud microphysics, CanAM4 127–9, **128**; autoconversion and accretion processes 128

cloud optical depth 138, 162, *163*

cloud radiative effects (CREs) 138

cloud-resolving model (CRM) 127, 128, 129

clouds: and cumulus parameterization 160–3, *161–2*, 166–7; mean distribution, CanAM4 135–9; optical properties of 130; overlap of precipitation with 141; parameterization of 231

Clouds and Earth's Radiant Energy System (CERES): Atmospheric Radiation Measurement (ARM) Program Validation Experiment (CAVE) network 114; Energy Balanced and Filled (EBAF) dataset 138, *139*

cloud vertical overlap 132

cloud water content 109, 149, 162

cloudy boundary layer, GEM15 109

cluster analysis 4

cold start problem 172, 220

conductivity-temperature-depth (CTD) rosette system 7–8

confidence interval 45

Continental Ice Dynamics Model (CIDM) 182–3

continuity equation 130, 181

convective available potential energy (CAPE) 109, 129, 148, 150, 151; for large-scale ascent case 155, *155*; for stronger downdraft experiment 152, *154*

convective drying 151, *152*

convective heating 151, *152*

convective-radiative equilibrium test 151–3, **151**, *152–4*

convective scale momentum budget equation 129

correlated-*k* distribution (CKD) method 130, 131, 135

coupled models, initialization of 172

coupling strategy, UVic Earth System Climate Model 182–3

cross-validation 43

cumulative probability space (CPS) 130; band spectrum ranges, absorbers, and number of intervals for **131**

cumulus cloud fraction 125, 129

cumulus parameterization 146–7; closure condition 150; clouds, radiation, and surface energy budget 160–5; convective effects on large-scale temperature and moisture fields 147; convective-radiative equilibrium test 151–3, **151**, *152–4*; downdraft ensemble 149–50; effect on GCM climate simulations 157–65; effects of imposed large-scale vertical motion 155–6, *155–6*; entrainment rates, determining 167–8; moist convective adjustment 156–7, *157*; precipitation and tropical circulation patterns 159–60, *160*; results from single column tests 151–7; updraft ensemble 147–9; zonal mean temperature and moisture

fields 158–9, *158*, *159*; *see also* Canadian Climate Centre general circulation model (CCC GCM)

cyclogenesis 17

cyclone tracking, CFL system study 11, *14*, 17

Darcy theory 134

Deacon Cell 202

deep convection: Canadian regional forecast system, 15-km 107, 109, 113; CanAM4 129; clouds 148; and cumulus parameterization 153, 155, 156, 157, 165, 167; in-cloud oxidation in 125; Kain-Fritsch deep convective scheme 109; vertical momentum transfer in 129

depth-integrated steric height (DISH) 203, 206, 207–8, *213*

detrainment: cumulus parameterization 148, 149, 155, 156, *156*; updraft 151, *152*

diffusive moisture transport (DIF) 202, *203*; plus advective moisture transport (ADVDIF) 202, *203*; UVic Earth System Climate Model 203, *211*, *213*; with wind feedback 203, *211*, *213*

digital filter, in GEM15 108

dimethylsulphide (DMS) 125

dissolved inorganic carbon (DIC) 198

diurnal temperature range (DTR) 41, 58, 63, 66, 70–1, 75; 1900–98 47–8, *48*; 1950–98 51–2, *53*

downdrafts: and cumulus parameterization 149–50; mass detrainment *152*; mass flux *152*; strong 150, 152, *154*, 155, 166

Drake Passage 189, 197

drift stations, CFL system study 5, *6*, *7*

dry static energy 149; for moist convective adjustment *157*

Durbin-Watson test 32

Dust Indicators and Records of Terrestrial and Marine Paleoenvironments (DIRTMAP) II database 126

dynamic-thermodynamic sea-ice model 173

Earth System Climate Model (ESCM) 171, 172

eddies 13–14, 19; mesoscale 183, 185, 187, 189

Eddington approximation 132

elastic viscous plastic constitutive law 179

El Niño-Southern Oscillation (ENSO) 52, 135

energy-moisture balance model 171, 172, 173, 174

entrainment 129; cumulus parameterization 151; maximum entrainment rate 148; rates, determining 167–8; updraft 151, *152*

Environment Canada, National Climate Data Archive 24, 79

Euler angle grid rotation 173, 232–3, *233*

Europe: precipitation indices of 73; temperature indices of 73

European Centre for Medium-Range Weather Forecasts (ECMWF) reanalysis 175

European Climate Assessment & Dataset (ECA&D) 62

evaporation 147, 174; in downdrafts 150; UVic Earth System Climate Model 186, *190*, *192*, 202, *205*, *212*

explanatory variables 44

extreme climate indices 44, 62; 1900–98 55–7, **55**, *56–7*, **56**; 1950–98 **56**, 57; *see also* precipitation indices; temperature indices

extreme events, fraction of annual precipitation falling in 37–8, *37*

fast ice sampling, CFL system study 5, *6*

fibric peat 94; saturated hydraulic conductivity of **96**, 97, *104*; soil-water characteristic curves of 97, *98*, *104*; volumetric water content for *104*

Fickian diffusion 174

field programs, CFL system study 7, 9, 17

Fifth Coupled Model Intercomparison Project (CMIP5) 123, 134, 135

FIRE Arctic Cloud Experiment (FIRE.ACE) 109

flags *see under* trace measurements

flaw leads *see* Circumpolar Flaw Lead (CFL) system study

fresh snow density 28, *28*, 29, 81

gamma distribution 23, 31, 130, 132
Gandin optimal interpolation technique 42, 43
gases, optical properties of 130
Gaussian probability distribution 127, 132
GCM-Oriented Cloud-Aerosol Lidar and Infrared Pathfinder Satellite Observations (CALIPSO) Cloud Product (CALIPSO-GOCCP) 136, *136*
GEM15 *see* regional forecast system, 15-km version
General Circulation Models (GCMs) 24, 41, 56, 58, 59
Geophysical Fluid Dynamics Laboratory (GFDL), Modular Ocean Model (MOM) version 2.2 183
Geostationary Operational Environment Satellite-West (GOES-W) satellite 107
geostrophic approximation 175
geostrophic wind velocities 175, 177, *180*
Gibbs effect 125, 137
glaciers 182
Glen's flow law 182
Global Climate Observing System (GCOS), Surface Network (GSN) 80
Global Environmental Multiscale (GEM) model 106–7
Global Land Cover 2000 (GLC2000) global dataset 135
global mean surface temperature 41, 167
global oceanic circulation 189
global precipitation 41
global warming 4; and moisture advection 202–3, 206–8; simulations, UVic Earth System Climate Model 224–31
greenhouse climate response index (GCRI) 42, 56
grid points, in GEM15 108
grid rotation, Euler angle 173, 232–3, *233*

heat capacity of peat 97
heat flux: from atmosphere 178–9; latent heat flux 19, 151, 164, 174, 175; from ocean 179; sensible 18, 19, 151, 164, 174, 175
heat transport, atmospheric 174, 201
hemic peat 94; saturated hydraulic conductivity of **96**, 97; soil-water characteristic curves 97, *98*; volumetric water content for *104*
Historical Canadian Climate Database (HCCD) 23, 24
horizontal diffusion, in GEM15 108
horizontal ice concentration 181
horizontal volume flux, continental ice 182
hybrid variable transformation approach 125
hydraulic conductivity of peat *see* saturated hydraulic conductivity

ice albedo feedback 15, 221
ice areal fraction 179
ice clouds 128; optical properties of 130
Ice Crystal Ratio (ICR) 29
ice crystals 28; aggregation time scale, CanAM4 129; sedimentation, CanAM4 128; and solid precipitation trace events, ratio of *29*, 82; trace measurements 28–9
ice motion anomalies, CFL system study 8–9, *10*, 15
ice sheet thickness: last glacial maximum 217–18, *218*; UVic Earth System Climate Model 183, *184*
ice surface temperature 179
ice thickness 178
ice thickness distribution (ITD) 181–2
ice volume per unit area 181
incoming shortwave radiation 174
index of agreement 95
inorganic ocean carbon cycle, UVic Earth System Climate Model 198, *201*
Intergovernmental Panel on Climate Change (IPCC) 123
intermediate complexity climate models 171–2
International Arctic Buoy Program (IABP) 8
International Geosphere-Biosphere Programme (IGBP), Ocean Carbon-Cycle Model Intercomparison Project (OCMIP) 198

International Polar Year (IPY) *see* Circumpolar Flaw Lead (CFL) system study
International Satellite Cloud Climatology Project (ISCCP) 136; mean total cloud fraction for 137, *137*
Intertropical Convergence Zone (ITCZ) 186, 201, 202
isopycnal mixing 187, 198, 202

Joint Air-Sea Interaction (JASIN) experiment 126

Kain-Fritsch (KF) deep convective scheme, GEM15 109
Kendall's rank correlation tau 45, 59, 65, 86
kriging 28
Kuo-Transient shallow convection scheme, GEM15 109

Labrador Sea 212, 215
large-scale temperature/moisture fields, effects of cumulus convection on 147
last glacial maximum (LGM), climate of 171, 172, 208, 210–15, 217–19
latent heat flux 19, 151, 164, 174, 175
layer clouds, CanAM4: cloud microphysics 127–9; statistical cloud scheme 126–7
leapfrog time step scheme, flux coupling for *186*
least squares method 59, 126, 185, *187*, *213*
level of free convection (LFC) 129
linear trend analysis 31–8
liquid to solid precipitation ratio 36, *36*
longwave radiative flux 174
low pressure systems, GEM15 119–20, *119*, *121*

Mackenzie Shelf 5
McLane Moored profilers (MMP) 8
M'Clure Strait 5, *9*, 14, 17, 19
Manual of Surface Weather Observations (MANOBS) 26
mass balance equation 182
mass fixer 140
mass flux 146; downdraft 149–50, *152*; for large-scale ascent case 155, *156*; net cloud *152*; updraft 146, *152*
maximum entrainment rate 148
maximum likelihood estimator 31
maximum number of consecutive dry days (MCDD) 74, 75
maximum-random overlap rule 141
maximum temperature 58; daily, 1900–98 45–6, *46*; daily, 1950–98 50, *51*
mean precipitation, annual, 1961–90 48, *49*
mean sea level pressure (MSLP), CFL system study 11, 16–17
mean temperature 43, 46–7, 58; annual, 1950–98 51; annual, 1961–90 45, *45*
melting energy, for sea ice 181
meridional overturning streamfunction 187, *197*; last glacial maximum 212, *215*; UVic Earth System Climate Model 202, *210*, 223, *225*, 226–228, *228*, *229*
mesoscale eddies 183, 185, 187, 189
Meteorological Service of Canada (MSC) 79; rain gauge 26, 43, 80
mineral dust, CanAM4 126, 131
minimum temperature 58; adjustment in 43; daily, 1900–98 46, *47*; daily, 1950–98 51, *52*
missing value estimation 23; in daily precipitation time series 30–1
mixed distribution function 31
Model for Ozone and Related Chemical Tracers (MOZART) 125
moist convection 146; adjustment, cumulus parameterization 156–7, *157*; CanAM4 129–30
MoisTKE scheme 109, 110
moist static energy 151, *153*; for moist convective adjustment *157*; for strong downdraft experiment 152, *154*; updraft 168

moisture advection 185; and global warming 202–3, 206–8; UVic Earth System Climate Model 201–2, 203, *211, 213*; with wind feedback 203, *211, 213*
moisture balance equation 174
moisture fields, effects of cumulus convection on 147
momentum balance 179
momentum viscosity 233
Monte Carlo Independent Column Approximation (McICA) 131, 132, 135
Monte-Carlo tests 45
Moving Vessel Profiler (MVP) 8; temperature and fluorescence section across M'Clure Strait *9, 14*
multi-year (MY) ice 4; mobility, CFL system study 8–9, *10*

National Centers for Environmental Prediction (NCEP) reanalysis 10, 177, 185, 186, 202
National Climate Data Archive (NCDA) 24, 79
National Snow and Ice Data Center (NSIDC) 8
national time series: precipitation 31–8, 65–6, **68**, 69, 71, 72, *75*; temperature 65–6, **68**, 70, *73*
near-global analysis 62, 73, 74
near-surface air temperature 175, *177*
near-surface density 175, *177*
near-surface wind velocities 175
net heat, ocean 184
net radiative energy flux, cumulus parameterization 164, *164*, 165, 167
Newton-Raphson procedure 148, 168
95th percentile reference value 64
Nipher shielded snow gauge 24, 28, 33, 36, 64, 81
NOAA-Cooperative Institute for Research in Environmental Sciences (CIRES) Climate Diagnostics Centers 10
non-metric multi-dimensional scaling 4
North Atlantic Deep Water (NADW) 187, 189, 197, 202, 212; potential temperature in 215, *217*

Ocean Carbon-Cycle Model Intercomparison Project 172
ocean circulation model 171
ocean component model 171
Oceanic General Circulation Model (OGCM) 172
oceanic stress 179
ocean model, UVic Earth System Climate Model 183–5, 186–7
ocean-sea-ice-atmosphere (OSA) interface 4; coupling *15*, 18, 19
ocean solar surface albedo, CanAM4 131
ocean surface emissivity, CanAM4 131
open-water sampling, CFL system study 5, *6*
optimal interpolation technique 42, 43, 58
organic soils: *vs.* mineral soils 94; saturated hydraulic conductivity of *94*; soil wetness of *94*
orography, UVic Earth System Climate Model 175, *176*
out-going longwave radiation (OLR), cumulus parameterization 163, *164*, 167
outgoing planetary longwave radiation 174

peatland environments, hydraulic parameterization for: CLASS hydrology 94–5; impacts of new parameterization 97–101; method 95; research problem 93–4; saturated hydraulic connectivity 95, 97; sensitivity 100–1, **102**; site descriptions 95; soil moisture and water table level 97, 99–100, *99*; soil thermal regime in CLASS 95; specific yield 97; suction and wetness curves 97; temperature of peat 100, *103*; thermal conductivity and heat capacity 97
penetrative cumulus convection 146, 149, 158, 159, 167
percentage departure 32
percentile-based climate change indices 63
pit gauge 26

planetary albedo: cumulus parameterization 163, *164*, 167; UVic Earth System Climate Model 169, 174, 185
planetary boundary layer (PBL) 129, 146, 201
polynyas 4
porosity of peat 97
Portable In-situ Laboratory for Mercury Speciation (PILMS) 5
potential temperature: last glacial maximum 214–5, *217*; UVic Earth System Climate Model 187, *195*, 202, *209*, 224, *226*
Potsdam Institute, Climate and Biosphere Model (CLIMBER) group 171
precipitation 174; mean distribution, CanAM4 135–9; overlap with clouds 141; patterns, cumulus parameterization 159–60, *160*; production, limitation of 149; UVic Earth System Climate Model 186, **189**, *190*, 201, 202, *204, 212, 213*; *see also* second generation Adjusted Precipitation for Canada - Daily (APC2-Daily) dataset
precipitation indices 61–5; annual averages over 1961–90 **64**; data and methodologies 63–5, **64**; discussion 73–5; national time series 65–6, **68**, 69, 71, 72, *75*; probability density function 66; six stations representing different climatic regimes *65*; trend estimation 65; trends, 1900–2003 **67**, 71–2, *74–5*; trends, 1950–2003 66, **67**, **68**, 69, *69, 70*
precipitation time series, daily 22–4; adjusted annual total precipitation *34*; adjusted autumn total precipitation 35, *35*; adjusted winter total precipitation 35, *36*; annual precipitation 33, *33*; annual precipitation for north of 55°N 34–5, *34*; automation issues in 38; autumn precipitation 33–4, *33*; autumn precipitation for north of 55°N 35, *35*; climate regions and stations used in studies *23*; effect of adjustments on 29–30, *30, 31*; extreme events, fraction of annual precipitation falling in 37–8, *37*; fresh snow density adjustments *28*; homogenous, developing 24, 26–31; list of stations **25**; rain gauge measurements, adjustments associated with 26; regional and national comparison 31–8; region and station selection 24; snow ruler measurements, adjustments associated with 28–9; station joining and missing value estimation 30–1; summer precipitation for north of 55°N *34*, 35; trace measurements, adjustments associated with 26–8; trends **32**
precipitation totals 43, 58; 1950–98 53, *54*; 1961–90 48, *49*, 50
precipitation trends 40–2; data 42–4; extreme and abnormal climate indices, 1900–98 **55–6**, 55–7, *56–7*; extreme and abnormal climate indices, 1950–98 **55**, **56**, 57; major political boundaries *42*; methods 44–5; problems in data 41; southern Canada, 1900–98 48–50, *49–50*; station locations used for gridding *42*; whole of Canada, 1950–98 53–5, *54–5*
precipitation verification, REM15 119, *120*; for summer cycles 113, *115*; for winter cycles 112–13, *114*
pressure gradient force 129
primitive equation oceanic general circulation model 173
probability density function (PDF) 31; for precipitation indices 66, 72, *76*; for temperature indices 66, 70, *71*
proportionality factor 150

quantitative precipitation forecasts (QPF) 107, 110, 113, 115, 119
quasi-equilibrium assumption 146, 155

radiation, CanAM4: optical properties for gases, clouds, aerosols, and surface 130–1; radiative transfer 131–2, 138
radiation budget, cumulus parameterization 163–4, *164*, 167
radiation flux verification, REM15 114, **115**
radiative-convective model 153
radiative forcing, UVic Earth System Climate Model 172, 174, 219, 221, *222*, 224
Radiometrics MVP-3000A microwave profiling radiometer *13*
Radiometrics tP/WVP 3000 microwave radiometer profiler (MWRP) 11, 17
rainfall gauge measurement 79; adjustments associated with 26; APC2-Daily dataset adjustments 80, **81**, 82

rainfall trace measurement 28, 29, 82

rainfall trends, APC2-Daily dataset: Canada, 1950–2009 87, *88, 89*; southern Canada, 1900–2009 87, *88, 91*

rain/snow ratios 36, *36*

Rayleigh damping 175

red noise 44

Reference Climate Stations (RCS) 80

regional climate trends 44

regional data assimilation system (RDAS) 107

regional forecast system, 15-km version (GEM15) 106–7; bias and equitable threat scores *118*; changes to dynamical configuration 107–8; changes to physics package 108–10; cloudy boundary layer 109; and GEM24, changes between **107**; geopotential height bias and rms errors *116*; Kain-Fritsch deep convective scheme 109; Kuo-Transient shallow convection scheme 109; low-pressure system, 48-h forecast of 119, *119*; objective evaluation 110–17; parallel run 114–17; precipitation, 24-h accumulation of *120*; subgrid-scale orographic drag 109–10; subjective evaluation 117–20; summary of **107**; temperature bias and rms errors *117*; winter and summer cycles 110–14, *111–12*

regional precipitation time series 31–8

region selection, for precipitation time series 24

relative humidity, CFL system study 11, *13*, 17

relative vorticity, CFL system study 9, *10*, 15, 19

Representative Concentration Pathway (RCP4.5) scenario 134

residual circulation 11–12, 19

resolution of GEM15 107–8

Richardson number 133

RM Young 05103 anemometer 11

root mean square errors (RMSE) *42*, 43; GEM15 110, 114–15, *116, 117*

Rotronics MP 101A temperature and relative humidity sensor 11

ruler *see* snowfall ruler measurements

salinity, UVic Earth System Climate Model 184, 187, *196*, 202, *209*

sapric peat 94; saturated hydraulic conductivity of **96**, 97; soil-water characteristic curves 97, *98*; volumetric water content for *104*

saturated hydraulic conductivity: of peat *94*, 95, **96**, 97, *104*; sensitivity of saturation ratio, water table level and peat temperature to **102**

saturated soil volume 95

saturation specific humidity 126, 127, 174

Scanning Multichannel Microwave Radiometer (SMMR) 8

Schmidt number 198

sea ice: CFL system study 4, 5, 8; and cyclones 18; minimum 8–9, *10*

sea-ice boundary condition 134

sea ice concentration: CFL system study 9, *11*; UVic Earth System Climate Model 187, **198**, *199*, 226, *227*

sea-ice dynamics 179

sea ice extent 4, 9; last glacial maximum 212, *215*; UVic Earth System Climate Model 185, 186, 226

sea-ice model, UVic Earth System Climate Model 181–2, 187; dynamics 179; thermodynamics 178–9

sea level pressure (SLP) 175; CFL system study 10, 11, *12, 13*, 16–17; and sea ice 15

sea level rise, UVic Earth System Climate Model 226, *228*

sea salt aerosols 126, 131

sea surface salinity (SSS), UVic Earth System Climate Model 186–187, *194*, 202, *208*

sea surface temperature (SST) 52, 134, 175; last glacial maximum 213, 214, 215, *216*, 217, *218*; UVic Earth System Climate Model 186, *193*, 202, *207*, 208, 210, 211, 222–3, *224*, *227*, 228

second generation Adjusted Precipitation for Canada - Daily (APC2-Daily) dataset 78–9; adjustments for flags 81–2, *83*; adjustments for rainfall from rain-gauges 80, **81**; adjustments for snowfall from ruler measurements 81, *81*; annual total rainfall/ snowfall trends before and after adjustments 84–5, *86, 87*; data 79–80; impact of adjustments 82–5, **83**, *84, 85, 86*, 88–9; improvements made to 88; inhomogeneities 90; joining station observations 85, *86, 87*; locations of stations *80*; magnitude of adjustments for rain/snow measurement issues *85*; rainfall/snowfall trends for Canada, 1950–2009 86, 87, *88, 89–90*; rainfall/snowfall trends for southern Canada, 1900–2009 86, 87, *88, 91*

Self Contained Autonomous MicroProfiler (SCAMP) 8

semi-implicit method 128

sensible heat flux 18, 19, 151, 164, 174, 175

sensitivity studies, of total precipitation adjustments 29, *30*

shallow clouds, parameterization of 149, 156

shallow convection 14; CanAM4 129–30, 135; and cumulus parameterization 152, 153, 156, 158, 167; Kuo-Transient shallow convection scheme 109

similarity theory 133

simple climate models 171

simple dynamic/thermodynamic ice model 171

simple energy-balance atmospheric model 171

snow cover, UVic Earth System Climate Model 187, **198**

snowfall ruler measurements 24, 33, 79; adjustments, APC2-Daily dataset 81, *81*, 82; adjustments associated with 28–9

snowfall trends, APC2-Daily dataset: Canada, 1950–2009 87, *88, 90*; southern Canada, 1900–2009 87, *88, 91*

snow model, UVic Earth System Climate Model 179, 181

snow to precipitation ratio 58; 1950–98 54, *55*; 1961–90 50, *50*; precipitation indices 64, 66

snow water equivalent 24, 28

snow water equivalent adjustment factor 64, 81, *81*, 82, 83, 91

soil moisture of peat 97, 99–100, *101*

soil temperature, and moisture 95

soil-vegetation-atmosphere transfer scheme (SVAT) 93

soil-water characteristic curves of peat 94, 97, *98*

soil water retention curve 944

soil water suction 94, 97, *98*

soil wetness of organic soils *94*

solar forcing, UVic Earth System Climate Model 221–2, *223, 224, 225, 227–8*

solid precipitation: and ice crystal trace events, ratio of *29*, 82; liquid to solid precipitation ratio 36, *36*; solid to liquid precipitation 74; trace measurements 28, 29, 82; *see also* precipitation

southern oscillation index 44

spatial interpolation 28

Special Sensor Microwave Imager (SSM/I) 8

specific heat capacity 95

specific humidity: fields, cumulus parameterization 158, *159*; saturation 126, 127, 174

specific retention 95

specific yield of peat 94–5, 97

spherical geometry, momentum and tracer equations in 183

sponge layer, in GEM15 108

standardized departure 32

station joining: in APC2-Daily dataset 85, *86, 87*; in daily precipitation time series 30–1

station selection, for precipitation time series 24

statistical cloud scheme, CanAM4 126–7

statistical-dynamical approach 171

strong downdraft experiment, cumulus parameterization 150, **155**, 166; CAPE for 152, *154*; moist static energy for 152, *154*; water vapor mixing ratio for 152, *154*

subgrid-scale orographic drag, GEM15 109–10

sulphate aerosol 125

sulphur dioxide (SO_2), gas phase 125

summer cycles, REM15: precipitation verification of 24–48-hr forecasts for 113, *115*; radiation flux verification 114, **115**; upper-air verification of 48-hr forecasts for 110, *112*

summer precipitation 35; for north of 55°N *34*, 35
surface air temperature (SAT) 175; CFL system study 10, 11, *12, 13*, 17; last glacial maximum 212, *214*, 218; UVic Earth System Climate Model 185, *188*, **189**, 202, *206, 210, 211, 220*, 221–2, *223, 224, 227*, 228, *230*
surface energy, cumulus parameterization 164–5, *165*, 167
surface fluxes, CanAM4 133
surface modelling system 107
surface temperature: anomalies, CFL system study 17; cumulus parameterization 165, *166*, 167; global mean 41, 167; ice 179; *see also* surface air temperature (SAT)
surface verification for winter cycles, REM15 110, *113*
surface wind velocities 175

Taylor series 126, 168
temperature fields: effects of cumulus convection on 147; zonal, cumulus parameterization 158, *158*
temperature indices 61–3; annual averages over 1961–90 **64**; data and methodologies 63, **64**; discussion 73–5; national time series 65–6, **68**, 70, *73*; probability density function for 66; six stations representing different climatic regimes *65*; trend estimation 65; trends, 1900–2003 **67**, 70–1, *72, 73*; trends, 1950–2003 66, *67*–8, **67, 68**
temperature of peat 100, *103*
temperature trends 40–2; data 42–4; extreme/abnormal climate indices 44; extreme/abnormal climate indices, 1900–98 **55**–6, 55–7, *56*–7; extreme/abnormal climate indices, 1950–98 **55, 56**, 57; major political boundaries *42*; methods 44–5; southern Canada, 1900–98 45–8, *45*–8; station locations used for gridding *42*; whole of Canada, 1950–98 50–3, *51*–3
thermal conductivity of peat 95, 97
thermal diffusive transport coefficient 201–2
thermodynamics: advection, CanAM4 124–5; multilayer 181–2; sea-ice model 178–9, 181–2
thermodynamic sea-ice model 171
thermohaline circulation 171, 175, 203, 206, 207, 231
time steps, in GEM15 108
trace: classification of 27; definition of 26
trace measurements 43; adjustments associated with 26–8; correction results **27**; ice crystal 28–9; precipitation adjustments, APC2-Daily dataset 81–3, *83*; precipitation indices 64; rainfall 28, 29, 82; snow 28; solid precipitation 28, 29, 82
trace occurrence ratio 27, 29, 82
tracer diffusion 233
transect mode sampling, CFL system study 5
tropical cooling at last glacial maximum 172, 211, 213–14, 219
t-test 32
turbulence, CanAM4 132–3
Type-A gauge *see* Meteorological Service of Canada (MSC), rain gauge
Type-B gauge 26, 29, 36, 43, 79, 80

University of Victoria (UVic) *see* UVic Earth System Climate Model
updrafts: and cumulus parameterization 147–9; mass detrainment/entrainment 151, *152*; mass flux *152*

upper-air verification, REM15: for summer cycles 110, *112*; for winter cycles 110, *111*
upwelling 19; CFL system study 13; diffusion-energy balance climate model 171
US Standard Hydrometeorological Exchange Format (SHEF) surface observation network 110, 113, *114*, 115
UVic Earth System Climate Model 169–70; atmospheric model 173–8, 185–6; climate of last glacial maximum 208, 210–15, 217–19; Continental Ice Dynamics Model and coupling strategy 182–3; coupling between models 185; evaluation 219–21; experimental design 221, *222*; flux coupling for leapfrog time step scheme *186*; geographic latitude and longitude in rotated frame of reference *173*; geostrophic wind velocity anomalies for doubling of CO_2 177, 180; global warming and moisture advection 202–3, 206–8; global warming simulations 224–31; grid rotation 232–3, *233*; inorganic ocean carbon cycle 198; land elevations and ocean depths *176*; long timescale memory of ocean 219; model drift relative to total energy and fresh water 185, *187*; moisture advection 201–2; ocean model 183–5, 186–7; operators in spherical coordinates 233; parameterization of wind stress anomalies 175–8; river drainage basins *176*; sea-ice model 181–2, 1877; snow 179, 181; standard sea-ice model 178–9; transient *vs.* equilibrium climatologies at 1850 and 1998 221–4; water mass formation 189, 197–8

Vaisala PtB210 sensor 11
van Genuchten formulation 94, 97, 101
vertical heat conduction/storage in sea ice 181
volcanic aerosols 125, 131

water masses 13; formation, UVic Earth System Climate Model 187, 189, 197–8, 202
water retention curves of peat 97
water table level of peatland 95, 97, 99–100, *99*, **100**, *102*
water vapor mixing ratio 149, *153*; for moist convective adjustment *157*; for strong downdraft experiment 152, *154*
wetting loss 26, 29, 43, 79
white noise 44
Wilcoxon Rank Sum test 66
wind feedbacks 177–8; advective/diffusive moisture transport with 203, *211, 213*; and last glacial maximum 218–19
wind speed 133; CFL system study 10, 11, *12, 13*, 17
wind stress anomalies, UVic Earth System Climate Model 175–8, 219, *220*
wind stress vector 175
wind undercatch 29, 38, 43, 79
winter cycles, REM15: precipitation verification of 24–48-hr forecasts for 112–13, *114*; radiation flux verification 114, **115**; surface verification 110, *113*; upper-air verification of 48-hr forecasts for 110, *111*
winter total precipitation 35, *36*
World Meteorological Organization (WMO): CCL/CLIVAR Expert Team on Climate Change Detection, Monitoring and Indices (ETCCDMI) 62; Solid Precipitation Measurements Intercomparison project 79

zooplankton 4, 7